W9-BMJ-877

## ABOUT THE Cover

***The cover of*** The Design of Life ***features an artist's portrayal of human brain circuitry as it might appear if magnified many thousands of times. The portrayal illustrates an intriguing discussion of the human brain in Chapter One, quoted here in part:***

During the first eighteen months from conception, the brain's neurons are formed, deployed, and connected in a tsunami of activity, *at the rate of 250,000 per minute*, until 100 billion neurons are arrayed in a powerful, organized matrix. Each neuron may have tens of thousands of finger-like appendages, or dendrites, which connect with other neurons and dendrites in a bafflingly complex circuitry. No two neurons are exactly the same, with the result that the circuitry of each brain is unique. That circuitry is more complex than all the telephone circuitry on the face of the earth. Three decades ago science-writer Isaac Asimov was so impressed with the densely organized complexity of the human brain that he wrote: "In Man is a three-pound brain, which, as far as we know, is the most complex and orderly arrangement of matter in the universe."

# the Design of Life

*Discovering Signs of Intelligence in Biological Systems*

*William A. Dembski | Jonathan Wells*
*Editor, William A. Dembski*

*The important thing in science is not so much to obtain new facts as to discover new ways of thinking about them.*

- Sir William Lawrence Bragg

Foundation for Thought and Ethics, Dallas

**The Foundation for Thought and Ethics**
**Dallas, 75248**



Published in the United States

Cataloging-in-Publication data will be held on file with the Library of Congress

ISBN-13: 978-0-9800213-0-1
ISBN-10: 0-9800213-0-8

Typesetting, design and production *Characters, Buell Design,* Dallas, Texas

# Dedication

In honor of Drs. Paul (1914 – 2003) and Margaret (1919 – ) Brand. Convinced that we live in an intended world, that humans are not accidental happenings, and that the stunning design of the human body rightly inspires awe and gratitude, these medical pioneers dedicated their lives to alleviating the devastating effects of Hansen's disease (leprosy). Today millions of people around the globe with this disease enjoy a quality of life once unimaginable. At enormous personal cost to themselves and in opposition to settled medical practices and superstitions, this amazing couple pursued decades of painstaking research, developing orthopedic and ophthalmologic techniques that revolutionized the medical treatment of leprosy. Their lives attest to the vast potential of science to illumine the world and enhance human life.

# TABLE OF Contents

# Foreword

In 1999, the Kansas Board of Education established new science standards for Kansas public schools. These standards advocated a sober assessment of Darwin's theory of evolution and left the responsibility for how to teach the science of biological origins to local school districts. This was, by any reasonable account, a modest change to the Kansas Science Standards. Yet critics saw the new standards as the next salvo in the ongoing "evolution wars." They responded by misrepresenting the new standards and accusing the Board of eliminating the teaching of evolution and replacing it with religious dogma. The ensuing controversy quickly escalated until the change in Kansas science standards became a "shot heard 'round the world." National and international media joined major scientific organizations across the globe in heaping ridicule and scorn upon Kansas.

Earlier in the decade I had read the seminal works of Phillip Johnson (*Darwin on Trial*, Regnery Gateway, 1991) and Michael Behe (*Darwin's Black Box*, Free Press, 1996). These and other writings familiarized me with the idea of intelligent design and convinced me that its proponents were onto something big—that standard evolutionary theory was not nearly as well confirmed as it was widely touted and that intelligent design (ID), as an alternative approach to biological origins, had real scientific and intellectual merit. Nonetheless, it took the Kansas controversy to bring me "out of the closet."

As a native Kansan engaged in full-time medical research in the Kansas City area, I became actively involved in the debate, writing letters to the editor and speaking at a public forum sponsored by the Board of Education. It was here that I met John Calvert and Jody Sjogren, fellow travelers in the ID community. Together we formed Intelligent

Design Network (IDnet). The goal of IDnet was to foster institutional objectivity in the teaching of origins science.

Why should such an organization be necessary at all? The Kansas controversy made it clear that the "institutions of science"—from the national academies to textbook writers and publishers to the public schools—were uniformly committed to a materialistic, reductionistic view of origins. In consequence, they reflexively opposed the dispassionate consideration of any evidence that does not fit with that perspective.

After losing the majority in 2001 (which resulted in a reversion to the Darwin-only standards), a new Kansas Board, whose majority favored a free and open discussion of evolution, won in the 2004 elections. Thus, the time came for another round of discussions on how the Kansas Science Standards should treat biological origins.

The Board therefore organized a Review Committee, on which I was invited to serve. Working in that capacity gave me the opportunity to bring my scientific perspective (twenty-five years in medical research producing over a hundred scientific publications and many funded research projects) directly to the Board. Nonetheless, the two individuals who chaired the Committee favored a "Teach Darwin Selectively" approach, focusing only on the positive evidence that confirms Darwinian evolution. Moreover, the Committee itself was similarly inclined (by a ratio of three to one).

It therefore promised to be an uphill battle for the eight of us on the Committee (the "Minority") who favored a "Teach Darwin Comprehensively" approach, focusing not only on the positive evidence that confirms Darwinian evolution but also on the negative evidence that disconfirms it. Thanks to a majority on the Board that favored the latter approach, the Minority Position received a sympathetic hearing and was ultimately accepted (at least in large part) by the Board.

What was the Minority Position? What were its main tenets? Remarkably, in laying out the evidence for and against Darwinian evolution, this book by William Dembski and Jonathan Wells outlines the Minority Position in exquisite detail and with substantial scientific support. To be sure, in also presenting the theory of intelligent design, *The Design of Life* goes well beyond the Committee's mandate, which was concerned exclusively with the teaching of evolution and not with the teaching of intelligent design. (Contrary to widespread reports, the Board did not mandate the teaching of ID; in fact, it explicitly stated in the new Standards that it was not mandating ID.)

This book both provides the critical analysis of Darwinian evolution and also reflects the attitude of free and open inquiry that we hoped would become the norm for everyone with an interest in biological origins, first in Kansas and then, ideally, worldwide. In chapter seven, the authors make a statement that crystallizes the problem we were trying to remedy in Kansas, and then they go further to offer a solution that even Charles Darwin would likely have found acceptable:

> Evolutionary biology, by unfairly privileging Darwinian explanations, has settled in advance which biological explanations must be true as well as which must be false apart from any consideration of the empirical evidence. This is not science. This is arm-chair philosophy. . . . [I]n the *Origin of Species*, he [Darwin] wrote: "A fair result can be obtained only by fully stating and balancing the facts and arguments on both sides of each question."

In 2005, the Kansas Board agreed with the Minority: To be informed citizens, students need to be informed regarding scientific controversies. They need to know the multiple definitions of "evolution" and understand to what extent the scientific evidence backs up each meaning. They must be aware that the great mysteries of life remain just that—mysteries awaiting a satisfactory explanation.

The scientific community continues to wrestle with deep and fundamental questions: Where did the universe come from? How did life originate? How did a coded language (i.e., DNA) come to form the basis of life? How could multicellular life have originated from unicellular life? What is the origin of complex molecular machines that are inside every cell and that are necessary for life?

These and other problems have stubbornly resisted the standard materialistic, reductionistic approach to science, and students need to know this. In particular, students need to realize that old lines of evidence historically used to support Darwin's theory have come under significant scientific criticism in recent years and that entirely new lines of evidence have seriously challenged the theory (especially evidence from molecular biology).

Scientific claims are, by their very nature, tentative and always subject to change in light of new evidence. Students of science therefore need to be encouraged to keep an open mind and to let evidence speak for itself, not only in regard to biological origins but in regard to all other scientific issues. Indeed, this attitude stands as the bedrock of all true scientific inquiry, and it, above all others, needs to be nurtured in students. Especially in Kansas!

As of this writing (February 2007), a new Kansas Board of Education has shifted back to a teach-only-the-evidence-that-supports-Darwin approach to the study of biological origins. In fact, they have gone further, mischaracterizing science as a reductionistic enterprise that "describes and explains the physical world in terms of matter, energy, and forces." By so defining science, the new board has not only defined intelligent design out of existence but has also redefined what it means to be human. In particular, human free will and consciousness, which science studies, must, according to these new standards, be described only by reference to matter, energy, and physical forces. Far from being objective and neutral, the new standards now endorse a materialistic philosophy and worldview.

Despite this political ping pong over the Kansas Science Standards and despite the increasing stridency of those who would promote and enforce an ideologically charged conception of science, significant progress in framing the relevant questions over biological origins has been made, and an enduring record of what objectivity in science education might look like has been created. I am confident that others will pick up where we left off in Kansas, and, with the help of volumes like this, will make Darwin's hope of achieving a "fair result" by "balancing the facts and arguments on both sides" increasingly a reality.

*The Design of Life* gives all interested parties in the debate over biological origins the hard scientific evidence they need to assess the true state of Darwin's theory and of the theory of intelligent design. But it does much more: it carefully fosters the attitude of open inquiry that science needs not only to thrive but also to avoid becoming the play-thing of special interests. The authors, William Dembski and Jonathan Wells, are to be commended for writing a sparklingly clear book that empowers readers to navigate the captivating and controversial waters of biological origins.

William S. Harris, Ph.D.
*Director of Nutrition and Metabolic Diseases Research,*
*Sanford Research*
University of South Dakota
Sioux Falls, South Dakota

# Preface

Of *Pandas and People* was the first book to propose intelligent design as a scientific alternative to Darwinian evolution. In fact, it marked the first use of the term "intelligent design" as the scientific investigation into the effects and products of intelligent causes within biology. The scientific status of intelligent design remains to this day hotly debated. Yet the fledgling case for it advanced in *Pandas* looked to the very same methods of testing used throughout the sciences. These methods assess hypotheses in light of evidence and thus ensure that all scientific hypotheses, however well established they may look at the moment, are subject to refutation in light of novel evidence. Accordingly, these methods keep science honest, ensuring that the outcome of any scientific investigation is not predetermined. *Pandas*, far from prejudging the case for or against intelligent design, sought to let the evidence for design in biological systems speak for itself, unimpeded by either religious or materialistic ideology.

More than a decade has passed since the Foundation for Thought and Ethics commissioned Percival Davis and Dean Kenyon to write and later update *Pandas*. When the second edition of *Pandas* appeared in 1993, intelligent design consisted of sporadic criticisms of Darwinism and offered only glimmers of what a positive science of intelligent design might look like. Since then, intelligent design (or ID) has grown from a small and marginalized protest against Darwinian evolution to a comprehensive intellectual program for reconceptualizing biology. Intelligent design has now laid the foundations for a general biology whose fundamental organizing principle is not blind material forces but intelligently devised information.

The impact of intelligent design is being felt in both the scientific community and the culture at large. Front page stories in major newspapers such as the *New York Times* have given intelligent design respectful treatment in their science section.[1] Periodicals such as *Time* and *Newsweek* have featured it on their front covers.[2] Television programs, movies, and popular novels are exploring the theme of intelligent design.[3] Talk shows and news programs—everything from ABC's *Nightline* to Jon Stewart's *The Daily Show*[4]—regularly discuss the topic. The *Nova*-style science documentary *Unlocking the Mystery of Life* argues forcefully for intelligent design and has been broadcast in all major PBS markets (from New York to Los Angeles). At the same time, the BBC counterpart to *Nova*, called *Horizon*, has produced a documentary titled *A War on Science* challenging intelligent design.

On the scholarly and educational fronts, intelligent design is also making deep inroads. Peer-reviewed articles supporting intelligent design have begun to appear in the mainstream biological literature (e.g., *Protein Science, Proceedings of the Biological Society of Washington,* and *Journal of Molecular Biology*). Research scientists have begun to found labs devoted to intelligent design research. For instance, Douglas Axe, formerly a molecular biologist at Cambridge University, has founded The Biologic Institute; and Robert J. Marks II, Distinguished Professor of Electrical and Computer Engineering at Baylor University, has founded The Evolutionary Informatics Lab.[5]

Universities such as Cornell, Stanford, and Cal Berkeley now have student chapters known as IDEA Clubs that support intelligent design (IDEA = Intelligent Design and Evolution Awareness).[6] School boards, state legislatures, and the courts are weighing in on whether intelligent design may legitimately be taught in the public school science curriculum (the most notable instance being *Kitzmiller v. Dover*—see the epilogue). As a result, intelligent design is now being vigorously debated throughout the academic and scientific communities. It is high time, therefore, to issue a sequel to *Pandas* that reflects the progress of intelligent design over the last decade.

Darwinian theorists have long acknowledged that biological organisms "appear" to be designed. Oxford zoologist Richard Dawkins, a leading Darwinian spokesperson, has admitted, "Biology is the study of complicated things that give the appearance of having been designed for a purpose."[7] Statements like this echo throughout the biological literature. The late Francis Crick, Nobel laureate and co-discoverer of the structure of DNA, wrote, "Biologists must constantly keep in mind that what they see was not designed, but rather evolved."[8] Darwinists insist that the appearance of design is illusory because evolutionary mechanisms such as natural selection entirely suffice to explain the observed complexity of living things.

Over the last forty years, however, many evolutionary theorists have acknowledged fundamental difficulties with the Darwinian explanation for apparent design.[9] As a result, an increasing number of scientists have begun to argue that organisms *appear* to be designed because they actually *are* designed. These scientists (known variously as *design proponents* or *design theorists*) see impressive evidence of actual intelligent design in biological systems. As their numbers have grown, their work has sparked a spirited scientific controversy over this central question of biological origins. They argue that, contrary to Darwinian orthodoxy, biology displays abundant evidence of real, not just apparent, design.

Biologist Jonathan Wells is a case in point. He has found persuasive evidence for design in embryological development and in the molecular biology of the cell.[10] Moreover, through his book *Icons of Evolution* (Regnery, 2000), Wells has also become the leading spokesperson for correcting textbook errors in the teaching of biological evolution. In addition, mathematician William Dembski has published an important work on the theoretical underpinnings for detecting design. In *The Design Inference: Eliminating Chance Through Small Probabilities* (Cambridge University Press, 1998), he shows how design is empirically detectable and therefore properly a part of science.

The Foundation for Thought and Ethics is therefore extremely fortunate to have Dembski and Wells author this sequel to *Of Pandas and People.* Though originally planned as a third edition of *Pandas, The Design of Life* quickly took on an identity all its own. More than two-thirds of the material is completely new, and what remains of the original material has been thoroughly reworked and updated. Though there is continuity with the old book, *The Design of Life* is essentially a new book. As a standalone volume aimed at the general reader, *The Design of Life* provides the evidence and conceptual tools necessary to understand the scientific case for intelligent design.

Despite the progress that this volume represents, the Foundation for Thought and Ethics remains extremely grateful to Percival Davis and Dean Kenyon for laying the groundwork for it. In writing *Pandas* under the editorial eye and learned pen of Charles Thaxton (himself a seminal thinker in the intelligent design movement),[11] Davis and Kenyon drew from a wealth of experience and expertise. Davis had coauthored with Eldra Solomon and Harvard biologist Claude Villee what at the time was the best-selling college biology textbook for biology majors (originally titled *The World of Biology* and later retitled simply *Biology*).[12] Kenyon, a professor of biology at San Francisco State University, was one of the top authorities in the world on the origin of life. Not only did he coauthor a seminal text on the subject (*Biochemical Predestination*), but he also contributed to the prestigious Festschrift volumes for both Alexander Oparin and Sidney Fox (when the first edition of *Pandas* appeared, Fox was the most frequently cited origin-of-life scientist in high school biology textbooks).[13]

Davis and Kenyon have left their imprint throughout *The Design of Life*, but especially in the chapters on macroevolution (2), fossils (3), biological similarity (5), and the origin of life (8). Most of their insights remain valid. Yet, with the passage of time, their work has had to be updated. For instance, the origin-of-life chapter that Davis and Kenyon wrote for *Pandas* has a wonderful treatment of spontaneous generation, Oparin's hypothesis, and the work of Stanley Miller and Sidney Fox. But since the last edition of *Pandas* was published, there have been many new proposals for the origin-of-life, including the RNA world and various self-organizational scenarios. The present volume thoroughly critiques these more recent scenarios. Moreover, with the evidence and theoretical insights that have emerged since the publication of *Pandas*, this volume demonstrates far more convincingly than its predecessor could that the origin of life requires an intelligent cause.

The need for a book like this is more urgent than ever. Whenever the topic of evolution comes up, many scientists and educators give the impression that all fundamental debate about biological origins has long since ceased.[14] Nor do the media have the information necessary to correct this impression, perpetuating instead the stereotype that any challenge to Darwinian evolution is a challenge to science and must be religiously motivated (see the epilogue). But evolution, in its contemporary neo-Darwinian form, is not the only scientific account of biological origins. There is in fact a substantial scientific literature that critiques the adequacy of Darwinian explanations for the complexity and "apparent design" of biological organisms.[15] Thus the debate—*the scientific debate*—over Darwinian evolution remains very much alive. *The Design of Life* provides readers with an up-to-date overview of intelligent design and its contribution to that debate.

The Foundation for Thought and Ethics is grateful to the many people who helped bring this project to fruition. Dean Kenyon and Percival Davis deserve enormous thanks in laying the groundwork for this volume. The fellows and staff of Discovery Institute's Center for Science and Culture provided invaluable assistance always: from reading and proofing drafts to offering key biological insights; from digging up references to marketing the book. This help is perhaps not surprising since the authors, William Dembski and Jonathan Wells, are themselves Senior Fellows of the Center for Science and Culture. But it was gratifying to see such an outpouring of support across the board. Among the Center's fellows and staff who contributed significantly to the content of this book, Michael Behe, Scott Minnich, Stephen Meyer, Paul Nelson, and Casey Luskin stand out.

William Harris, Denyse O'Leary, James Barham, and Jonathan Witt read the manuscript in its entirety and offered detailed, helpful comments that greatly improved it. William Harris went even further and graciously wrote the foreword. Edward Peltzer vetted the origin-of-life chapter. Finally, Edward Sisson did the spadework for the epilogue, teasing apart the actual Scopes Trial from the mythology that has developed on account of its dramatization in the play and movie *Inherit the Wind*. To all the individuals who helped on this project, named and unnamed, the Foundation for Thought and Ethics owes a great debt and expresses its heartfelt thanks.

Jon A. Buell, *President*
The Foundation for Thought and Ethics
Dallas, Texas

# THE MEANINGS OF "Evolution"

Some meanings of "evolution" are uncontroversial, such as that organisms have changed over time, that organisms can adapt to changing environmental conditions, or that gene frequencies may vary in a population. If this is all that evolution meant, the general public would leave it well enough alone. Thus, when school boards and biology teachers must answer what they are teaching about biological origins, they often provide an innocuous version of evolution: *Of course you believe that organisms have changed over time . . . Surely you've heard of bacteria developing antibiotic resistance . . . This is evolution in action.*

Such depictions of evolution may alleviate public fears and sidestep controversy, but only for the moment. In fact, they hide what is really at stake in the debate over evolution. Bacteria developing antibiotic resistance do indeed exemplify evolution in action. But this is small-scale evolution (microevolution), which no one disputes and which is irrelevant to the really big claims of evolutionary biology.

Evolutionary biology makes two big claims:

1. The bacteria that develop antibiotic resistance and you, the human whose immune system cannot fend off the bacteria, are, along with all other organisms, descendents from a common ancestor in the distant past; and

2. The process that brought the bacteria and all other organisms into existence by descent from a common ancestor operates by chance and necessity and thus without any discernible plan or purpose.

The first of these is a claim about natural history and is known as "common descent" or "universal common ancestry." According to it, there is a common ancestor to which all living organisms trace their lineage. The second asserts that evolutionary change proceeds by purely material mechanisms and thus requires no intelligent guidance. Intelligence, on this view, is a product of evolution rather than something that guides it.

These twin pillars of evolutionary biology may rightly be credited to Charles Darwin. In proposing his mechanism of natural selection acting on random variations, Darwin seemed to remove any need for intelligence in accounting for biological systems. Instead, he made chance (in the form of random variations) the raw material for biological innovation, and necessity (in the form of natural selection) the driving force that separates among those variations, preserving organisms whose variations confer reproductive advantage while eliminating the rest.

This is the Darwinian mechanism of evolutionary change, and most biologists look to some version of it to explain biological diversification and to justify the first of Darwin's pillars, common descent. For instance, University of Chicago evolutionary geneticist Jerry Coyne writes,

> There is only one going theory of evolution, and it is this: organisms evolved gradually over time and split into different species, and the main engine of evolutionary change was natural selection. Sure, some details of these processes are unsettled, but there is no argument among biologists about the main claims. . . . [W]hile mutations occur by chance, natural selection, which builds complex bodies by saving the most adaptive mutations, emphatically does not. Like all species, man is a product of both chance *and* lawfulness. [*"Don't Know Much Biology,"* June 6, 2007, www.edge.org]

Throughout this book, we use the terms "evolution" and "Darwinism" interchangeably to denote this view of evolution.

# CHAPTER ONE Human Origins

## 1.1 WILLIAM JAMES SIDIS

William James Sidis (1898–1944) was perhaps the smartest person who ever lived. Estimates of his IQ range between 250 and 300. At eighteen months he could read the *New York Times.* At two he taught himself Latin. At three he learned Greek. At four he was typing letters in French and English. At five he wrote a treatise on anatomy and stunned people with his mathematical ability. At eight he graduated from Brookline High School in Massachusetts. He was about to enter Harvard, but the entrance board suggested he take a few years off to develop socially. He complied, and entered Harvard at eleven. At sixteen he graduated cum laude, and then became the youngest professor in history. He inferred the possibility of black holes twenty years before Subrahmanyan Chandrasekhar did. As an adult, he could speak more than forty languages and dialects.

Yet the stress of possessing such an amazing intellect took its toll on Sidis. Instead of being appreciated and admired for his intellectual gifts, he was regarded as a freak—an intellectual performer to be stared at rather than a fellow human being to be esteemed. As a teenager at Harvard, he suffered a nervous breakdown. As a professor at Rice University, he was unable to bear the constant media attention. In his early twenties, he resigned his professorship and withdrew from all serious intellectual pursuits. In 1924, a reporter found him working at a low-paying job in a Wall Street office. Sidis told the reporter that all he wanted was anonymity in a job that placed no demands on him. He spent the rest of his life working menial jobs.[1] What does the story of William James Sidis have to do with human origins?

Evolutionists believe that humans evolved from ape-like ancestors and therefore share many features with modern apes. Many evolutionists go further and claim that human capacities merely extend capacities already present in evolutionary ancestors. Darwin himself took this view in *The Descent of Man:*

> The difference in mind between man and the higher animals, great as it is, certainly is one of degree and not of kind. We have seen that the senses and intuitions, the various emotions and faculties, such as love, memory, attention, curiosity, imitation, reason, etc., of which man boasts, may be found in an incipient, or even sometimes in a well-developed condition, in the lower animals.[2]

*Harold Morowitz*

Some evolutionists, on the other hand, claim that humans exhibit capacities that are genuinely novel and cannot be explained in terms of the capacities of evolutionary ancestors. These include "emergentists" like Harold Morowitz.[3] They acknowledge that although important similarities between humans and apes exist, there are also far-reaching differences, especially differences in intellectual and moral capacities. For them, extravagant abilities like those of William James Sidis indicate that the difference between humans and other animals is radical, and represents a difference in kind and not, as Darwin held, merely a difference in degree.[4]

Did humans evolve from ape-like ancestors? Did those ape-like ancestors evolve from small furry mammals? Did those small furry mammals evolve from reptiles, which in turn evolved from fish? If we go back in time far enough, is there an evolutionary ancestor of all the organisms that we see? Is that common ancestor a single-celled organism? Did biological evolution from this last universal common ancestor proceed without any intelligent guidance but simply as the result of blind material forces? And did the first life arise through a process of chemical evolution in which non-living matter organized itself spontaneously, again without intelligent guidance?

According to the grand story of evolution, the answer to all these questions is Yes. Notwithstanding, as scientists and critical thinkers, how do we determine whether this story is true? To answer this question, we must examine the processes in nature by which biological complexity and diversity could emerge. Some processes in nature are blind, operating without goals, ends, or purposes. Other processes in nature are intelligent, operating with goals, ends, and purposes. How do we tell the difference, and how do we do so in the case of biological systems? In particular, what sorts of processes must operate in nature to bring about someone like a William James Sidis? Are purely material forces enough or is intelligence also required? These are the questions we will examine in this book.

THREE KEY DEFINITIONS

**Intelligent Design.** *The study of patterns in nature that are best explained as the product of intelligence.*

**Intelligence.** *Any cause, agent, or process that achieves an end or goal by employing suitable means or instruments.*

**Design.** *An event, object, or structure that an intelligence brought about by matching means to ends.*

## Is Intelligent Design Scientific?

In reflecting on the significance of Darwin's theory, evolutionary biologist Francisco Ayala remarked, "The functional design of organisms and their features would therefore seem to argue for the existence of a designer. It was Darwin's greatest accomplishment to show that the directive organization of living beings can be explained as the result of a natural process, natural selection, without any need to resort to a Creator or other external agent." To this Ayala immediately added, "The origin and adaptation of organisms in their profusion and wondrous variations were thus brought into the realm of science."[5]

*Charles Darwin*

*Dr. Francis Ayala University of California, Irvine*

With this last comment, Ayala clearly suggests that prior to Darwin the study of biological origins was not properly a part of science. And since the study of biological origins prior to Darwin focused heavily on intelligent design, Ayala is in effect claiming that to explain biological complexity and diversity with reference to design cannot properly be regarded as scientific. Philosopher of biology David Hull makes this point explicitly: "He [Darwin] dismissed it [design] not because it was an incorrect scientific explanation, but because it was not a proper scientific explanation at all."[6] *Continued on next page*

But this cannot be right. Many special sciences employ the concept of design. Indeed, many of those sciences would be inconceivable without it. Archeology assumes that humans of past ages have left evidence of their lives and cultures, and that that evidence is distinguishable from the effects of blind material forces. Forensic science assumes that humans, when committing crimes, try to cover their tracks; yet, when they try to cover their tracks, they often fail, and the tracks lead back to them and not, as they would like, to "natural causes." Other special sciences that require the concept of design include artificial intelligence, cryptography, and random number generation.

*The Arecibo radio astronomy dish is located near the northern coast of Puerto Rico. Its 300-meter diameter makes the giant SETI instrument the world's largest.*

Nor does design always have to refer to human design. Some psychologists study animal learning and behavior. Animals display intelligence and can design things. For instance, the dams that beavers build are designed. Nor does design have to be confined to Earth. The Search for Extraterrestrial Intelligence (SETI) looks for signs of intelligence in radio signals from outer space. SETI's underlying assumption is that we can sift out naturally occurring radio signals to make out those that are designed.

*Leslie Orgel* *Francis Crick*

Biologists Francis Crick and Leslie Orgel have even proposed that life is too complex to have arisen here on planet Earth and so must have been seeded by intelligent space aliens (traveling to our solar system in spaceships).[7] Though regarded as wildly implausible by some, their theory of *directed panspermia,* as it is called, is nonetheless regarded by the scientific community as falling within the bounds of science. The Crick-Orgel theory proposes a design-based view of life on Earth.

Science itself needs to employ the concept of design to keep itself honest. Plagiarism and data falsification are, unfortunately, far more prevalent in science than anyone would care to admit.[8] *The Chronicle of Higher Education* reports a striking case in point:

> Raymond G. De Vries, an associate professor of medical education at the University of Michigan at Ann Arbor, and three colleagues last year reported surveying more than 3,000 scientists about whether they had ever engaged in misbehavior, such as changing a study because of pressure from a source of funds, or failing to present data that contradict one's own research. One-third of the scientists acknowledged they had committed some form of research misbehavior.[9]

A crucial factor in keeping such abuses in check is our ability to detect them. In all these cases, what is being detected is design.

If design is so readily detectable within various special sciences, and if its detectability is one of the key factors keeping scientists honest, why should design be barred on a priori grounds from biology? What if biological systems exhibit patterns that clearly reveal design? The point of this book is to show that such patterns do exist in biological systems and that there are no good reasons for barring design from biology.

## 1.2 OUR FOSSIL ANCESTORS?

Let us start by considering why evolutionists think that humans evolved from ape-like ancestors. Evolutionary accounts of the history of the human race take for granted two things: that humans and apes evolved from an earlier common (ape-like) ancestor, and that their evolution did not require any guidance by intelligence. Does the fossil record support this view? Does it support other interpretations?

Humans are classified as belonging to the genus *Homo* and the species *sapiens.* The genus *Homo* in turn falls within the family *Hominidae*, which includes the apes, and, in particular, the chimpanzees (genus Pan). Among extant apes, chimpanzees are thought to be the closest evolutionary cousin of humans. Thus, if humans evolved from ape-like ancestors, their evolution would be entirely at the genus level. Compare this to the evolution of reptiles into mammals, which represents a class-level transition (see Chapter 4). Since evolutionists think it plausible that reptiles evolved into mammals (which represents a much higher-level transition), it is hardly surprising that they think it even more plausible that ape-like creatures evolved into humans.

Nevertheless, when one examines the actual data and arguments, the case for human evolution becomes less obvious. The fossil record contains several extinct species within the genus *Homo:* most recently *Homo neanderthalensis* (the Neanderthals, formerly considered a subspecies of *Homo sapiens,* but now increasingly considered a separate species); then *Homo erectus;* and, going even further back, *Homo habilis.* Each of these species had many distinctly human characteristics (for instance, the ability to make tools whose sophistication far exceeds any tools employed by apes).

And yet, there is no clear genealogical evidence demonstrating the evolution from *Homo habilis* into *Homo erectus* into *Homo neanderthalensis* into ourselves, *Homo sapiens.* To be sure, there are similarities. *Homo neanderthalensis* is, by any criterion (anatomical, physiological, cultural) closer to *Homo sapiens* than is *Homo erectus,* and similarly *Homo erectus* is closer to us than is *Homo habilis.* At best, this shows that if humans evolved, then the common ancestor of *Homo sapiens* and *Homo neanderthalensis* is more recent than the common ancestor of *Homo sapiens* and *Homo erectus.* And this common ancestor, in turn, is more recent than the common ancestor of *Homo sapiens* and *Homo habilis.* But such an inference presupposes rather than establishes that humans evolved.

The same problem recurs when we try to argue for human evolution at the genus level. The generally accepted date for the formation of our genus, *Homo,* is about 2.5 million years ago (*Homo habilis* and *Homo rudolfensis* are considered the first true members of our genus). The line leading to our genus, *Homo,* is said to have diverged from the line leading to our closest ape cousins, the chimpanzees, at least 5 million years ago. In the interim are the *Australopithecines,* an extinct genus within the *Hominidae.* They include *Australopithecus anamensis* (circa 4 million years ago), *Australopithecus afarensis* (circa 3.5 million years ago), and *Australopithecus africanus* (circa 2.5 million years ago).

As before, one can argue on the basis of structural similarity in the fossil record that our common ancestor with *Australopithecus africanus* is more recent than our common ancestor with *Australopithecus afarensis,* and that this common ancestor, in turn, is more recent than our common ancestor with *Australopithecus anamensis.* But again, this reasoning is based on the assumption that the *australopithecines* and we humans in fact share a common ape-like ancestor. As we shall see in Chapter 5, structural similarity, as exhibited in the fossil record, is not enough by itself to establish such evolutionary connections. What's needed, instead, is independent evidence for the temporal ordering being proposed and for the genealogical connections.

## 1.3 THE NINETY-EIGHT PERCENT CHIMPANZEE?

Scientists look increasingly to genetic data for independent evidence that humans evolved from ape-like ancestors. The underlying assumption is that life forms that have very similar genetic structures are closely related. In recent years, genome mapping has enabled detailed comparisons to be made between the DNA of humans and chimpanzees. Indeed, the most widely cited evidence for human evolution outside the fossil record is genetic.

The base sequences in human and chimpanzee DNA are 98 percent similar. This fact is taken as decisive confirmation of ape to human evolution. But what does this genetic similarity really mean? Consider, first, that because there are only four nucleotide bases, whenever one lines up distinct strands of DNA, even entirely random strands will, on average, be 25 percent similar. Any claim of similarity faces this discount at the outset.

Consider, further, that humans and chimpanzees don't have exactly the same number of DNA base-pairs. In the 1980s, when the 98 percent similarity figure was first proposed, researchers also thought that the genome of chimpanzees was 10 percent larger than that of humans.[10] But in that case, if one lined up all of human DNA with all of chimpanzee DNA, 10 percent of the chimpanzee DNA would have no human counterpart. Looked at in this way, initial reports of the similarity between human and chimpanzee DNA should have noted at least a 10 percent difference, but they did not. This difference in genome size has largely vanished: current estimates for the length of human and chimpanzee genomes are much closer, with 3.1 billion base-pairs for chimpanzees and 3.2 billion base-pairs for humans.[11]

Where, then, does the "98 percent" figure come from? In 1984, Charles Sibley and Jon Ahlquist performed a DNA–DNA hybridization experiment in which the DNA of each species was heated in order to separate the individual strands, and the strands from the two species were mixed and allowed to recombine.[12] Human DNA combined with chimpanzee DNA, and vice versa. The degree of matching between the strands was measured by heating the human–chimp DNA combination and measuring the temperature at which the combined strands separated. Thus, on thermodynamic grounds, Sibley and Ahlquist found a 1.63 percent difference between the two species, and thus a 98.4 percent identity.

Genetic similarities between humans and chimpanzees parallel other similarities between the two. For instance, humans and chimpanzees share gross morphological similarities. In the eighteenth century, before the universal common ancestry of living forms was widely accepted, Linnaeus classified the chimpanzee as *Homo troglodytes* ("primitive man"). According to Jonathan Marks, "When the chimpanzee was a novelty in the 18th century, scholars were struck by the overwhelming similarity of human and ape bodies. And why not? Bone for bone, muscle for muscle, organ for organ, the bodies of humans and apes differ only in subtle ways."[13] With so many obvious physical similarities, genetic similarities between humans and chimpanzees are hardly surprising.

Even so, to say that human and chimpanzee DNA are 98 percent similar can be seriously misleading. That's because we tend to think of DNA in terms of written language. DNA strands form sequences from a four-letter alphabet (usually represented by A, T, C, and G). Likewise, books written by humans in English form sequences from a 26-letter alphabet. Yet, there is a crucial difference between the way humans read written texts and the way cells make sense of DNA. If two books written by humans are 98.4 percent similar, they are essentially the same book. That's because such texts are written to be deciphered not by computers or machines but by competent readers who can recognize random errors and skip past them.

On the other hand, if two sequences of DNA are 98 percent similar, their functions may be vastly different. That's because the cell does not possess a capacity for deciphering DNA comparable to that of humans for deciphering texts. Written language incorporates redundancy and contextual cues that enable us to determine the words and meaning

### PHYSICAL DIFFERENCES BETWEEN HUMANS AND CHIMPANZEES[14]

How similar are humans and chimpanzees when we look not at the level of genes but at the level of gross morphology? Consider the following differences:

1. *The feet of chimpanzees are prehensile, in other words, their feet can grab anything their hands can. Not so for humans.*
2. *Humans have a chin and protruding nose whereas apes do not.*
3. *Human females experience menopause; no other primates do (the only known mammal besides humans to experience menopause is the pilot whale).*
4. *Humans are the only primate in which the breasts of the female are apparent when not nursing.*
5. *Humans have a fatty inner layer of skin as do aquatic mammals like whales and hippopotamuses; apes do not.*
6. *Male apes have a bone in the penis called a baculum (10 millimeters in chimpanzees); humans do not.*
7. *Humans are mostly right-handed. Chimpanzees show no handedness preference.*
8. *Humans sweat; apes do not.*
9. *Humans can consciously hold their breath; apes cannot.*
10. *Humans are the only primates that weep.*

These are just a few of the more obvious physical differences between humans and chimpanzees. But the key difference, of course, resides in the intellectual, linguistic, and moral capacities of humans.

of a text even when it has been substantially altered. On the other hand, random errors in DNA (as with random errors in computer code), even if isolated and few, often introduce radical changes in function that can be disastrous if not fatal.

Because of the complex ways that cells use genetic information, very small genetic changes can critically alter biological function. Proteins, which are specified by genes, interact to form higher-order networks that are not evident from nucleotide or amino-acid sequences alone, and thus cannot be discovered from sequence analyses. Consequently, two organisms might have nearly identical sets of genes, and even situate those genes in roughly the same order along a chromosome; and yet utilize those genes so differently as to produce markedly different organisms.

The lesson here is that small changes can have very significant effects on biological systems *if those changes are just the right changes.* In particular, because the gene expression system operates holistically, large-scale reworking of it would require more than the trial-and-error tinkering characteristic of standard evolutionary theory. Rather, its reworking would require multiple coordinated changes. Such changes indicate the activity of a designing intelligence.

## 1.4 THE BENEFITS OF BIGGER BRAINS

In explaining human evolution, evolutionists emphasize the evolution of the human brain. They maintain that the greater size and more complex organization of the evolving hominid brain explains key differences in behavior and accomplishment between humans and other animals (notably, the apes). In particular, they see a strict correlation between cognitive capacities and brain size. Thus, cognitive capacities such as those demonstrated by William James Sidis are said to require organisms with sizable brains.

Evolutionists have two competing explanations for how the human brain evolved to its present size and complexity: One is that it evolved through natural selection because bigger brains made hominids smarter and therefore more likely to survive and reproduce. The other, championed by Stephen Jay Gould, argues that bigger hominid brains were at first an accidental byproduct of the evolutionary process, which, only after bigger brains had been present for some time, would make hominids smarter. The first view sees bigger brains as an adaptation—something that confers an immediate benefit. The second view sees bigger brains as a preadaptation—something that is not of immediate benefit but can be turned to advantage later.

No one doubts that the human brain has remarkable capacities. Even so, evolutionists have no detailed scientific explanations of how it evolved. Take a recent report in *Nature* by Michael Hopkin titled "Jaw-Dropping Theory of Human Evolution: Did Mankind Trade Chewing Power for a Bigger Brain?" According to Hopkin,

> Researchers have proposed an answer to the vexing question of how the human brain grew so big. We may owe our superior intelligence to weak jaw muscles, they suggest. A mutation 2.4 million years ago could have left us unable to produce one of the main proteins in primate jaw muscles. . . . Lacking the constraints of a bulky chewing apparatus, the human skull may have been free to grow, the researchers say.[15]

Think of what is being argued here. Evolutionists are not simply arguing that a very modest mutation affecting jaw muscles gives brains room to grow. Rather, they are arguing that—given room to grow—brains will in fact grow, getting bigger and bigger till—*presto!*—intelligence, language, culture, and amazing people like William James Sidis emerge. This isn't so much an argument as it is wishful speculation. How would we know whether it was true?

Evolutionists rarely rise above such speculation when accounting for human cognitive capacities in terms of brains size. Usually they don't even get that far. Usually they can't even identify a concrete biological feature that might be implicated in the distinctly human aspects of cognition. That's why the jaw-dropping theory of bigger brains aroused so much excitement among evolutionary biologists—here, at least, was an actual genetic mutation that might be implicated in bigger brains and, thus, in human cognition.

A brief survey of the facts concerning the brain's development and complexity suggests that something more than mere brain size is required to explain human intelligence. During the first eighteen months from conception, the brain's neurons are formed, deployed, and connected in a tsunami of activity, at the rate of 250,000 per minute, until 100 billion neurons are arrayed in a powerful, organized matrix. Each neuron may have tens of thousands of finger-like appendages, or dendrites, which connect with other neurons and dendrites in a bafflingly complex circuitry. No two neurons are exactly the same, with the result that the circuitry of each brain is unique. That circuitry is more complex than all of the telephone circuitry on the face of the earth.

Three decades ago science-writer Isaac Asimov was so impressed with the densely organized complexity of the human brain that he wrote: "In Man is a three-pound brain, which, as far as we know, is the most complex and orderly arrangement of matter in the universe."[16] In the intervening years since Asimov offered this insight, the complexity of the human brain has, in light of further scientific investigation, become even more impressive.

Nevertheless, Asimov also held that "there is nothing magic about the creative ability of the human brain, its intuitions, its genius. It is made up of a finite number of cells of finite complexity, arranged in a pattern of finite complexity." Indeed, he saw the human brain as the product of a purely materialistic evolutionary process. Thus, he continued, "When a computer is built of an equal number of equally complex cells in an equally complex arrangement, we will have something that can do just as much as the human brain can do to its uttermost genius."[17]

Asimov said this in 1975. Such a computer has never been built and is not on the horizon. His remarks form not an argument but more wishful speculation. Asimov thought that if a sufficiently powerful computer ran suitable programs—*voila!*—human consciousness and thought would snap into place. But the human brain is nothing like a computer. There is no evidence that consciousness and intelligence can be reduced to computation and complexity. All that neuroscientists have observed is a correlation between complex neural circuitry and intelligent agency. What they lack is any theory of how, if at all, neural circuitry makes intelligent agency happen.

To sum up, evolutionists simply assume that evolution produces bigger brains. And why not? Evolution, after all, is said to have produced everything else of biological significance. Attributing bigger brains to evolution is therefore hardly a stretch. And once bigger brains have evolved, spectacular cognitive abilities are supposed to follow as a matter of course. The complex neurological organization simply occurs, of itself, through chance events and natural forces. But how, exactly? Unfortunately, evolutionists have no answers here. But this lack of answers and uncertainty raises another question: namely, to what extent are bigger brains really necessary for our cognitive abilities?

## 1.5 THE BENEFITS OF SMALLER BRAINS

It is natural to think that bigger brains equal more intelligence, but this is a misleading simplification. In discussions relating brain size to cognitive capacities, it is important to consider brain size not merely in absolute terms (e.g., weight or volume of brain) but also in relation to body size. Elephants, for instance, have bigger brains than humans. Another crucial factor related to intelligence is the brain's inherent organizational complexity. For instance, compared with the rat, each neuron in the human brain makes between ten and 100 times more synaptic connections.

In the evolutionary literature, all of our spectacular cognitive abilities—mathematical genius, musical genius, poetic genius—are tied, whether directly or indirectly, to our large complex brains.[18] Now, it's certainly true that large complex brains are correlated with increasing intelligence. But correlation, as every scientist will admit, is not causation. Moreover, the correlation is far from perfect. Humans with small or damaged brains have often shown normal or above-normal mental powers. This suggests that human mental powers cannot simply be equated with brain size. Indeed, an evolutionary case can be made for the utility of smaller brains.

*An African Grey Parrot*

For instance, the expression "bird-brain," in suggesting that someone has a small brain and therefore low intelligence, is a misnomer. Some birds possess remarkable cognitive abilities far beyond anything we might expect on the basis of brain-size. Consider Irene Pepperberg's research with Alex, one of four African Grey parrots that she has trained:

> Alex, the oldest, can count, identify objects, shapes, colors and materials, knows the concepts of same and different, and bosses around lab assistants in order to modify his environment. [The researchers] have begun work with phonics and there is evidence to suggest that, someday, Alex may be able to read.[19]

Given such anomalies as Alex, why should we think that big brains are required for higher cognitive functions? In fact, there are reliable reports of people exhibiting remarkable cognitive function with very much reduced brain matter. For instance, anthropologist Roger Lewin reported a case study by John Lorber, a British neurologist and professor at Sheffield University:

> "There's a young student at this university," says Lorber, "who has an IQ of 126, has gained a first-class honors degree in mathematics, and is socially completely normal. And yet the boy has virtually no brain." The student's physician at the university noticed that the youth had a slightly larger than normal head, and so referred him to Lorber, simply out of interest. "When we did a brain scan on him," Lorber recalls, "we saw that instead of the normal 4.5-centimeter thickness of brain tissue between the ventricles and the cortical surface, there was just a thin layer of mantle measuring a millimeter or so. His cranium is filled mainly with cerebro-spinal fluid."[20]

Or consider the case of pioneer microbiologist Louis Pasteur. As historian of science Stanley Jaki remarks,

*Louis Pasteur*

> A brain may largely be deteriorated and still function in an outstanding way. . . . A famous case is that of Pasteur, who at the height of his career suffered a cerebral accident, and yet for many years afterwards did research requiring a high level of abstraction and remained in full possession of everything he learned during his first forty some years. Only the autopsy following his death revealed that he had lived and worked for years with literally one half of his brain, the other half being completely atrophied.[21]

Evolutionists, when confronted with such anomalies, will often remark that the brain contains lots of redundancy. Lorber himself concludes that "there must be a tremendous amount of redundancy or spare capacity in the brain, just as there is with kidney and liver."[22] But that raises another problem. If much of the brain is redundant, then why didn't we evolve the same cognitive abilities without developing larger brains? Redundancy carries hidden costs. Big brains make it difficult for human babies to pass through the birth canal, which, historically, has resulted in heavy casualties—many mothers and babies have died during delivery. Why should the selective advantage of bigger brains with lots of redundancy outweigh the selective advantage of easier births due to smaller brains that, nonetheless, exercise the same cognitive functions, though with lowered redundancy?

There are many deep questions here. Evolutionists may be right that large complex brains have an inherent selective advantage. But that has yet to be established. It remains an open question how our higher mental capacities (such as composing a symphony or proving a deep mathematical theorem) relate to the size and structure of our brains. Evolutionists generally regard mind as simply a function of electro-chemical activity in the brain. But this materialist assumption (that mind is reducible to brain) remains for now without empirical support. What we have are correlations between brain images and conscious mental states. What we do not have is a causal mechanism relating the two.

Quite the contrary. There are now good reasons for thinking that no such causal mechanism exists and that mind is inherently irreducible to brain.[23] This is good news for intelligent design, which treats intelligence as irreducible to material entities and the mechanisms that control their interaction. At the same time, it does not mean that intelligence should be regarded as something "supernatural." Supernatural explanations invoke miracles and therefore are not properly part of science. Explanations that call on intelligent causes require no miracles but cannot be reduced to materialistic explanations. Indeed, design theorists argue that intelligent causation is perfectly natural, provided that nature is understood aright.

## 1.6 LANGUAGE AND INTELLIGENCE

When evolutionists look to the fossil record, genetic similarity, and brain size to substantiate human evolution, they are arguing that humans evolved from ape-like ancestors because these share similar physical structures (e.g, bones, cranial capacity, and DNA sequences). But evolutionists also look to cognitive-behavioral similarities between humans and presumed ape-like ancestors to substantiate human evolution. Thus, for instance, some evolutionary theorists will argue that human language is a straightforward evolutionary development from animal communication systems. The evidence is unconvincing.

Take the capacity of apes for simple symbol manipulation. Apes are capable of acquiring a rudimentary communication system. For instance, Barbara King, a biological anthropologist at the College of William and Mary, describes an ape that developed a taste for champagne and learned to refer to it symbolically.[24] King interprets this capacity as further confirmation of our common ancestry with the apes.[25] But what does this ape really know about champagne other than "that bubbly yellow liquid that tastes good"? And even this goes too far, tacitly attributing linguistic practices to apes that they give no evidence of possessing.[26]

Does the ape have any concept of what champagne actually is, namely, an alcoholic beverage made by fermenting grapes, turning it into wine, and then carbonating it? Can the ape acquire this concept as well as the related concepts needed to understand it? Can the ape deploy this concept in an unlimited number of appropriate contexts, the way humans do? Not at all. The difficulty confronting evolution is to explain the vast differences between human and ape capacities, not their similarities. The communication systems of apes and other animals are not on a continuum with human language. The premier linguist of the 20th century, Noam Chomsky, explained this clearly:

> When we study human language, we are approaching what some might call the "human essence," the distinctive qualities of mind that are, so far as we know, unique to man and that are inseparable from any critical phase of human existence, personal or social. . . . Having mastered a

> language, one is able to understand an indefinite number of expressions that are new to one's experience, that bear no simple physical resemblance and are in no simple way analogous to the expressions that constitute one's linguistic experience; and one is able, with greater or less facility, to produce such expressions on an appropriate occasion, despite their novelty and independently of detectable stimulus configurations, and to be understood by others who share this still mysterious ability. The normal use of language is, in this sense, a creative activity. This creative aspect of normal language use is one fundamental factor that distinguishes human language from any known system of animal communication.[27]

Chomsky is here responding to a standard maneuver in the evolutionary literature: many evolutionists, upon identifying a similarity between humans and apes (or other animals more generally), use this similarity not to elevate apes but, rather, to lower humans. In particular, such evolutionists downgrade the feature of our humanity that is the assumed basis for the similarity. We've just seen this in the case of human language: because humans and apes both have communication systems, human language is said to be just a more sophisticated (more highly evolved) version of ape communication. Not so. Human language, with its infinite adaptability to different contexts and its ability to generate novel concepts and metaphors, has no counterpart in the communication systems of other animals. Jonathan Marks summarizes the situation as follows:

> For all the interest generated by the sign-language experiments with apes, three things are clear. First they do have the capacity to manipulate a symbol system given to them by humans, and to communicate with it. Second, unfortunately, they have nothing to say. And third, they do not use any such system in the wild.[28]

In the same way, evolutionists tend to downgrade human intelligence when comparing it with ape and animal intelligence. From the vantage of contemporary evolutionary theory, intelligence is not a fundamental feature of reality but a product of evolution acquired by us and other animals because of its value for survival and reproduction. But is that all intelligence is? Might not intelligence, instead, be a fundamental feature of the world, a principle that animates the whole of reality, responsible for the marvelous patterns we see throughout the biophysical universe and reflected in the cognitive capacities of animals—and preeminently so in humans? The very fact that the world is intelligible and that our intelligence is capable of understanding the world points to an underlying intelligence that has adapted our intelligence to the world.

Darwinian evolutionists resist this conclusion by attributing the fit between our intelligence and the world to natural selection. Accordingly, they suggest there is a selective advantage to accurately understanding the world. But this is far from clear. Accurate representations of reality need not enhance, and in fact can be detrimental, to survival and reproduction. Suppose you are accosted by a dog that you don't think is dangerous. Because you don't think the dog is dangerous, you don't exhibit fear and thus are actually less likely to be attacked by the animal. Nevertheless, the reality may be that the dog is extremely dangerous. Thus, by misconstruing the reality of the situation, you actually improve your chances of survival and reproduction.

Intelligence, when viewed as a product of natural selection, is merely a tool for survival and reproduction. Such a tool is under no obligation to give us an accurate understanding of the world. The evolutionary process, as Darwin conceived it, places no premium on accurately representing reality. The process by which our minds evolved, according to Darwin, places a premium solely on survival and reproduction. Since misrepresentations of reality could facilitate survival and reproduction better than accurate representations, there is no reason to think that our minds are adapted to know the actual state of the world. Indeed, our minds are, on standard evolutionary principles, more likely to operate at the expense of truth, preferring expedience and gratification.

Darwin himself felt the force of this objection: "With me the horrid doubt always arises whether the convictions of man's mind, which has been developed from the mind of the lower animals, are of any value or are at all trustworthy."[29] To appreciate the full significance of Darwin's remark, apply the doubt he expresses here to evolutionary theory itself: On what basis can we have confidence in evolutionary theory if it is the product of a human mind that "developed from the mind of the lower animals"? Darwin's theory, as an explanation of how the human mind arose, is therefore self-referentially incoherent—in other words, the theory logically defeats itself. Thus, to the degree that we place confidence in it as an accurate account of our human origins, to that degree we have no basis for placing confidence in it. Alternatively, unless a designing intelligence specifically fitted our conceptual apparatus to the world around us, the convictions of our mind are inherently untrustworthy and can provide us with no reliable understanding of human origins. To sum up, when evolutionists note some similarity between humans and animals, they tend not to elevate animals by seeing in them a partially developed trait that finds its full expression in humans. Rather, they tend to demote humans by dismissing their marvelous gifts as products of a blind evolutionary process that merely embellishes capacities already present in animal ancestors. This is especially the case for language and intelligence. Instead of stressing human distinctiveness, they stress commonality with animals. From an intelligent design perspective, the study of human origins needs to pay proper attention to both human distinctiveness and commonality with animals. Intelligent design is a new science, so how best to do this is an open field of inquiry.

## 1.7 MORALITY, ALTRUISM, AND GOODNESS

The human characteristic that poses the greatest difficulty for evolutionary theory is not extraordinary cognitive ability. Cognitive ability is usually (though not always) rewarded, at least to some extent. So, even though the evidence for the evolution of cognitive ability may be weak or nonexistent, an evolutionary story can still be told that extraordinary cognitive ability arose because it was useful to our hunter-gatherer ancestors. But what about ethics and, in particular, altruism? What about the willingness of some human beings to risk or sacrifice themselves for others, without reasonable hope of reward? How does evolution explain such acts?

According to evolutionary psychology (currently one of the hottest evolutionary subdisciplines), the story runs as follows: We, and other primates, live in societies structured by moral norms. Those norms facilitate cooperation. They get us to help each other—to behave altruistically. On evolutionary principles, altruism must therefore be a strategy for facilitating survival and reproduction. In particular, altruism does not reflect a designer's intention for us, nor does it reflect any benevolence underlying the universe. According to evolutionary psychology, altruism comes in two versions. In one version, altruism, even though it may require sacrificing oneself, nonetheless may also benefit the survival of kin (blood relatives), thus promoting one's genes, and therefore is likely to be favored by evolution. In the other version, altruism is not really a sacrifice at all but a form of exchange: you scratch my back and I'll scratch yours. The first of these is known as *kin selection,* the second as *reciprocal altruism.*

Photographer: Jim Harrison

*Michael Ruse* *E. O. Wilson*

The point to realize is that altruism, the kindness we display toward others at a cost to ourselves, is, according to evolutionary psychology, merely grease that keeps evolutionary skids running smoothly. Indeed, evolutionary psychologists and evolutionary ethicists reinterpret all our moral impulses in this light. Michael Ruse and E. O. Wilson are remarkably straightforward in this regard:

### DO CHIMPANZEES HELP OTHER CHIMPANZEES?

Yes, if they are related to them or otherwise know them. But in a letter to the science journal *Nature* (October 27, 2005), researchers revealed that chimpanzees will not help unknown chimps, even if helping would cost nothing. They noted,

> Experimental evidence indicates that people willingly incur costs to help strangers in anonymous one-shot interactions, and that altruistic behaviour is motivated, at least in part, by empathy and concern for the welfare of others (hereafter referred to as other-regarding preferences). In contrast, cooperative behaviour in non-human primates is mainly limited to kin and reciprocating partners, and is virtually never extended to unfamiliar individuals. Here we present experimental tests of the existence of other-regarding preferences in non-human primates, and show that chimpanzees (Pan troglodytes) do not take advantage of opportunities to deliver benefits to familiar individuals at no material cost to themselves, suggesting that chimpanzee behaviour is not motivated by other-regarding preferences.[30]

> The time has come to take seriously the fact that we humans are modified monkeys, not the favored Creation of a Benevolent God on the Sixth Day. In particular, we must recognize our biological past in trying to understand our interactions with others. We must think again especially about our so-called "ethical principles." The question is not whether biology—specifically, our evolution—is connected with ethics, but how.[31]

> As evolutionists, we see that no [ethical] justification of the traditional kind is possible. Morality, or more strictly our belief in morality, is merely an adaptation put in place to further our reproductive ends. Hence the basis of ethics does not lie in God's will. . . . In an important sense, ethics as we understand it is an illusion fobbed off on us by our genes to get us to cooperate. It is without external grounding. Like Macbeth's dagger, it serves a powerful purpose without existing in substance.[32]

> Ethics is illusory inasmuch as it persuades us that it has an objective reference. This is the crux of the biological position. Once it is grasped, everything falls into place.[33]

This ethics-as-illusion view of morality makes perfect sense within an evolutionary worldview. Even so, how do Ruse and Wilson know that ethical principles are merely an illusion? As will become clear in subsequent chapters, the actual evidence for evolutionary theory (especially the grand claim that natural selection is the principle force driving evolution) is slender at best. So to base evolutionary psychology on conventional evolutionary theory is like building a house of cards on a castle of sand.

Equally problematic for Ruse and Wilson is that their evolutionary view of morality cannot be squared with the facts of our moral life. Within traditional morality, the main difficulty is to come to terms with the problem of evil. For evolutionary ethics, by contrast, the main difficulty is to come to terms with the problem of good. Evolutionary theorizing regards reproductive advantage as lying at the root of ethics. Yet it is a fact that people perform acts of kindness that cannot be rationalized on evolutionary principles. Altruism is, as a matter of human practice, not confined simply to one's in-group (those to whom one is genetically related). Nor is altruism outside one's in-group always simply a quid pro quo. People do, in fact, often transcend their drive for reproductive advantage (of their own genes or of their kins'[34]).

Holocaust rescuers, who aided the escape of Jews and others persecuted by the Nazis at great cost and risk to themselves, provide a particularly striking example of genuine altruism. Biologist Jeffrey Schloss, who studies this area, writes:

> Holocaust rescuers exhibited patterns of aid that uniformly violated selectionist [i.e., evolutionist] expectations. Not only was the risk of death clear and ongoing, but it was not confined to the rescuer. Indeed, the rescuer's family, extended family, and friends were all in jeopardy, and recognized to be in jeopardy by the rescuer. Moreover, even if the family escaped death, they often experienced deprivation of food, space, and social commerce; extreme emotional distress; and forfeiture of the rescuer's attention. What's more, rescuing was unlikely to enhance the reputation of the rescuer: Jews, Gypsies, and other aided individuals were typically despised, and assisting them so violated the laws and prevailing social values that the social consequences included ostracism, forfeiture of possessions, and execution. While it is possible to speculate that reputation and

USHMM, Courtesy of Sytske Huisman-Bakker

*Dutch rescuers, Berend Philip Bakker and Jeltje Bakker-Woudsma, two among many honored for their selfless courage.*

group cohesion within subcultural enclaves could been enhanced by rescuing, there is little evidence that such enclaves existed, and most rescuers do not testify to belonging to, or knowing of a group that would have extended support or approval, much less reward or esteem for their actions. Moreover, the overwhelming majority were absolutely secretive about their behavior, not even disclosing it to closest friends or family members outside their immediate dwelling. Finally, the "most unvarying" feature of the behavior and attitudes of all the rescuers was the complete absence of group or individual connections to those aided.[35]

*Mother Teresa*

How does evolutionary ethics make sense of people who transcend their selfish genes? Genuine human goodness, which looks to the welfare of others even at one's own (and one's genes') expense, is an unresolvable problem for evolutionary ethics. Its proponents have only one way of dealing with goodness, namely, to explain it away. Mother Teresa is a prime target in this regard: If Mother Teresa's acts of goodness on behalf of the poor and sick can be explained away in evolutionary terms, then surely so can all acts of human goodness.

For the prominent proponent of evolutionary psychology E. O. Wilson, goodness depends on "lying, pretense, and deceit, including self-deceit, because the actor is most convincing who believes that his performance is real."[36] Accordingly, Wilson attributes Mother Teresa's acts of goodness to her belief that she will be richly rewarded for them in heaven. In other words, she was simply looking out for number one, acting selfishly in her own self-interest, looking to cash in on the Church's immortality. As Wilson puts it, "Mother Teresa is an extraordinary person but it should not be forgotten that she is secure in the service of Christ and the knowledge of her Church's immortality."[37]

In fact, after Mother Teresa's death in 1997, her published letters revealed that she suffered from depressive episodes throughout her life in which she experienced grave crises of faith, though she remained faithful to her mission to the end.[38] But this wrinkle presents no insuperable difficulty for evolutionary ethics. If Mother Teresa's goodness cannot be dismissed as self-serving, it can be dismissed as maladaptive. Thus, evolutionary ethics can always argue that Mother Teresa's genetic program misfired, so distorting her ethical sensibilities as to make her an evolutionary dead-end, one that evolution would be sure to weed out because it has no use for such extreme do-gooders. To clinch their case, they need merely note that as a Catholic nun, Mother Teresa took a vow of celibacy, left no offspring, and therefore failed to pass on her genes. Thus, instead of treating Mother Teresa as a model of goodness to which we should aspire, evolutionary ethics regards Mother Teresa as either a self-serving hypocrite or a freak of nature with no future.

Such rationalizations of human goodness are now standard fare in the evolutionary psychology and evolutionary ethics literature.[39] Certainly, they denigrate our moral sensibilities. More significantly, however, they don't square with the facts. There is little evidence that those who are motivated to risk or sacrifice themselves for others are, in general, less well adapted than others or that they seek a reward, such as personal comfort, increased status, or more offspring, any of which might be explained by evolutionary psychology. Apart from clear countervailing evidence, their own testimony that they are doing what they think is morally right should be accepted at face value. In that case, however, the question remains: what is the origin of the morality that motivates them? Here an intelligent design approach connects most readily with the approach to ethics known as "natural law" (not to be confused with what evolutionists typically mean by "laws of nature").[40] Within this approach, ethics represents conformity of behavior to the design constraints according to which humans were intended to operate.

## 1.8 MODIFIED MONKEY OR MODIFIED DIRT?

In responding to criticisms of evolution based on the Bible (which portrays God as creating humans from the earth beneath our feet), Thomas Henry Huxley once remarked, "It is as respectable to be modified monkey as modified dirt."[41] From an intelligent design perspective, the crucial issue is not the respectability of humanity's material precursors (monkeys vs. dirt), but what was producing the modifications that made us what we are. In particular, is the source behind those modifications intelligent or simply the outworking of blind material forces?

Regardless of whether one is a biblical creationist or an atheistic Darwinist or anything in between, all are agreed that humans did not magically materialize out of nothing. Humans arose from preexisting material stuff. Indeed, the very word "human" refers to the earth (humus) that lies beneath our feet. In this respect, monkeys and humans are both modified dirt, and that is true regardless of whether humans are, in addition, modified monkeys. ID is compatible with each of these possibilities, and there are ID proponents who hold to each. Nonetheless, even those ID proponents who accept that humans descended from primate ancestors do not accept that we evolved in the ordinary sense.

Evolution, as the term is typically used, refers to a process by which organisms change apart from any need for intelligent guidance or intervention. It follows that evolution by intelligent design is not what most people mean by evolution. Nevertheless, once intelligence is allowed as a possible factor in the emergence of humanity, it becomes an open question whether humans are both modified monkeys and modified dirt (as with evolution) or merely modified dirt (as with biblical creation). We can ask the same sort of question about an archeological artifact. For instance, is an engraved metal bowl the result of reworking an existing bowl or was it made from scratch by casting liquid metal in a mold?

There may be good reasons for thinking that humans are redesigned monkeys (shared error arguments described in Chapter 5 provide one line of evidence). Even so, a design-theoretic perspective does not require that novel designs must invariably result from modifying existing designs—some designs could just be built from scratch. Hence, there may also be good reasons for thinking that a redesign process didn't produce humans and that, instead, humans were built from the ground up (for instance, what appear to be shared errors might not be errors at all).

Design theorists have not reached a consensus about just how humans emerged. Nevertheless, they have reached a consensus about the indispensability of intelligence in human origins, regardless of the process by which humans emerged. Thus, in particular, they argue that an evolutionary process unguided by intelligence cannot adequately account for the remarkable intellectual gifts of a William James Sidis or the remarkable moral goodness of a Mother Teresa.

*The general notes on the CD included with this book expand on the material in this chapter and throw light on the following discussion questions.*

## 1.9 DISCUSSION QUESTIONS

1. Briefly summarize the fossil evidence for human evolution. Are there any nonhuman fossils (e.g., the Australopithecines) that have been conclusively shown to be ancestral to modern humans? If not, on what grounds can the fossil record be taken as supporting human evolution? Is further independent evidence required? Assuming that the fossil record supports evolution, does it also reveal the mechanism by which evolution operates?

2. What does it mean to say that humans and chimpanzees share 98 percent of their genes? Does this mean that humans and chimpanzees are 98-percent similar? Does the genetic similarity between humans and chimpanzees indicate that they are descended from a common evolutionary ancestor? Support your answer.

3. List some ways in which humans and chimpanzees differ at the level of gross morphology (anatomy and physiology). How can such differences be squared with the genetic similarity between humans and chimpanzees? Are there differences between humans and chimpanzees that are surprising given their genetic similarity?

4. How is the brain size of organisms related to their intelligence? Is there a strong correlation or are there examples of smaller brains that exhibit remarkable cognitive abilities? What is the significance of the case study by John Lorber in which he describes a young man with a high IQ who "has virtually no brain"? Does appealing to redundancy in the brain, as Lorber does, adequately explain such anomalies? Why or why not?

5. Are our cognitive abilities simply a product of brain function? Or, are those abilities not reducible to brain function? Together, these two questions summarize the famous mind-body problem. What light, if any, does the relation between brain size and intelligence throw on the mind-body problem?

6. Is human language unique among animal communication systems? How so? Summarize Noam Chomsky's view that human language ability is fundamentally different from the communication systems of other animals. Is Chomsky's view widely accepted among evolutionary anthropologists such as Barbara King? What is Barbara King's view? Which view about the nature of human language and animal communication systems do you find more compelling? Why?

7. What are the three main evolutionary hypotheses for explaining the emergence of higher cognitive abilities such as mathematics in humans? How, for instance, do Darwinists employ these hypotheses to explain human mathematical ability? What, if any, evidence is there to support these hypotheses? [See general notes to section 1.6.]

8. Define morality and altruism. From a Darwinian perspective, can there be anything like a truly selfless act? Why are reciprocal altruism and kin selection, as developed within evolutionary theory, incompatible with altruism in its ordinary sense of selfless acts of kindness? Who were the holocaust rescuers? Who was Mother Teresa? How does Darwinism explain the altruistic acts of people like holocaust rescuers and Mother Teresa?

9. What does E. O. Wilson mean when he describes morality as an "illusion fobbed off on us by our genes"? Can this view of morality be squared with the facts of our moral life? Comment on the following remark from section 1.7: "Within traditional morality, the main difficulty is to come to terms with the problem of evil. For evolutionary ethics, by contrast, the main difficulty is to come to terms with the problem of good."

10. Comment on T. H. Huxley's famous claim that "it is as respectable to be modified monkey as modified dirt." Did humans evolve from monkeys? Are there compelling reasons for thinking that humans did evolve from monkeys? Are there compelling reasons for thinking that they did not? Which of these positions is compatible with intelligent design? Are both compatible? Support your answer.

# CHAPTER TWO 2 Genetics and Macroevolution

## 2.1 DARWIN'S THEORY

Darwin's *Origin of Species,* published in 1859, did more to establish evolution as a scientific theory than any other single scientific work. In that book Darwin contended that no existing species was individually created. Instead, he argued that every species descended from preexisting species as a result of natural selection. Moreover, he argued that all species trace their ancestry back to one or a few original beings or forms. According to Darwin's theory, organisms exhibit new traits when natural selection chooses among numerous natural genetic variants. Certain variants, the theory says, will be passed on to future generations if they give their possessor a competitive edge or advantage over other organisms in the population and therefore leave more offspring. Darwin argued that as evolution progressed, advantageous new traits would accumulate until a new species was formed.

*Charles Darwin*

Darwin began the *Origin of Species* by looking at animal breeding. As an animal breeder himself, Darwin was amazed at how much animals could change through selective breeding. By selecting animals with particular traits and allowing them to reproduce, breeders have produced differences within a single species that exceed differences between separate species in the wild. For instance, a Chihuahua and a Great Dane belong to the same species (see figure 2.1). By all appearances, the difference between these two dog breeds exceeds the

*Figure 2.1* *Chihuahua and Great Dane*

difference between, say, a fox and a coyote. Yet the fox and coyote, though members of the Canidae family, belong to different species (see figure 2.2). The Chihuahua and the Great Dane are the result of selective breeding; the fox and coyote, by contrast, exist in the wild. Could not nature, Darwin asked, produce the differences we see in selective breeding, only much more, given enough time?

Darwin noted that in nature many more offspring are born than can survive and reproduce. Since offspring vary, some may possess traits to a degree that exceeds that of other members in the population—such as longer legs. If this trait improves the ability of offspring to adapt to their ecological niche, then they stand a better chance of surviving and passing the trait on to their own offspring. If this trend continues over successive generations, eventually animals that possess longer legs will outnumber those that do not. Once that happens, the trait becomes established in the population.

*Figure 2.2* *Coyote and fox*

Darwin dubbed this process *natural selection* to emphasize its parallel with what breeders do when they select for given traits. Unfortunately, the term suggests that nature is capable of consciously selecting among organisms, foreseeing what traits will benefit them in the future and adjusting the course of evolution to bring about those traits. Yet, according to Darwin, evolution, as controlled by natural selection, is a blind mechanical process that cannot act teleologically by adjusting means to ends. The Darwinian mechanism of natural selection that controls the evolutionary process is simply the environment's way of weeding out harmful traits and giving beneficial traits free play. Moreover, what the environment deems harmful or beneficial depends on the organism's immediate needs and not on what the organism needed in the past or might need in the future. It follows that in Darwin's theory of natural selection, evolution proceeds without plan or purpose.[1]

Nonetheless, Darwin argued that natural selection, though not following any actual plans or purposes, is capable of producing biological traits that appear to be the result of purposive action. Indeed, he ascribed remarkable skill to natural selection: "Natural selection picks out with unerring skill the best varieties."[2] Or again: "Natural selection is daily and hourly scrutinising, throughout the world, every variation, even the slightest; rejecting that which is bad, preserving and adding up all that is good; silently and insensibly working, whenever and wherever opportunity offers, at the improvement of each organic being in relation to its organic and inorganic conditions of life."[3]

Ever since Darwin wrote these words, the biological literature has ascribed to natural selection numerous honors and awards. Accordingly, evolutionary biologists have ascribed to it artistry and skill. They have compared it favorably with a composer of music, a master of ceremonies, a poet, and a sculptor. Richard Dawkins has identified natural selection with a watchmaker, albeit a blind one.[4] Steven Pinker compares natural selection to a software engineer.[5]

Despite natural selection's immense popularity to this day,[6] a growing number of scientists now question its effectiveness as a creative force in biology. Some, like Stuart Kauffman, Brian Goodwin, and Robert Laughlin, take a self-organizational approach in which laws of self-organization and complexity take precedence over natural selection.[7] Others, like the authors of this book, take a design-theoretic approach in which living things appear to be designed because they actually are designed. Both the self-organizational and the design-theoretic approach demote natural selection, attributing the creative potential for building new organisms and new biological structures to forces other than natural selection.

Darwin did not originate the idea of natural selection. Several earlier naturalists had observed this process taking place in nature. They took, however, a more measured view of it than Darwin. For instance, Edward Blyth, one of Darwin's predecessors and a proponent of design in biology, saw natural selection as a conservative force for maintaining species within settled boundaries and for weeding out unfit individuals and thereby aiding the survival of existing species. He recognized that if all organisms were designed according to basic blueprints, then there must be some kind of quality control—some way to keep the original types from drifting away from their original designs. He believed that natural selection played this role by eliminating those organisms whose traits were too deviant.

Darwin's revolutionary idea was to reconceptualize natural selection as a force capable of producing new species. For Darwin, species were fluid, and natural selection could exploit that fluidity and drive large-scale evolutionary change. For Blyth, by contrast, species varied only within settled boundaries, and natural selection was merely a regulator that kept species within those boundaries. Prior to Darwin, most scientists believed in design and saw species as essentially fixed (including the father of taxonomy, Carolus Linnaeus).

The fixity of species found support in Plato's theory of ideal types, which proposed an ideal fixed blueprint to which each organism conformed. Contemporary proponents of intelligent design reject two aspects of Plato's theory. First, design theorists acknowledge that organisms are capable of significant change. The strict fixity or rigidity of species entailed by Plato's theory is therefore relaxed within the contemporary theory of intelligent design. Second, design theorists reject the idea that the design of every species had to be perfect or ideal. Design can be real without being ideal (for instance, the car you drive is actually designed though not ideally designed—you can imagine ways its design could be improved).

Even though natural selection is the linchpin of Darwin's theory, Darwin recognized that any creative potential inherent in the evolutionary process depends not simply on natural selection but also on the sources of variation on which natural selection operates. That's because natural selection does not actually produce new traits. It can only act on traits that already exist. Darwin put it this way: "Unless profitable variations do occur, natural selection can do nothing."[8]

What, then, is the source of evolutionary novelty—of the profitable variations that lead to new traits and structures upon which natural selection can act? Take the giraffe. How could such a creature have evolved? Its oversized limbs, stretched-out neck, and ungainly posture suggest that everything is precariously out of proportion. And yet its parts are marvelously coordinated with each other. It moves with graceful ease and delivers such a powerful kick that it has few natural enemies. The outlandish body-shape of the giraffe is therefore a puzzle for evolution.

Jean Baptiste de Lamarck (1744–1829), an evolutionist who was active prior to Darwin, thought the giraffe's long neck resulted from its constant stretching upward to eat leaves high up in tree branches. In Lamarck's theory of evolution, the giraffe's neck changed in response to its continual need to reach higher, and this change was passed on to its offspring. As a result, the giraffe's neck gradually became longer. Lamarck's theory has been discredited. Scientists now recognize that body structures do not change in heritable ways in response to an organism's needs or habits. If it did, Olympic racers should give birth to yet faster racers, and the children of geniuses should be even smarter than their parents. That is generally not the case.

Darwin's theory of natural selection turned Lamarck's explanation on its head. Instead of the environment eliciting new and improved traits from the organism, Darwin maintained that something within the organism itself gives rise to new traits, which are then either preserved or weeded out by the environment. Lamarck emphasized an organism's need for a longer neck to survive as well as its ability to meet that need and pass the benefit to future generations. Darwin emphasized the environment's role in

Jean Baptiste de Lamarck (1744–1829), a French biologist, believed that altering an organism's environment would alter its needs and behavior, and that this in turn would alter its structure. In 1809 he published a book titled *Zoological Philosophy,* in which he wrote:

*Jean Baptiste de Lamarck*

> The giraffe lives in places where the ground is almost invariably parched and without grass. Obliged to browse upon trees it is continually forced to stretch upwards. This habit maintained over long periods of time by every individual of the race has resulted in the forelimbs becoming longer . . . and the neck so elongated that a giraffe can raise his head to a height of eighteen feet without taking his forelimbs off the ground.[9]

favoring organisms with longer necks and then preserving any that happened to come along—the giraffe, as it turned out.

Thus, in addition to the environment's role in natural selection, Darwin also needed something within organisms to give rise to new traits. Even so, to maintain that organisms either survive or perish depending on whether they possess traits that suit them to their environment was no new insight. Darwin's key insight—contested to this day—was that the same source within organisms capable of producing slight variations could be vastly extrapolated. Thus, over time, this source of variation, when sifted by natural selection, would produce new traits which, in turn, would accumulate to produce entirely new organisms.

But what was this source of variation within organisms? Darwin himself did not know (neo-Darwinism, the contemporary form of Darwinism, regards the source of variation as random changes in DNA—more on this later in the chapter). Yet regardless of the source of variation and his ignorance of it, Darwin held a definite view on how variations were transmitted. Heredity, for Darwin, was a matter of mixing or blending parental traits. According to his theory of blending inheritance, offspring should be intermediate between their parents not only in physical appearance but also in whatever hereditary material they receive and pass on. For example, according to the theory of blending, if a red morning glory is crossed with a white one, the resulting flowers should all be blended and therefore pink.

But what happens if these pink flowers are then crossed with each other? According to Darwin's theory of blending, the offspring from this cross should also be pink, and so on throughout successive generations. Red and white flowers would thus be lost forever. Does this in fact happen? Whether such blending actually occurs can be tested experimentally. When red and white flowers are crossed, the first generation consists of pink flowers. But when these pink flowers from the first generation are crossed, red and white flowers return in the second generation. It's as though red and white, instead of being blended, go underground and then reemerge. This shows that the theory of blending or mixing inheritance is false. The theory (incorrectly) predicts that pink flowers, when crossed, should stay pink.

Furthermore, blending inheritance produces sameness, not differences. It cannot explain how new variations arise. Therefore, Darwin's theory could not be reconciled with his view of heredity. Not until the sixth edition of his *Origin of Species* (published in 1872, thirteen years after the first edition) did Darwin give any indication that his theory of natural selection was incompatible with his view of heredity. According to natural selection, an advantageous new trait would have to be preserved and passed on undiluted to future generations. But blending inheritance does not allow for traits to pass undiluted from one generation to the next. Darwin's theory of blending inheritance rendered evolution by natural selection effectively impossible, though for some years he failed to acknowledge the inconsistency.

*Gregor Mendel*

## 2.2 MENDEL ON INHERITANCE

At the same time that Darwin was formulating his theory, an Austrian monk named Gregor Mendel was conducting experiments on how traits are inherited from parents to offspring. Mendel discovered that traits could be lost in one generation only to reappear in a later generation. For example, when he crossed a pea plant yielding wrinkled seeds with one yielding round seeds, all the offspring in the first generation had round seeds. Was the wrinkled trait therefore lost? No. It reappeared in the next generation of pea plants.

Mendel concluded that heredity is governed by "factors" or "particles" (later called genes) passed from parent to offspring. A trait might disappear temporarily, but the gene responsible for the trait remains present within the organism and may be passed to its offspring. When a breeder, for example, causes some characteristic to appear or disappear, this represents neither a true gain nor a true loss. It represents merely the interplay of dominant genes (whose associated traits are manifest) and recessive genes (whose associated traits are, for the moment, concealed). A trait that is seemingly lost may therefore still be present and reappear. Conversely, a trait that seems to appear out of nowhere may not be new at all but simply the expression of a recessive gene that existed all along. When breeders produce new show dogs or meatier cattle, they are in fact merely shuffling genes to bring formerly recessive genes to expression (see figure 2.3).

In crossing different varieties of garden peas and analyzing the results, Mendel inferred six principles of inheritance:

***Figure 2.3*** *Dog breeding produces widely differing varieties of dogs not by adding new genetic material to the gene pool but by selecting smaller sets of genes from the larger and richer store of genetic material already in the gene pool.*

1. The inheritance of traits is determined by (what were later called) genes that act not like a fluid that can be blended but rather like individual particles that retain their identity.

2. Genes come in pairs for each trait, and the genes of a pair may be alike or different.

3. When genes controlling a particular trait are different, the effect of one is observed (dominant) in the offspring while the other one remains hidden (recessive).

4. In gametes (eggs and sperm) only one gene of each pair is present. At fertilization gametes unite randomly, which results in a predictable ratio of traits among offspring.

5. The genes controlling a particular trait are separated during gamete formation; each gamete carries only one gene of each pair.

6. When two pairs of traits are studied in the same cross, they sort independently of each other.

Although Mendel's principles have since been expanded and refined, they still remain basically sound today. It was unfortunate for Darwin that he did not learn about Mendel's work. Ironically, Darwin had a copy of the crucial paper by Mendel outlining Mendel's theory of inheritance in his library. Yet Darwin never read Mendel's paper:

examination of the paper after Darwin's death showed that its pages were bound together and left uncut (formerly, in the manufacture of books, flat printed sheets were folded and sewn together but not cut apart). It is a matter of speculation whether Darwin would have given up blending inheritance in favor of Mendel's theory had he studied Mendel's six principles.

Whereas Darwin was developing a theory of far-ranging organismal change, Mendel was demonstrating that living things are remarkably stable. Perhaps because of Darwin's success in focusing attention on change, Mendel's theory was not taken seriously until the first decade of the twentieth century. In Mendel's theory, genes behave like individual particles and can be inherited essentially without change. When Mendel's work was rediscovered at the beginning of the twentieth century, it was welcomed enthusiastically by Darwinists and integrated into their theory of natural selection—this modification of Darwinism is referred to as *neo-Darwinism.* Mendelian genetics promised to revitalize Darwinism. For instance, it explained how a single new advantageous trait could survive and eventually become dominant in a population.

Yet, Mendelian genetics has proved a mixed blessing for Darwin's theory. On the one hand, it provides the stability necessary for a trait to become established in a population. On the other hand, stability is just what Darwinism doesn't need if evolutionary change is to be so far-ranging as to produce the whole complex tree of life from a single-celled organism. Natural selection has a much easier time of it working with and taking advantage of hereditary factors that are stable (as occur in Mendel's theory). But this very stability stands in the way of these hereditary factors changing sufficiently to induce genuinely novel traits (as required by Darwin's theory).

Darwin proposed a theory of massive evolutionary change. By evolution (or "descent with modification" as he called it), Darwin meant that all organisms trace their lineage back to one or a few common ancestors. In this grand view of evolution, all organisms take their place along a great tree of life. According to Darwin's theory, given vast amounts of time, vast amounts of evolutionary change are possible, far more than can be observed directly.

Mendelian inheritance, by contrast, suggests a much more limited view of evolution. Mendelian inheritance accounts for breeders producing sweeter corn or fatter cattle through the shuffling of existing genes. At the same time, because of the stability of genes, it also explains the inability of breeders to turn corn into another kind of plant or cattle into another kind of animal. What breeders accomplish is diversification within a given species, a limited form of change known as *microevolution.* Mendelian inheritance underwrites microevolution.

In contrast, Darwinism requires wholesale changes in the physical and behavioral characteristics of organisms, the entry of novel information leading to increases in biological complexity, and, ultimately, the origin of novel types of organisms. In other words, it requires a massive form of change known as *macroevolution.* The occurrence of microevolution is not a matter of debate between Darwinists and intelligent design proponents. Microevolution can be observed, and scientists acknowledge it. What is at issue is macroevolution.

Darwin's theory claims that over long enough stretches of time microevolution is just macroevolution. In effect, the theory claims that there is no fundamental difference between these two types of evolution. According to the theory, large-scale changes are merely a gradual accumulation of small-scale changes. The following sections in this chapter examine the genetic basis for this extrapolation from micro- to macroevolution. As we will see, neither Mendelian genetics nor contemporary molecular genetics nor more recent work on evolutionary developmental biology ("evo-devo") supports the neo-Darwinian view that known sources of genetic change facilitate macroevolution.

## 2.3 GENETIC DIVERSITY

Mendel distinguished dominant from recessive genes. In his original experiments with peas, both forms of a gene (known as *alleles*) are present all along, but only one is expressed for a given trait. Not all traits, however, are associated with a single gene. Some are associated with multiple genes. For instance, many genes determine human skin color: a number of dark and light genes working in combination affect this trait's expression. Thus, two hybrid individuals, each with a full complement of light and dark genes, could in principle produce offspring exhibiting the complete range of possible skin colors (see figure 2.4). In such a distribution, the extremes of light and dark would be rarest. Most of the population would center around the average between the extremes. This "blending" of skin colors differs from the mistaken idea of blending inheritance: with skin color, genes are not losing their identity by being blended; rather, they are maintaining their identity, but being combined in different ways.

***Figure 2.4*** *Skin color in humans is determined not by just one but by a combination of genes. Thus, the skin colors of a couple's children may be very different from their own depending on which of their skin color genes are expressed in each.*

If a population with genes for the entire spectrum of skin shades moved into a geographical area where dark skin was an advantage, what might happen? Clearly, natural selection would favor the darker skinned individuals. Thus, the combination of genes producing the darkest color would be established in that geographical area. Would there be a change from one species to another? No. The only change would be in the prevalence of certain gene combinations.

The English sparrow illustrates this point. It was introduced to North America in 1850. Early attempts to establish this species in America were unsuccessful. Eventually, however, the birds obtained a limited foothold in a few localities where they remained relatively few in numbers for several years. Ultimately, they adapted to their new habitat and their population underwent runaway growth (probably due to the widespread use of the horse, which provided a food source of insects that bred in the horse droppings and of grain used for horse feed). Today English sparrows live throughout most of the continental United States.

When samples of the sparrow population were taken from several geographic areas in the United States, the birds from colder climates were, on average, found to be larger and have shorter extremities than those from warmer climates. The ideal sparrow body-type thus seems to vary according to geographic region (see figure 2.5). These differences in size and extremity were long known for distinct species of birds living at different latitudes. But with the sparrow these differences were observed to develop within a single species.

***Figure 2.5*** *The distribution pattern of the English sparrow by body-types illustrates that species adapt to environment as advantageous combinations of genes already present are utilized.*

How does one explain the success of the English sparrow in adapting to different geographic locales within the United States? The answer may lie in the way certain genes express themselves. The first of these transplanted birds—known as the founders—presumably possessed all the genes necessary to produce the range of body-types and sizes observed today. But the founders had yet to develop the specialized combinations of genes that we observe in the American forms of today. This could explain why the sparrows were at first established only in small populations: the sparrows needed time for specialized combinations to emerge before the birds could become vigorous and abundant.

When such combinations eventually did occur, they could have placed individuals possessing them at an advantage, who were then favorably selected for by regional environmental factors. Yet, it was the combination of existing genes that proved advantageous, not the origination of novel genes. Obviously, then, genetic diversity in itself constitutes an advantage to a population. For instance, the diverse range of antibodies (immunoglobulins) in the human gene pool hinders pathogenic bacteria from becoming established in the human population as a whole.

The sparrow example shows that much of the variety we see in species has nothing to do with, as neo-Darwinism supposes, novel small-scale changes in genes. In such examples, the genes themselves are not changing. Rather, they are merely being shuffled around, expressing novel combinations of genes that were already present. Combinations of both expressed and unexpressed genes (in the genetic reserve) can give biological populations adaptive potential. Even among a relatively small population of founders, there is much greater *polymorphism* (i.e., potential diversity in body form) than appears to exist at the beginning.

As another example of polymorphism, consider that humans living in cold climates tend to be bulkier and to have shorter extremities than those living in warm climates. This difference reflects the relationship of surface area to mass. A body will radiate heat more rapidly if its ratio of surface area to volume is high (increased, for example, by a slender body form with less mass and long and skinny appendages). It will conserve heat if its ratio of surface area to volume is low (decreased by a round body form with short stocky appendages).

An Eskimo from Alaska and an African from the Nile region belong to the same species. Yet the contrast in body builds is striking. The long arms and legs of the African, an advantage for radiating excess body heat in hot climates, would be a disadvantage in the arctic, where they would cause excessive cooling and be more susceptible to frostbite than the shorter extremities of the Eskimo. Natural selection favored different body types by adapting the same species to a wide range of climatic conditions. This must not be confused with the emergence of novel genes, much less the transformation of one species into another. It is combinations of fixed genes, not the individual genes themselves, that in this case are subject to natural selection.

If a population is unable to expand into new environments or adapt to changing conditions, it is likely to remain small and may even go extinct. Small population size is a danger to any species. When an organism mates, it contributes a sperm or an egg to the offspring. The gametes contain only half of the organism's genes. Thus, when mating occurs, the partners contribute only half of their total complement of genes (excluding sex-linked alleles). By having a large number of offspring, organisms ensure that most of their genes will be expressed through mating (although only half of their total genetic complement is contributed to each offspring, it is a different half each time).

Consequently, the larger the number of offspring, the greater the number of gene combinations and the greater the percentage of the gene pool preserved. It follows that a low reproductive rate will increase the probability that genetic information will be lost. Such loss of information reduces the polymorphism of the population. If it continues, the species' ability to adapt to changing environments will be lost and the species itself may become extinct. Rare or endangered species often become extinct due to a loss of genetic diversity and an attendant reduction in polymorphism. The best way to achieve genetic diversity is for species in a population to reproduce randomly. This ensures that any adaptive potential residing in a population's genes is preserved.

Intensive breeding may produce interesting and useful varieties, but it tends to diminish the adaptive gene pool of the lineage, leading to increased susceptibility to disease and environmental insults. It also tends to concentrate defective traits through inbreeding. The farther morphology shifts from the species norm (average), the more it produces developmental discordance, stress, and decreased fertility. Such stressed populations tend to rebound toward the species norm, which works against large-scale evolutionary change.

In conclusion, natural selection helps a species to flourish by favoring gene combinations that allow it to adapt to new and changing conditions. When confined to combinations of existing genes, natural selection is therefore a force for preserving rather than transforming species. But that raises a question: does natural selection merely preserve existing genes or does it also help to create new ones (as it must if it is to bring about the novel genetic information required to originate new species)? To answer this question, we turn next to the physical structure of genes.

## 2.4 THE MOLECULAR BASIS FOR GENES AND EVOLUTION

According to Darwin's theory, new features must arise before natural selection can act. Only then can competitive advantages occur, and only then can natural selection act to shape the course of evolution. Macroevolution therefore requires that something in the organism change to enable natural selection to take effect. But what exactly is it that changes in the organism? Following Mendel, biologists focused on genes as the "change factor" driving biological evolution. But what exactly are genes?

Although Mendel laid the groundwork for our modern understanding of genetics, it took another hundred years before science discovered the molecular basis for inheritance. In the 1940s some scientists concluded that DNA was somehow responsible for genetic inheritance. But it wasn't until 1953 that two scientists, Francis Crick and James Watson, discovered the structure of the DNA molecule—the now famous double helix. DNA molecules encode genetic information via four nucleotide bases, which function as letters of an alphabet. The four bases come in two groups: the purines guanine (G) and adenine (A), and the pyrimidines thymine (T) and cytosine (C) (see figure 2.6). DNA molecules consist of two chain-like strands consisting of a sugar-phosphate backbone to which the nucleotide bases are attached. The two strands run in opposite directions and spiral around each other. Weak hydrogen bonds connect complementary paired bases, A to T and G to C.

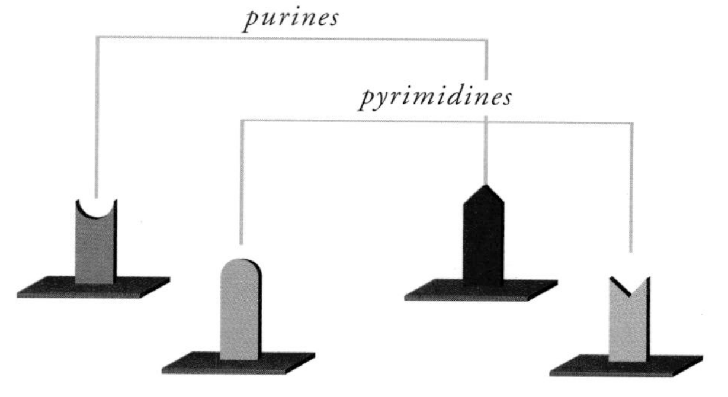


***Figure 2.6*** *Bases. There are two types of bases, purines (guanine and adenine) and pyrimidines (thymine and cytosine).*

Because sequences of DNA are able to code for proteins, it is accurate to describe such coding sequences of DNA as *messages.* In fact, there is a deep connection between sequences of nucleotide bases in DNA that code for proteins and sequences of alphabetic characters written on a page that convey meaning. Just as we can tell the difference between gibberish and meaningful text, so the cell can tell the difference between a random sequence of DNA bases and a message. Former University of London cell biologist E. J. Ambrose makes this point as follows: "There is a message if the order of bases in DNA can be translated by the cell into some vital activity necessary for survival or reproduction."[10]

Most possible sequences of DNA code for no biologically significant structures. Likewise, with human language, most alphabetic sequences are gibberish. Only an extremely small proportion of the astronomical number of alphabetical sequences possible on this page would convey the same message as this one. So too, an extremely small proportion of the astronomical number of possible DNA sequences can code for a functional protein of a given type. Cambridge paleontologist Simon Conway Morris (though not himself a proponent of intelligent design) recognizes this fact and its significance: "Isolated 'islands' provide havens of biological possibility in an ocean of maladaptedness. No wonder the arguments for design and intelligent planning have such a perennial appeal."[11]

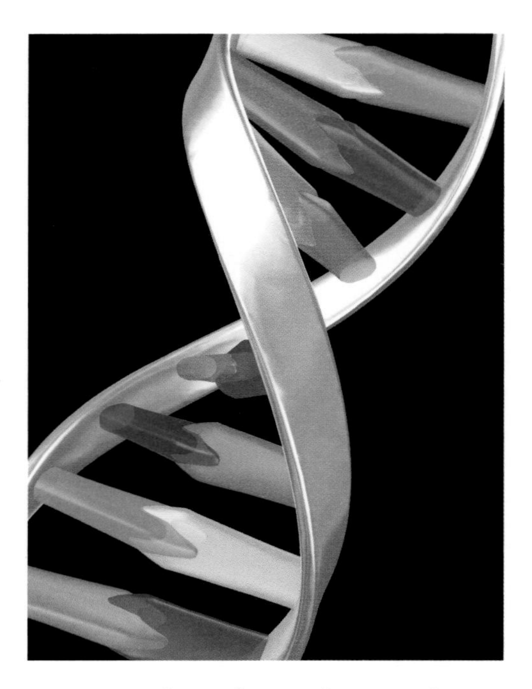

Nonetheless, neo-Darwinists think they have an answer for DNA's language-like ability to code for biologically significant structures, one that does not require intelligent design. Their answer looks to natural selection acting on gene mutations. Mutations are molecular changes in DNA, and they fall into two classes. First, there are point mutations. Here the changes occur in the individual nucleotide bases of the DNA.[12] Second, there are chromosome mutations. These involve not individual nucleotide bases but entire sections of DNA. Chromosome mutations can result in the duplication of a segment of DNA, its loss or recombination at another place on the same or a different DNA molecule, or even its inversion within the molecule.

Point mutations are rare. How rare? It's been known for some time that a gene, on average, changes only once in every 100,000 to 1,000,000 replications (reproductions).[13] Another way to express the point mutation rate is to find out how many gametes contain at least one mutant gene. Studies show a mutant gene will, on average, occur once in every 10 to 100 gametes. While the underlying causes for this basic rate of point mutation are not well understood, the rate can be increased by certain environmental factors such as heat, chemicals, and radiation.

A point mutation in a coding gene can be viewed as a random change in functional information. As a unit of functional information in the cell, a coding gene is much like a word (a unit of meaningful information) in a book. What would happen if the letters in some of the words in this book were randomly changed? Would the book be improved? Generally not. Such changes would in all likelihood decrease rather than increase the meaningful information conveyed by this book. If enough random changes occurred, the result would be gibberish.

Mutations have the same effect in the biological world. Most are harmful, while most that are not harmful are merely neutral, neither helping nor hindering the organism. In fact, except for extreme conditions in which harsh environmental insults enormously increase selection pressure (as when bacteria exposed to antibiotics develop antibiotic resistance), no beneficial point mutations are known.

Such beneficial mutations, however, involve small-scale changes in single protein molecules and provide no evidence for macroevolution. Moreover, when environmental pressure is reduced, the benefit conferred tends to be lost. For instance, antibiotic resistance tends to reduce reproduction rates, so that when the antibiotic is removed, the original bacteria (the "wild type" with higher reproduction rates) reemerge and again dominate the population. This isn't evolution in any full-blown sense. This is one step forward and one step back.

In the overwhelming majority of cases, point mutations in functioning genes are deleterious or even lethal. In other words, they produce structural impairments and genetic diseases or, worse yet, death. As a consequence, most mutations are actually selected against by natural selection. Some lethal mutations, in fact, terminate the development of an organism as early as the zygote stage.

Because of these limitations on point mutations, some neo-Darwinists look to chromosome mutations to bring about macroevolutionary change. They consider gene duplication especially important in this regard. This is because once duplicated, a gene is potentially available to serve some new function. Normally a gene is dedicated to an existing function (such as making a protein or regulating the production of proteins). But if it is duplicated, the duplicated gene is redundant and therefore not required to serve the existing function, which is adequately being carried out by the original gene. The duplicated gene is therefore free from the constraints of selection pressure (i.e., the pressure to perform the existing function) and can go off by itself to explore other genetic possibilities. By "navigating" through "genetic hyperspace," the duplicated gene is thus, according to neo-Darwinists, likely to achieve some new function, which then can provide the basis for macroevolutionary change.

Although this story has a certain initial plausibility, it quickly falls apart when probed.[14] For one thing, when a duplicated gene is decoupled from selection pressure, how it changes is determined entirely by chance. Unlike the original gene, which remains functional, the duplicated gene just sits there, waiting to be transformed into some new functioning gene, whether by other chromosome mutations (e.g., inversion) or by point mutations. But, as we just saw, point mutations are exceedingly rare. Moreover, any additional chromosome mutation to a duplicated gene is merely the further action of chance rearranging and repositioning the gene.

But perhaps we have been focusing unduly on point and chromosome mutations without giving proper due to natural selection. Perhaps, mutations, when suitably sifted by natural selection, can supply the raw material for macroevolution. Let's now consider why skepticism is in order about the claim that mutations are the source of the highly complex structures we find in biology.

## 2.5 THE ADAPTATIONAL PACKAGE

To determine what sorts of genetic changes macroevolution requires, one first needs to be clear what the key feature of biological organisms is that macroevolution must explain. A biological organism is more than the sum of its individual structures. In discussions of biological evolution, this point is often missed because evolution is thought to proceed by cumulating advantages. But organisms are not just bundles of accumulated advantages. An organism's ability to function successfully requires an entire adaptational package, that is, a set of structures that are carefully coordinated with one another to help the organism make a living. The challenge for macroevolution is to bring about such adaptational packages.

An excellent example of an adaptational package is the giraffe. What impresses people most about the giraffe is its long neck. Darwin himself drew attention to the giraffe's neck. In the *Origin of Species* he wrote:

> The giraffe, by its lofty stature, much elongated neck, forelegs, head and tongue, has its whole frame beautifully adapted for browsing on the higher branches of trees. It can thus obtain food beyond the reach of the other *Ungulata* or hoofed animals inhabiting the same country and this must be a great advantage to it. . . .[15]

***Figure 2.7*** *A drinking giraffe. When a giraffe bends its head to the ground to graze or drink, only an adaptational package of sophisticated blood pressure controls keeps the blood vessels in the giraffe's brain from bursting.*

The advantage of the giraffe's long neck for "browsing on the higher branches of trees" is, however, not nearly as obvious as Darwin makes out. Consider that the neck of the female giraffe is two feet shorter, on average, than that of the male. If a longer neck were truly needed to reach above the existing forage line, then the females would have soon starved to death and the giraffe would have become extinct.

Darwin was correct when he called the giraffe "beautifully adapted," but he did not have enough information to appreciate the full extent and refinement of the adaptations. Observe some giraffes eating and drinking in the zoo, and you will notice that they don't just raise their heads to eat leaves high up in trees but also bend their heads to the ground to eat grass and drink water (see figure 2.7). Given their long legs, giraffes could be said to need a long neck less to reach up into the trees (which are not the only source of vegetation in many terrains) than to reach the ground to drink water.

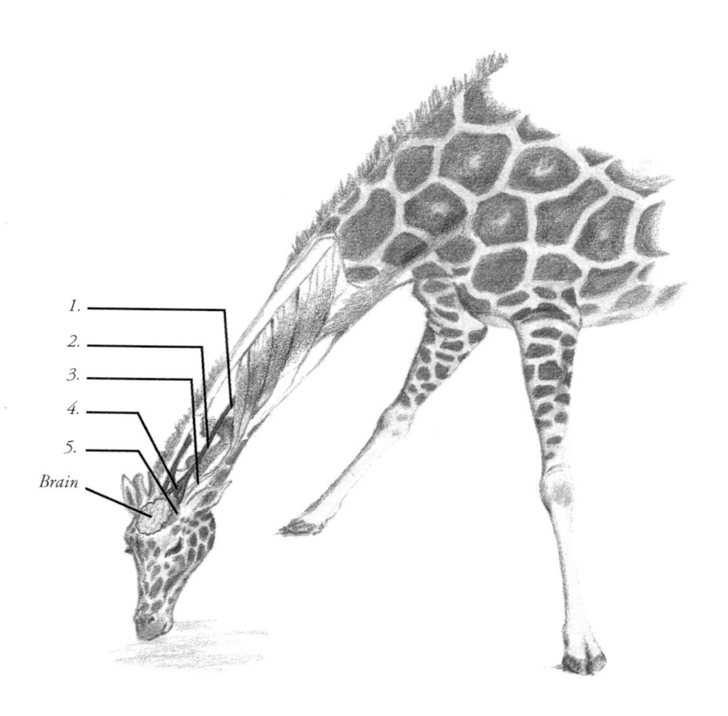


*Figure 2.8* As the giraffe drinks, an adaptational package protects its brain from hemorrhaging. These include: (1) pressure sensors along arteries that detect a rise in blood pressure; (2) increased muscle fiber in artery walls toward the head that allows greater control through artery constriction; (3) heavily valved veins that control the return of blood to the heart; (4) arteries approaching the head branch into the (5) rete mirabile and others that bypass the brain.

The giraffe is an integrated adaptational package whose parts are carefully coordinated with one another. To fit successfully into its environmental niche, the giraffe presumably needed long legs. But in possessing long legs, it also needed a long neck. And to use its long neck, further adaptations were necessary. When a giraffe stands in its normal upright posture, the blood pressure in the neck arteries will be highest at the base of the neck and lowest in the head. The blood pressure generated by the heart must be extremely high to pump blood to the head. This, in turn, requires a very strong heart. But when the giraffe bends its head to the ground it encounters a potentially dangerous situation. By lowering its head between its front legs, it puts a great strain on the blood vessels of the neck and head. The blood pressure together with the weight of the blood in the neck could produce so much pressure in the head that without safeguards the blood vessels would burst.

Such safeguards, however, are in place. The giraffe's adaptational package includes a coordinated system of blood pressure control (see figure 2.8). Pressure sensors along the neck's arteries monitor the blood pressure and can signal activation of other mechanisms to counter any increase in pressure as the giraffe drinks or grazes. Contraction of the artery walls, the ability to shunt arterial blood flow bypassing the brain, and a web of small blood vessels between the arteries and the brain (the *rete mirabile,* or "marvelous net") all control the blood pressure in the giraffe's head. The giraffe's adaptations do not occur in isolation but presuppose other adaptations that all must be carefully coordinated into a single, highly specialized organism.

In short, the giraffe represents not a mere collection of isolated traits but a package of interrelated traits. It exhibits a top-down design that integrates all its parts into a single functional system. How did such an adaptational package arise? According to neo-Darwinian theory, the giraffe evolved to its present form by the accumulation of individual, random genetic changes that were sifted and preserved piecemeal by natural selection. But how could such a piecemeal process, in which mutation and selection act on the spur of the moment with no view to the future benefit of the organism, bring about an adaptational package, especially when the parts that make up the package are useless, or even detrimental, until the whole package is in place? That's the trouble with integrated packages—they are package deals that offer no benefit until the entire package is in place[16] (see figure 2.9).

***Figure 2.9*** *Some parts of the adaptational package fail to arrive on schedule.*

To be sure, random genetic changes might adequately explain changes in a relatively isolated trait, such as an organism's color. But major changes, such as the evolution of a giraffe from an animal with short legs and short neck, would require an extensive suite of coordinated adaptations. The complex circulatory system of the giraffe must appear at the same time as its long neck or the animal will not survive. If the various elements of the circulatory system appear before the long neck, they are useless or even detrimental. This interdependence of structures strongly suggests a top-down design that is capable of anticipating the total engineering requirements of organisms like the giraffe.

The biological literature is filled with examples of adaptational packages. Some organisms, such as arthropods (a group that includes modern crabs and lobsters), even appeared with their adaptational packages intact during the Cambrian explosion. The Cambrian explosion marks the sudden appearance in the fossil record of numerous multicellular animals exhibiting diverse body plans (see Chapter 3). For most of these animals, evidence of fossil ancestors is completely lacking (with but one or two exceptions, there are no known Precambrian precursors). And yet these organisms arrive fully formed in the fossil record as integrated adaptational packages.

As always, microevolution is not the issue here. Moth populations that over generations shift in color from light to dark or mosquitoes that exhibit resistance to DDT are often cited as examples of evolution by natural selection. But such examples only illustrate small changes in the gene frequency of populations. A shift in the dominant moth coloring requires no new genetic information because the *alleles* (variant genes) are already present in the population. In contrast, major changes require major coordinated adaptations, which in turn require impressive amounts of new functional and genetic information. When we fully appreciate the informational requirements for the origin of even a modest new biological structure, much less the origin of a major adaptational package, we can see what a tall order it is for blind mechanisms such as mutation and natural selection to account for them.

According to E. J. Ambrose, selection pressure from the environment is too general for the demands of evolution: "The sort of message which the physical or biological environment can transmit to the organism in the way of new information is an extremely simple one, of the yes or no type such as 'Can I find food higher up the hill or not?'"[17] Simple information like this, however, even when cumulated over time, is not the tightly integrated information needed to coordinate the numerous changes that must occur to build novel complex biological structures and body types. To evolve novel adaptational packages, populations therefore face an information hurdle.

One way to see this hurdle is in the phenomenon of phylogenetic inertia. *Phylogenetic inertia* denotes the tendency of populations to maintain an average morphology as well as a limited degree of variability around the population average. How can mutations overcome phylogenetic inertia to evolve new adaptational packages? It's not clear that they can. Chromosome mutation may exchange parts of gene sequences. But there is no evidence that such "new" genes can provide the steady accumulation of novel traits (to say nothing of their coordination) that natural selection needs for Darwinian evolution to be effective. Chromosome mutation merely reshuffles existing genes.

The only known way to introduce genuinely new genetic information into the gene pool is by mutations that alter the nucleotide bases of individual genes. This is different from chromosome mutation, in which sections of DNA are duplicated, inverted, lost, or moved to another place in the DNA molecule. Point mutations do not merely rearrange but fundamentally alter the structure of existing genes. Such mutations typically result from random copying errors of DNA and are intensified through exposure to heat, chemicals, or radiation.

Could chromosome and point mutations working in tandem provide the raw material for macroevolutionary change? As the primary source of evolutionary novelty in the neo-Darwinian theory, mutations have been studied intensively for the past half century.

*Figure 2.10* *Rapid reproduction and abundant supply make fruit flies, especially* Drosophila, *excellent subjects for experiments that investigate mutations.*

The fruit fly is a case in point. Its genome is easily manipulated and its short lifespan and reproductive cycle allows scientists to observe and track many generations (see figure 2.10). As a result, it has been the subject of numerous experiments. By bombarding it with radiation to increase the rate of mutations, scientists now have a pretty clear idea what kind of mutations can occur.

There is no evidence of mutations in fruit flies creating new structures. Mutations merely alter existing structures. For instance, mutations have produced crumpled, oversized, and undersized wings. They have produced double sets of wings (one set of which doesn't work and thus is deleterious to the organism). But they have not created a new kind of wing. Mutations have also created monstrosities, like fruit flies with legs growing where they should have antennae, a condition known as *Antennapedia* (see figure 2.11). But even such monstrosities merely rearrange existing structures, albeit in bizarre ways. Nor have mutations transformed the fruit fly into a new kind of insect. Experiments have simply produced variations of fruit flies.

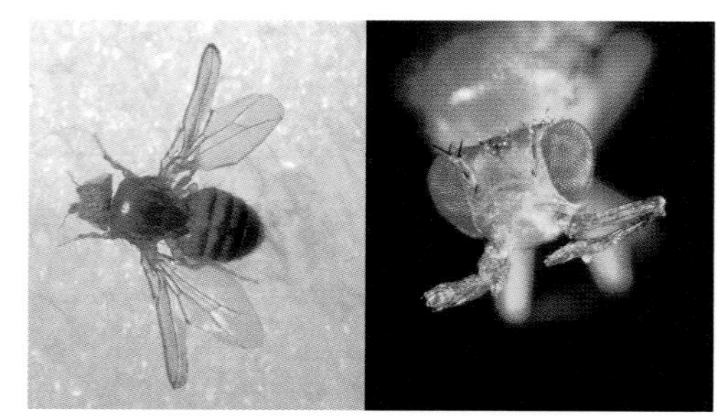

*Figure 2.11* *Four-winged fruitfly and fruitfly with legs growing in place of antennae. These mutants are evolutionary dead-ends—they are not promising intermediates on the way to some new species.*

In conclusion, to generate an adaptational package requires not piecemeal change but integrated, systematic change. Moreover, the source of such change must impart massive amounts of new functional information into an organism. Such information, however, gives no evidence of resulting from the interplay of mutation and selection. Indeed, it gives no evidence of being reducible to matter and energy at all. As Norbert Wiener, one of the founders of information theory, remarked: "Information is information, not matter or energy. No materialism which does not admit this can survive at the present day."[18] Just as the information on this printed page is distinct from the ink and paper that make up the page, so the information in biological systems is distinct from its material constituents. What is the source of the information needed to build adaptational packages? As with the information in written messages and engineered systems, the only source known to be capable of generating information such as we see in biological systems is intelligence.

## 2.6 HOW MANY GENES MUST CHANGE?

Even though mutation and selection give no evidence of generating novel adaptational packages, one can still ask how many genes would have to change for macroevolution to be possible at all. Cell biologist E. J. Ambrose estimated that it "is most unlikely that fewer than five genes could ever be involved in the formation of even the simplest new structure previously unknown in the organism."[19] Ambrose then showed that even with only five genes involved, it would be incredibly improbable that the functional information needed for a new structure could arise by chance mutations. He began by noting the rate of nonharmful mutations (mutations that are either favorable or neutral). By conservative estimates, no more than one new nonharmful mutation occurs per generation in a population of 1,000 (most genes have a mutation frequency smaller that one in 100,000, and most of those mutations are harmful).

The probability, then, of two nonharmful mutations occurring in the same organism is one in 1,000,000. (The probability of two independent events occurring is the product of their independent probabilities; thus 1/1,000 x 1/1,000 = 1/1,000,000.) The odds of five nonharmful mutations occurring in the same individual are one in one thousand million million! (To calculate this probability, multiply 1/1,000 times itself five times; the result is 1/1,000,000,000,000,000.) For all practical purposes, there is no chance that these five mutations could occur within the life cycle of a single organism.

But suppose five nonharmful mutations occurred within the gene pool of a single species (thus in a population of organisms rather than in a single organism). Suppose further that these mutations occurred over time and were preserved in the heterozygous state. Given the recombination potential from extensive interbreeding, couldn't the five new genes eventually come together in a descendant? The Hardy-Weinberg Law states that in the absence of selection or other outside forces, random mating ensures that the proportions of genes in a population remain the same from generation to generation. Accordingly, the proportions of these five mutated genes to their nonmutated counterparts in the rest of the species' population will remain constant.

It follows that the mere production of more offspring doesn't improve the odds of these genes coming together through recombination. Their number will only increase if the individuals carrying these mutated genes are favored by selection or perhaps by the operation of chance in small populations (as with genetic drift; see the next chapter). If the mutations occur in noncoding regions of the genome, then natural selection won't eliminate them since a gene that doesn't code for a trait won't contribute any advantage or disadvantage for survival to the organism and thus won't be selectable.

The neutrality of such noncoding regions would thus give the five mutated genes a better chance to come together through recombination in reproduction. But there would still be huge probabilistic obstacles.

What are the odds that individual organisms carrying these separate genes would find each other among a population of, say, one million? The organisms would need to mate at the proper time and in the right order to combine all five genes in a single individual. Moreover, the resulting new set of five genes would have to code for a truly new structure to increase the organism's level of complexity over its parental species (as required by macroevolution). But such a scenario seems utterly implausible, requiring coincidence piled on top of coincidence.

Even if such an explanation were plausible, there's still more to explain. Suppose our set of five new genes becomes fixed in an individual chromosome, where they are not expressed in the heterozygous state because the complementary genes on the matching chromosome are dominant, thus screening off our five genes and preventing them from being expressed. Suppose, too, that all five are gathered in the same region of the chromosome, which would be highly improbable, though not impossible, by means of mechanisms that break and rejoin chromosomes. If they were indeed located in one area, an additional mutation of another gene could conceivably switch that region from recessive to dominant (acting as a switch gene for this gene cluster).

Could such a convergence of genes on a chromosome bring about the formation of a new structure in an individual organism? Such five- or six-gene clusters are now known to control wing coloration of mimetic butterflies such as *Papilio dardanus* in Africa.[20] Controlling color forms, however, falls far short of originating novel complex structures. No examples of five- or six-gene clusters are as yet known to account for complex structures.

Ambrose did not merely consider the improbability of developing a five-gene cluster of nonharmful mutants. He noted that even the simplest biological structures would require many more than the five genes he considered. He also pointed out that the improbability of getting the right genes into a cluster fades "into insignificance when we recognize that there must be a close integration of functions between the individual genes of the cluster, which must also be integrated into the development of the entire organism."[21] Ambrose therefore concluded that "hypotheses about the origin of species fall to the ground, unless it is accepted that an intensive input of new information is introduced at the time of isolation of the new breeding pair."[22]

This bleak picture, which Ambrose painted over twenty years ago, has not grown brighter in the interim. Evolutionary biologists increasingly look to *co-option* and *coevolution* to resolve the problem that Ambrose raised. According to these hypotheses, evolution is not supposed to require all five (or however many) genes required for a given structure to come into place all at once. Rather, one gene could be formed and have a selective advantage for some structure whose function is distinct from the one being evolved into. Then another gene might emerge that enhances the first gene with respect to another structure serving a still different function. In this way new genes could emerge gradually, be deployed for one function, and then be redeployed (i.e., "co-opted") for another function as an organism's structures and functions gradually evolve in tandem (i.e., "coevolve").

The chief difficulty with this story is that no evidence supports it. As merely a conceptual possibility, it has an initial plausibility. But there is no detailed step-by-step Darwinian pathway by which co-option and coevolution are known to produce any new complex biological structure. To be sure, it does happen that biological structures employed for one purpose are sometimes co-opted for other purposes.[23] But for a complex structure requiring numerous genes, what needs to be explained is not an isolated case of redeployment (the co-option of a single gene) but an orchestrated sequence of redeployments that continually maintains fitness as genes are gradually added and as functions and structures gradually change and converge on the structure and function in question. There is no evidence for such gradual interlocking sequences of redeployments. We shall return to this issue in Chapter 6.

The difficulty here goes beyond merely evolving some new biological structure. The degree of complexity in biological organisms is stunning—far greater than the complexity of gene clusters and the structures they induce. Organisms are organized into a hierarchy of nested systems each composed of multiple interdependent structures. For an organism to function properly, its structures must fit together or integrate within systems that are themselves structural components of higher-level systems. The DNA coding for such systems of structures would require hundreds, if not thousands, of genes. Biological structures never sit in isolation but always need to be coordinated, working together within larger systems that assist the organism in making a living. The Darwinian mechanism shows no ability to effect such coordination.

## 2.7 "EVO-DEVO"

In concluding this chapter, we turn to a recent proposal by Darwinists to account for macroevolution, namely, "evo-devo" (pronounced *EE-voh-DEE-voh).* This term was coined in the 1990s. It stands for *evolutionary developmental biology.* Evo-devo attempts to merge two subdisciplines of biology: evolutionary biology, which studies the mechanisms by which populations of organisms change over generations, and developmental biology, which studies the mechanisms by which individual organisms grow from conception to maturity. Evo-devo takes as its starting point that genetic mechanisms are the key to both evolutionary and developmental biology. The merger of evolutionary and developmental biology, therefore, looks to key genes that influence development and could in principle also influence changes in development and thereby lead to macroevolutionary change.

What if, for instance, a gene that controls development could somehow induce a change early in development? Even a small change early in development might have huge consequences for the organism's anatomy and physiology. Think of an arrow aimed accurately at a target. Left to fly unperturbed, the arrow will land in the target's bull's-eye. Yet the earlier in flight that the arrow is diverted from its trajectory, the wider off the mark it will be when it lands. The promise of evo-devo is that genetically induced changes early in development, though small and easily attainable in themselves, might nonetheless lead to macroevolutionary changes. In this way evo-devo seeks to do an end-run around the more traditional neo-Darwinian approach to macroevolution, which emphasizes the steady accumulation of microevolutionary changes leading to macroevolution. Evo-devo, by contrast, promises rapid evolutionary change at a small cost, namely, the cost of mutating a few key genes that control early development.

Yet despite this initial promise, evo-devo is now in a state of crisis. To be sure, its study of genes that control development continues apace. And the field is making some progress in understanding how genetic developmental mechanisms assist in microevolutionary change (like changes in butterfly eyespots). But William Jeffery, an evolutionary developmental biologist at the University of Maryland, concedes that evo-devo's attempt to understand how developmental genes induce macroevolutionary change, is "at a dead end."[24] The problem is that evo-devo looks to conserved genes (genes that are essentially the same across widely different organisms, often in different phyla) to study how macroevolutionary change might have occurred. But that raises a fundamental problem. Elizabeth Pennisi, in a 2002 report about evo-devo for the journal *Science*, stated the problem this way: "The lists [of conserved genes give] no insight into how, in the end, organisms with the same genes came to be so different."[25]

To understand the problem, let's back up. More than a century ago, biologists observed that parts of some animals occasionally develop like other body parts normally found elsewhere in the organism. For example, the antenna of an insect sometimes develops as a leg (see figure 2.12). Such transformations were dubbed "homeotic" by William Bateson in 1894 to indicate that the affected part has become "like" some other part of the organism.[26] With the rise of modern genetics, such transformations were traced to mutations in "homeotic genes" that specify the identities of certain groups of cells during embryonic development.

So, do homeotic genes provide a key piece of evidence for macroevolution? Simple logic shows that the answer is No.[27] Precisely because homeotic genes are universal, they cannot explain the differences in organisms supposedly due to macroevolutionary change. Here's why: if biological structures are determined by their genes (as supposed by neo-Darwinism), then different structures must be determined by different genes. If the same gene is implicated in structures as radically different as a fruit fly's leg and a mouse's brain, or an insect's eyes and the eyes of humans and squids, then that gene really isn't determining much of anything.

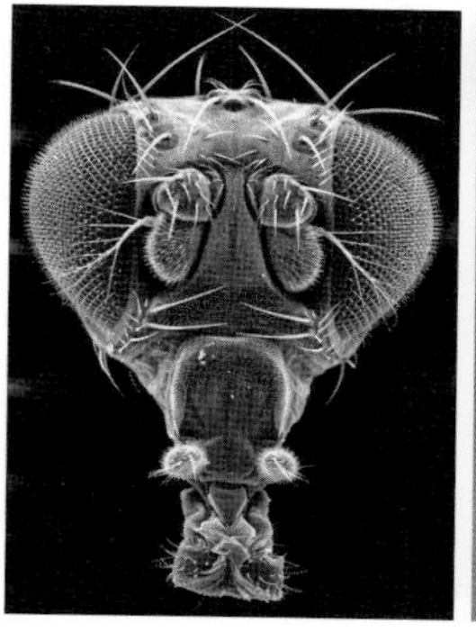
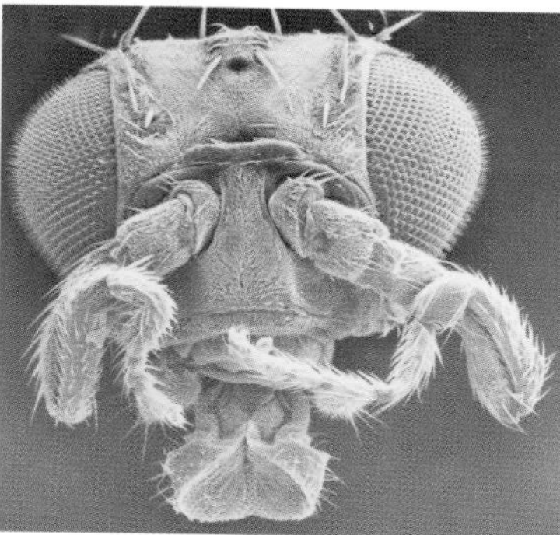

*Figure 2.12 A normal fruit fly and one with legs growing out of the head.*

Consider the analogy of an ignition switch in a vehicle. One might find similar ignition switches in vehicles such as automobiles, boats, and airplanes—vehicles which otherwise are very different from each other. Perhaps, in some sense, an ignition switch can be called a "master control"; but except for telling us that a vehicle can be started by turning on an electrical current, it tells us nothing about that vehicle's structure and function. Similarly, except for telling us how an embryo directs its cells into one of several built-in developmental pathways, homeotic genes tell us nothing about how the actual biological structures are formed. As homeotic genes turn out to be more and more universal, the "control" they exercise in development turns out to be less and less specific.

To sum up, developmental geneticists have found that the genes that seem to be most important in development are remarkably similar in many different types of animals, from worms to fruit flies to mammals. Initially, this was regarded as evidence for genetic programs controlling development. But biologists are now realizing that it actually constitutes a paradox: if genes control development, why do similar genes produce such different animals? Why does a caterpillar turn into a butterfly instead of a barracuda?

In the end, the problem with using developmental biology to underwrite evolutionary biology is that the changes needed in development to facilitate macroevolution simply do not occur. Previously, we used the analogy of an arrow diverted off its course early in its trajectory. The earlier it is diverted, the farther it will be from its intended target. But that's because there's nothing in the arrow to bring it back on course. A developing embryo, by contrast, takes heroic steps to get back on course and reach its developmental endpoint regardless of how early in development it is perturbed. Developmental trajectories are self-correcting in ways that trajectories of arrows are not.

Developmental biologists often study embryogenesis by perturbing embryological development. One of the most remarkable features of developing embryos is their resilience: even with intense experimenter interference, a surprising number of them develop to adulthood. Remarkably, although interference may introduce deformities, the basic endpoint of development never changes. If they survive, fruit fly eggs always become fruit flies, frog eggs always become frogs, and mouse eggs always become mice. Not even the species changes. Every embryo is somehow programmed to develop into a particular species of animal.

Significantly, such programs are not genetic programs. To be sure, genes do play a role in development. But to say that they control or determine development is a vast overstatement. There is now considerable evidence that genes alone do *not* control development. For example, when an egg's genes are removed and replaced with genes from another type of animal, development follows the pattern of the original egg until the embryo dies from lack of the right proteins. (The rare exceptions to this rule involve animals that could normally mate to produce hybrids.) The "Jurassic Park" approach of putting dinosaur DNA into ostrich eggs to produce a *Tyrannosaurus rex* makes exciting fiction, but it ignores scientific fact.

What about mutating the DNA instead of replacing it completely? Using a technique called "saturation mutagenesis," biologists have found that mutations in developmental genes often lead to death or deformity, but they have never produced anatomical changes that benefit the organism. Furthermore, DNA mutations never alter the endpoint of embryonic development; they cannot even change the species. An embryo needs the right genes to make new proteins, and its development suffers without them. But being dependent on genes is not the same as being controlled by them. A house under construction needs suitable building materials, but these materials do not determine its floor plan.

If DNA were in control of development, then you should be able to produce a replica of yourself by putting your DNA in a human egg which has had its own DNA removed. This is the reasoning behind the recent uproar over "cloning." But such a "clone" would not be an identical (albeit younger) copy of you. How it looked would depend in large measure on information in the enucleated egg that received your DNA.

But not even "identical" twins are replicas of each other. They frequently differ somewhat in physical characteristics, and they always differ—sometimes dramatically—in temperament and interests. Yet "identical" twins share not only the same DNA, but also the same egg cell and (usually) the same womb. Even the imperfect similarity exhibited by "identical" twins requires more than the same DNA.

Another piece of evidence against the idea that DNA controls development (as opposed to merely influencing it, which it certainly does) comes from the fact that adult cells contain the same DNA as a fertilized egg. Yet the cells of an adult animal differ markedly from each other in form and function. But if they have the same DNA, why are they so different? Part of the answer is that each cell type utilizes only a portion of its genetic repertoire, with factors outside of the DNA turning on the appropriate genes. But if development requires that DNA be controlled by factors outside of itself, then DNA does not control development.

Why, then, does the view that genetic programs control development continue to be so popular? The answer, to a large extent, depends on its logical connection with neo-Darwinian evolution. Genetic programs are a corollary of the neo-Darwinian synthesis of Mendelian genetics and Darwinian evolution. According to neo-Darwinism, genetic mutations provide the raw materials for evolution, and natural selection modifies organisms through changes in gene frequencies. Development is what turns a single cell into a worm instead of a mouse. Hence, if evolution can change worms into mice by modifying their genes, then it must do so by modifying genes that control development. Conversely, if development is controlled by something other than genes, then evolution must be due to something other than genetic mutations and changes in gene frequencies. Consequently, if the notion that genetic programs control development is false, then so is neo-Darwinism. Neo-Darwinism logically entails the control of development by genetic programs.

If the evidence for neo-Darwinism were compelling on other grounds, the centrality of genetic programs in evolution might be true in spite of the counterevidence described above. But the evidence for neo-Darwinism turns out to be surprisingly thin. Mutations that are supposed to provide the raw materials for neo-Darwinian evolution can do so only if they benefit the organism, and mutations in developmental or homeotic genes are always harmful. In fact, the only mutations that are known to be beneficial are those that affect immediate interactions between a mutated protein and other molecules. Such mutations in DNA can confer antibiotic and insecticide resistance, but they never lead to the large structural changes needed for full Darwinian evolution. Mutations cannot even produce a new species, much less change a fish into an amphibian or a dinosaur into a bird.

The universality of homeotic genes in development is supposed to be due to their presence in a common ancestor, but the preponderance of evidence suggests that the common ancestor lacked the features which those homeotic genes now supposedly control. From a Darwinian perspective, this is a serious problem. According to neo-Darwinism, complex gene sequences gradually evolve by conferring selective advantages on the organisms that possess them. But gene sequences confer selective advantages only if they program the development of useful adaptations. If a primitive animal possessed homeotic genes but lacked all of the adaptations now associated with them, then those genes must have originated prior to those adaptations. How, then, did homeotic genes evolve?

Neo-Darwinism maintains that such genes evolved by encoding primitive adaptations that remain to be discovered, but this is ad hoc speculation. The bottom line is that each new piece of evidence demonstrating the universality of homeotic genes (and thus their independence from any particular adaptation) makes their presence in a supposed common ancestor more difficult for neo-Darwinism to explain. Ironically, the very discoveries that evolutionary developmental biologists until recently have found so exciting are actually adding to the list of difficulties for their theory.

But what if homeotic genes are the product of intelligent design? In that case, finding the same homeotic genes across major divisions of organisms parallels what one finds in ordinary engineering practice: rather than reinvent the wheel, design engineers reuse existing designs. Just as an engineer would not be surprised to find similar ignition switches in different kinds of vehicles produced by the same manufacturer, so biologists who admit design need not be surprised to find similar homeotic genes in different types of animals. Given a design hypothesis, it makes perfect sense why such gene sequences would be incorporated into a wide variety of organisms that otherwise share few structural similarities. To be sure, the design hypothesis needs to be fleshed out before it can provide a detailed explanation of how the design was implemented. But this is not a problem unique to design: the Darwinian account of homeotic gene formation needs as well to be fleshed out before it can provide a detailed explanation of how those genes evolved. Despite this lack of detail in both approaches, intelligent design is more consistent than neo-Darwinism with recent discoveries in developmental genetics.

## EVIDENCE THAT GENETIC PROGRAMS DO NOT CONTROL DEVELOPMENT

1. *Placing foreign DNA into an egg does not change the species of the egg or embryo. (The rare exceptions to this rule involve animals that could normally mate to produce hybrids.)*
2. *DNA mutations can interfere with development, but they never alter its endpoint.*
3. *Different cell types arise in the same animal even though all of them contain the same DNA.*
4. *Similar developmental genes are found in animals as different as worms, flies, and mammals.*
5. *Eggs contain several structures (such as microtubule arrays and membrane patterns) that are known to influence development independently of the DNA. (See general notes.)*

*The general notes on the CD included with this book expand on the material in this chapter and throw light on the following discussion questions.*

## 2.8 DISCUSSION QUESTIONS

1. Discuss Edward Blyth's view of natural selection. Was he correct in thinking of natural selection as a conservative force that preserved organisms and maintained their integrity? Was Darwin justified in extending Blyth's view so that natural selection became not merely a conservative force but rather a creative force for fashioning new organisms and biological structures? Explain your answers.

2. Lamarck's theory of evolution is sometimes characterized by the phrase "the inheritance of acquired characteristics." Describe Lamarck's theory as well as his view of inheritance. What is the problem with Lamarck's theory? What sort of empirical data disprove Lamarck's theory?

3. Describe Darwin's view of blending inheritance. Why does Darwin's view of inheritance sit uneasily with his theory of evolution by natural selection? What must be true about inheritance if Darwin's theory is to be true?

4. Describe Mendel's theory of inheritance. What role do genes play in it? Is Mendel's theory still accepted today? How has Mendel's theory been modified as a result of the work by Watson and Crick on the structure of DNA? Does Mendel's theory support Darwin's theory of evolution, overcoming the problems resulting from Darwin's view of blending inheritance? Or does Mendel's theory raise additional challenges to Darwin's theory? If so, what are these challenges? In particular, does Mendel's theory make organisms too stable or too variable for Darwinian evolution to succeed?

5. How much genetic diversity does the gene pool of a population allow? What is the best way for a population to maintain its genetic diversity? What happens genetically when a small band of organisms breaks away from a population and tries to make a home in a new environment? Describe the case of the English sparrow when it was transplanted to the United States. Do the sorts of changes observed in that case lend support to Darwinism?

6. What is an adaptational package? How is the giraffe an adaptational package? Describe the giraffe's coordinated system of blood pressure control. What's wrong with claiming that the giraffe's neck got longer and longer as selection pressure kept selecting giraffes with longer necks?

7. What types of changes do genes undergo? A lot of genetic change is simply a matter of reshuffling preexisting genes. How do fundamentally new genes emerge? Do fundamentally new genes have to emerge for large-scale evolution to occur, say, for a reptile to evolve into a mammal? How much evolution can occur simply by reshuffling genes? When genes change, do the changes typically benefit or harm an organism? Genetic changes can be harmful, neutral, or beneficial. What proportion fall in each of these three categories? Give examples of harmful, neutral, and beneficial genetic changes. How reliable is the DNA copying mechanism? On the basis of its reliability, should we expect lots of genetic change? If not, what implications does this have for Darwinian evolution?

8. Why is so much genetic research done with fruit flies? What sorts of changes in fruit flies have geneticists observed in the course of experimenting with them? How do geneticists modify the genome of fruit flies? Since genetic research with fruit flies began 100 or so years ago, how much have fruit flies evolved? Does genetic research with fruit flies support Darwinism? (See the general notes.)

9. E. J. Ambrose describes the number of genetic changes that must occur for organisms to evolve novel structures under the supposition that genetic changes are indeed what drives macroevolution. Discuss Ambrose's work. Minimally, how many genetic changes would need to be involved in the evolution of a novel biological structure like an insect's wing? What sorts of changes would these be? How plausible is it that such changes are random and that natural selection can coordinate these changes to produce a novel biological structure?

10. What is "evo-devo" and what does it stand for? What is a developmental gene? In what sense are these genes universal? How does their universality undercut their use in explaining organismal diversity as a result of evolution? Do genetic programs control embryological development? What other factors might influence embryological development? Why is neo-Darwinism committed to genetic programs controlling development? Does developmental biology underwrite evolutionary biology? Why or why not?

# CHAPTER THREE The Fossil Record

## 3.1 READING THE FOSSIL RECORD

Fossils have long puzzled and fascinated us. Fossils used for ornamental or religious purposes have been found in Neanderthal graves. The philosopher Aristotle (384-322 BC) saw life as arising spontaneously from the earth and fossils as a botched attempt in this process. What could account for the shapes of animals and plants entombed in the earth and made of stone? In superstitious times, it might have seemed reasonable to attribute fossils to some mysterious force within the earth itself. But with the rise of modern science, it became clear that once-living plants and animals could turn to stone under the right conditions. If buried quickly enough (before being consumed by decay or scavengers), and if buried with the right mixture of minerals, any plant or animal could become a fossil.

Paleontologists, scientists who specialize in the study of fossils, read the fossil record as a chronicle of life in former ages (see figure 3.1). Skeletons, footprints, leaves, spores, animal tracks, feathers, worm burrows, and even bits of hide can all be found as fossils. By interpreting these clues, paleontologists attempt to reconstruct what living things were like in the past. What stories do the rocks tell? Like many questions in science, the answer depends in part on what background assumptions one brings to the data. If, for instance, one assumes the impossibility of intelligent design because of a prior commitment to materialistic explanations in science, then one will interpret the fossil record as a history of blind material forces operating without goal or purpose.

***Figure 3.1*** *A Trilobite. Second only to dinosaurs as the most widely-known fossils, trilobites are an index fossil, used to identify rocks from the Cambrian Period.*

Science has long been celebrated as immune to the subjectivity that affects other areas of human inquiry. Many people believe that the methodology of science provides a filter that removes the possible distorting effects of an individual scientist's philosophy or values. This ideal of scientific objectivity gained currency because many scientific theories—such as the germ theory of disease, the theory of gravitation, and Mendel's theory of heredity—are theories about the way things operate in the present. Such theories can be checked simply by comparing them with what actually happens. If, for instance, we have a theory about the Moon orbiting the Earth and the Earth orbiting the Sun, we can test it by predicting, say, a solar eclipse and then observing whether it occurs as predicted.

The possibility of such empirical checks is why science is widely considered to be value-neutral or objective. Regardless of their individual philosophies or points of view, scientists conducting the same experiment in the same way expect to get the same results. Many scientific theories describe these kinds of repeatable phenomena, which allow for direct evidence and admit clear proof or disproof. Such theories are independent of past observations. For instance, physical theory tells us that the Earth traces an elliptical orbit. Whether it in fact does so could be settled even if all past records of the Earth's orbit suddenly vanished—astronomers would simply need to make some new observations.

Unlike scientific theories that focus on things happening now, evolutionary theory focuses on singular events in the past. Fossils, in particular, which are so much the subject of evolutionary theorizing, represent singular historical events and not repeatable events such as the motion of planets. Regardless of how life originated in the first place (whether by intelligent design or by spontaneous generation) and regardless of how the giraffe or the aardvark originated (whether by intelligent design or by natural selection), such events are not reoccurring. The past has passed. As the geneticist Theodosius Dobzhansky remarked, "Evolutionary happenings are unique, unrepeatable, and irreversible. It is as impossible to turn a land vertebrate into a fish as it is to effect the reverse transformation."[1] So too, a biological origin by intelligent design might be unique, unrepeatable, and irreversible.

Paleontology is therefore a *historical science.* Like detective work, it looks for clues to reconstruct what might have happened or been the case in the past. Such reconstructions are often referred to as *historical narratives.* The clues paleontologists employ to construct their historical narratives come from fossils and molecular sequences (see Chapter 5). Such data are invariably incomplete and often ambiguous, allowing for multiple explanations and interpretations, though some explanations may be supported by more evidence than others.

Harvard's Ernst Mayr, the dean of American evolutionists until his death in 2005, made this point in his final book, *What Makes Biology Unique?* As he put it:

> A knowledge of history is . . . indispensable for the explanation of all aspects of the living world that involve the dimension of historical time. . . . To obtain its answers, particularly in cases in which experiments are inappropriate, evolutionary biology has developed its own methodology, that of *historical narratives* (tentative scenarios).[2]

Thus, according to Mayr, to sketch the shape of the past paleontologists propose tentative scenarios with varying degrees of plausibility.

Historical narratives, tentative scenarios, plausible hypotheses—these are hardly the raw materials for a rigorous, exact science. Even so, Darwinists treat the fossil record as a trump card. Leaving aside the question of how evolution occurred, Darwinists maintain that the fossil record nonetheless establishes that it occurred. Accordingly, fossils are said to demonstrate that the history of life on Earth is a gradually branching tree in which diverse organisms, over countless generations, meld imperceptibly one into another. In particular, they maintain that the fossil record overwhelmingly confirms large-scale evolutionary changes (macroevolution).

This conclusion, however, assumes that some materialistic form of evolution (like Darwinism) provides the only reasonable interpretation of the fossil record. That assumption is mistaken. Intelligent design also provides a reasonable interpretation of the fossil record. As we shall see in this chapter, it fits well with the empirical data. In particular, it agrees with the fact that certain fully-formed organisms appear all at once in the fossil record and are separated from other fossilized organisms by sizeable gaps.

## 3.2 THE "GRAVEST OBJECTION" TO DARWIN'S THEORY

Darwin proposed that all living things trace their lineage back to one or at most a few original forms. Contemporary Darwinists take Darwin's view to its logical extreme, holding that all organisms trace their lineage back to a single universal common ancestor. In these lineages, one organism gradually transforms into another—there are no sudden changes. According to Darwin, such transformations happened by "numerous, successive, slight modifications."[3]

To illustrate his theory, Darwin used the metaphor of a tree: The trunk represents the common ancestor, and the tips of the branches represent living organisms. If all organisms are modified descendants of a common ancestor, as Darwin thought, then the history of life would form a branching-tree pattern in which all species are part of what he called one "great Tree of Life" (see figure 3.2). Modern formulations of Darwin's theory are based on this assumption of universal common ancestry, or *common descent.*

***Figure 3.2*** *A generalized phylogenetic tree, showing the customary gradual pattern of evolution.*

In holding to common descent, how well does Darwin's theory match up with the fossil record? If Darwin's theory were true and the fossil record were reasonably complete, the fossils should reveal continuous lineages of creatures, with major groups of organisms blending smoothly and imperceptibly into one another via innumerable transitional forms (just as a color wheel blends one color imperceptibly into another). Indeed, the differences separating major groups in taxonomy (such as starfish and birds) are so great that Darwin's theory cannot avoid huge numbers of transitional forms. Darwin himself wrote in *The Origin of Species:*

> By the theory of natural selection all living species have been connected with the parent-species of each genus, by differences not greater than we see between the varieties of the same species at the present day; and these parent-species, now generally extinct, have in their turn been similarly connected with more ancient species; and so on backwards, always converging to the common ancestor of each great class. So that the number of intermediate and transitional links, between all living and extinct species, must have been inconceivably great. But assuredly, if this theory be true, such have lived upon this earth.[4]

Common descent requires that for any two organisms whatsoever, transitional forms connect the two. Moreover, because Darwin's theory is a theory of gradual evolutionary change, the greater the differences between two organisms, the greater the number of transitional forms required to connect the two. An abundance of such transitional forms in the fossil record would provide solid circumstantial evidence on which to build a theory of evolution. At the base of Darwin's great tree of life would be the last universal common ancestor (abbreviated LUCA), the most recent life form ancestral to all extant organisms. Over time, the tree would grow and branch out as new species appear. As more species evolve, clusters of genera would gradually emerge, forming families and eventually orders. As living things continue to branch and diverge, classes and phyla would form. In this way, all the major groupings of organisms would ultimately emerge, some near the base of the tree of life but most higher up in the branches (see figure 3.3).

Yet Darwin had no such fossil evidence. Scientists at the time of Darwin simply had not discovered the "missing links" his theory required. As Darwin noted in *The Origin of Species:* "The number of intermediate varieties, which have formerly existed on earth, [must] be truly enormous."[5] Yet this enormous number of intermediates could not be found in the fossil record, which did not reveal a continuous chain of creatures leading up to fish or reptiles or birds. Darwin conceded this fact: "Why then is not every geological formation and every stratum full of such intermediate links? Geology assuredly does not reveal any such finely graduated organic chain."[6] Indeed, this was, in Darwin's own words, "the most obvious and gravest objection, which can be urged against my theory."[7]

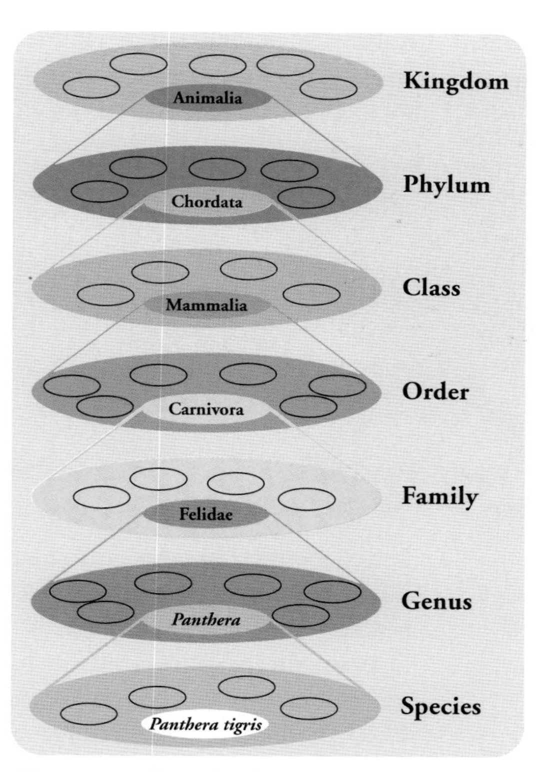


*Figure 3.3 A hierarchy of the major taxonomic groups used for biological classification*

To his credit, Darwin acknowledged that this absence of transitional forms created a serious problem for his theory. Nonetheless, despite this acknowledgment, many scientists continued to find his case for evolution compelling. Perhaps, they reasoned, the missing transitional fossils would eventually be found. In Darwin's day, fossil findings were patchy and the search itself was unsystematic. Darwin himself hoped that the missing transitional forms would turn up as scientists searched more deliberately and systematically. Thus began the search for "missing links."

What did paleontologists find? Many new fossils, to be sure. But what they didn't find were the numerous intermediates that, according to Darwin's theory, had once existed. Rather than plugging gaps in the fossil record, new fossils tended to create new gaps. Granted, a few oddball types did show up that combined features from very different organisms and thus failed to fit neatly within existing categories. Two notable examples are *Archaeopteryx*, an ancient bird with some reptilian features, and the duck-billed platypus, which has a bill like a duck but fur like a mammal (see figure 3.4). Nevertheless, even these oddities tended to fall primarily within one category rather than to share features equally from two or more categories.

Take *Archaeopteryx*. The feathers in *Archaeopteryx* appear to be identical to those in modern birds, having the structure of a genuine airfoil. Yet, in place of the standard adaptational package characteristic of birds, Archaeopteryx has several reptilian features, including a bony tail, teeth in its beak and claws on its wings. The case of the duck-billed platypus is similar. The platypus lays eggs and has a bill like a duck. But except for these bird-like features, it is like other mammals in possessing fur and suckling its young. Taxonomists classify it as a mammal, and it has never been considered a transitional form between birds and mammals. Most proposed missing links are like this—rather than merging taxonomic groups, they fall almost exclusively in one group or another.

***Figure 3.4*** *A duck-billed platypus.*

Today, almost 150 years after the publication of Darwin's theory, we know about thousands of fossil organisms that were unknown to Darwin. But still, the gaps between major groups of animals refuse to close. Ever since the publication of Darwin's theory, paleontologists have puzzled over this overwhelming scarcity of transitional fossils. It was one thing to hope to find missing links in Darwin's day when the science of paleontology was still in its infancy—perhaps scientists simply hadn't searched long or hard enough. But few still defend this explanation. Today, the number of fossils that have been unearthed is staggering, and new ones are being discovered faster than they can be catalogued (see figure 3.5).

The more fossils paleontologists have found, the clearer it has become that they form a pattern at odds with Darwin's theory. The pattern in the fossils is not a gradually branching tree but a collection of nested clusters separated by gaps. Perhaps that

***Figure 3.5*** *Many museums and research centers have back rooms so overstocked with fossils they can wait for years, or even decades, to be analyzed by scientists.*

should not be surprising—it is, after all, the same pattern we see among living organisms today. For example, there are many breeds of horses, but they are clearly separated from cattle. Likewise, there are many varieties of corn, but no one would confuse them with wheat. Varieties cluster around a basic morphological pattern (*morphe* comes from the Greek and means form, shape, or structure) rather than merging smoothly from one form to another.

## 3.3 MAJOR FEATURES OF THE FOSSIL RECORD

Apart from a video recording of life's history (which unfortunately does not exist), the fossil record provides the best evidence on which to base a reconstruction of the history of life. Three features of the fossil record stand out as scientists try to uncover how life began and diversified into its profusion of forms:

1. **The Cambrian Explosion.** A profusion of animal life forms appears in the rocks at the beginning of the Cambrian Period. During this period, the vast majority of the known animal phyla (over 95 percent) appeared within an exceedingly brief interval of geological time (lasting, according to current estimates, a maximum of 5 to 10 million years). Thereafter, apart from a few exceptions, new animal phyla stop appearing throughout the geological record. Phyla constitute the major groups of

animal life forms. They are distinguished by large differences in morphology and basic body plans. Any theory that attempts to account for the emergence of major taxonomic groups within the animal kingdom must explain how such a dramatic range of body plans appeared so abruptly.

2. **Stasis.** Once a life form makes its first appearance in the fossil record, it tends to persist largely unchanged through many strata of rocks. It may exist to the present, exhibiting virtually no change over tens or even hundreds of millions of years. Or it may have a long and unchanging history but then go extinct. This characteristic of organisms to remain largely unchanged throughout their duration in the fossil record is called stasis. Thus, instead of showing one species gradually transforming into another, fossils overwhelmingly exhibit minor variation within a given species. In this respect, the fossil record agrees with what breeders have known all along: breeding may produce interesting and unusual varieties of roses or dogs, but each retains the characteristics that make it a rose or a dog.

3. **Gaps.** Although the fossils appear to follow a rough progression (for example, fish precede reptiles which precede mammals), the fossil record doesn't support the Darwinian claim that the major taxonomic groups are connected to one another by biological descent. There is, for instance, no gradual series of fossils leading from fish to amphibians, or from reptiles to birds. Fossil types are fully formed and functional when they first appear in the fossil record. The earliest known fish fossils exhibit all the characteristics of today's fish. Likewise, reptiles in the fossil record have all the characteristics of present-day reptiles. This pattern holds across the board. There is a conspicuous scarcity of evidence for graded series of in-between fossils. Instead, numerous gaps exist throughout the fossil record.

Let us now consider these features of the fossil record more closely.

Easily the most dramatic feature is the sudden origin of animal phyla in the Cambrian Period known as the Cambrian Explosion. Phyla mark the fundamental divisions between animals. Distinct phyla differ radically in their basic body plans. As a consequence, the emergence of phyla signals where evolutionary modification would have to be most extensive. Yet, the great majority of living animal phyla (roughly thirty—the number varies because scientists disagree on details of how to classify them) appear in a remarkably brief period of time, geologically speaking. Current estimates place their emergence within a 5 to 10 million-year window near the Precambrian-Cambrian boundary. Given the usual 3.8 billion-year history of life, this is a blink of an eye. Even so, the fossil record provides no evidence that any of these extant phyla are connected by evolutionary intermediates.

The Cambrian Period started about 550 million years ago. To appreciate the significance of the Cambrian Explosion, let's back up and consider the fossils that appear prior to the Cambrian. Geologists have found evidence of life far earlier than the Cambrian. For instance, they have found sediments in Africa and Australia that contain fossilized single-celled organisms, which they estimate to be more than 3 billion years old. Sediments only slightly more recent have been found to contain fossils known as stromatolites, which are layered mats of photosynthetic bacteria. Precambrian fossils consist exclusively of single-celled organisms until just prior to the Cambrian.

Multicellular organisms slightly older than the Cambrian (no more than 620 million years ago) were first discovered in the Ediacara Hills of Australia in the 1940s. Since then, Ediacaran fossils have been found in many other parts of the world. The Ediacaran life forms differ substantially from those of the Cambrian (see figure 3.6). Many of them are so strange that it's not even clear if they can properly be regarded as animals. Ediacaran fossils consist of disc- and frond-like organisms. Some paleontologists think they are closer to lichens than to animals. Cambridge paleontologist Simon Conway Morris believes some of the Ediacaran fossils were animals. Yet he also maintains that, with at most a few exceptions, the Ediacaran organisms were not ancestral to the Cambrian phyla.

***Figure 3.6*** *A diorama exhibiting reconstructions of Ediacaran life forms based on fossils. Prominent in this exhibit are jellyfish (Ediacaria flindersi, "fried egg-shaped"; Mawsonites spriggi, blue fluted in center; Kimberella quadrata, bullet-shaped in center); sea pens (Charniodiscus arboreus, pink paddles; Charniodiscus oppositus, orange paddles); flat worms (Dickensonia costata, ovals on substrate); and algae (dark green fingers and filamentous tufts).*

There are three other examples of multicellular animals just prior to the Cambrian:

1. Tiny organisms with shells whose fossils are unlike any other group we know.

2. Trace fossils in which not the organism itself but the traces of its activity (like burrows or tracks) are fossilized.

3. Microscopic, soft-bodied bilaterians (i.e., worms with bilateral symmetry, or mirror-image left and right sides) 40 to 55 million years before the Cambrian.

Except for these examples, and perhaps for a few Ediacaran survivors, there is no fossil evidence of multicellular animals before the Cambrian. The Precambrian fossil record is now well documented. Yet at no point does the transition from the Precambrian to the Cambrian demonstrate the long history of gradual diversification required by Darwin's theory.

Nor do post-Cambrian rocks help much in elucidating the Cambrian Explosion. Paleontologists suspect that if additional animal phyla are found in post-Cambrian rocks, they too will trace back to the Cambrian Explosion. For example, no fossil record of the flat worms (phylum *Platyhelminthes*) is known until later, yet paleontologists believe they must have existed before most of the other phyla. Furthermore, another dozen possible phyla (the classification for these organisms is disputed) of now extinct soft-bodied animals have been found in the Middle Cambrian. But with more fossil evidence, the origin of these extinct groups might even date back to the narrower window of time ordinarily associated with the Cambrian Explosion. During this period, at least forty novel phyla suddenly emerged. Figure 3.7 illustrates how broad-ranging the Cambrian Explosion was. The Cambrian Explosion accounts for nearly all known animal phyla. Paleontologist James Valentine has speculated that some phyla remain to be discovered and estimates that at least sixty phyla originated at this momentous beginning:

> We consider these estimates [of the sixty or so phyla originating during the Cambrian Explosion] to be very conservative. If they are on the correct general order, then the evolutionary events near the Precambrian-Cambrian boundary not only occurred with unexpected rapidity within lineages but involved so many higher taxa as to form an evolutionary explosion without precedent, both rapid in pace and broad in scope.[8]

Not only does the Cambrian Explosion begin abruptly, but it also ends abruptly. The explosion of new animal phyla in the Early to Middle Cambrian is followed by a nearly complete silence—new animal phyla simply stop appearing throughout the

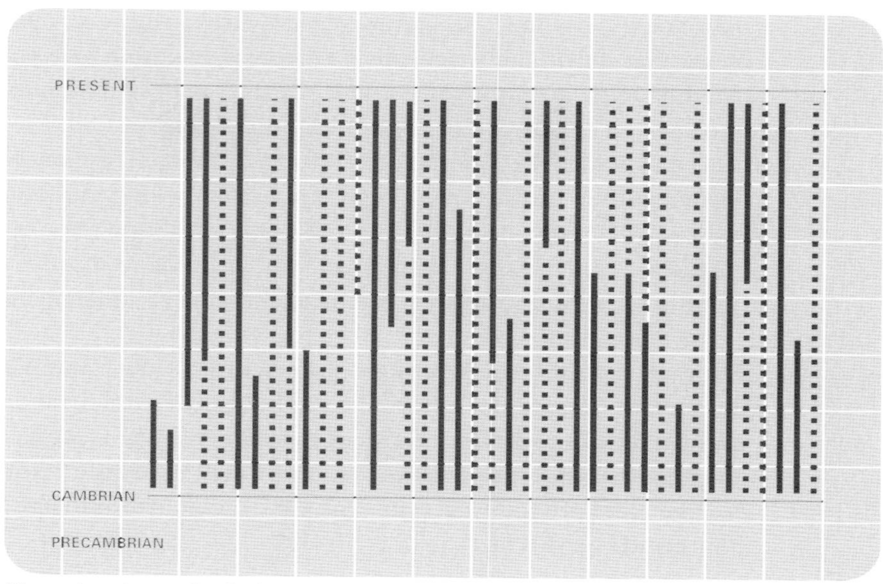


***Figure 3.7*** *A generalized schematic of the fossil record, showing the Cambrian origins of nearly all animal phyla in relation to the overall time scale of the history of animal organisms. Solid lines represent phyla confirmed in the fossil record. Dotted lines represent the presumed existence of phyla yet to be confirmed in the fossil record.*

remaining 500 million years or more of geologic time! In addition, the appearance of new classes (the level of classification just below phyla) drops off almost as dramatically. For instance, over half of the 62 well-defined and easily fossilized marine classes already appear by the close of the Cambrian, and another 29 percent first occur in the Ordovician (the geological period just following the Cambrian around 490 million years ago).[9]

According to Valentine and his coauthors, the number of new orders (the level of classification just below classes) also declines over time: "Appearances of marine orders are somewhat more dispersed but again show a pattern of major innovation during the early Paleozoic [the geologic era that starts with the Cambrian]."[10] This pattern is inconsistent with Darwin's great tree of life, in which the gradual diversification of existing organisms is supposed to precede fundamental disparities between them. Indeed, for higher levels of classification (such as phyla, classes, and orders) the fossil record shows the exact opposite.

The nearly simultaneous appearance of most known phyla becomes all the more remarkable when we consider that the variation within a phylum is quite small compared to the variation between phyla. Distinct phyla differ in their basic body plans whereas organisms from the same phylum share the same basic body plan. As a result, the morphological distance between two distinct phyla is greater than the morphological distance separating representatives within any given phylum. This means that the origin of new phyla represents evolution's greatest achievement in diversifying the forms of animal life. Yet, the origin of new animal phyla is crammed into the first one or two percent of the animal fossil record. Moreover, the fossils needed to bridge the morphological distance between distinct phyla and thereby to document their evolutionary origins are simply lacking. Although the extremely early and isolated appearances of the animal phyla constitute the most dramatic feature of the fossil record, this finding receives little attention in most biology textbooks. The Cambrian Explosion presents a profound challenge to contemporary evolutionary theory.

The same lack of fossil transitions characteristic of the Cambrian Explosion is found throughout the rest of the fossil record. Even if we pool fossil data from all over the world and from rocks irrespective of their proposed ages, we cannot form smooth, unambiguous transitional series. As more fossils have been discovered, the gaps have become more pronounced. University of Chicago paleontologist David Raup has commented on this fact:

> Knowledge of the fossil record has been greatly expanded [since the publication of Darwin's Origin of Species]. . . . Ironically, we have even fewer examples of evolutionary transitions than we had in Darwin's time. By this I mean that some of the classic cases of Darwinian change in the fossil record, such as the evolution of the horse in North America, have had to be discarded or modified as a result of more detailed information.[11]

Likewise, the late Harvard paleontologist Stephen Jay Gould has noted: "The extreme rarity of transitional forms in the fossil record persists as the trade secret of paleontology. The evolutionary trees that adorn our textbooks have data only at the tips and nodes of their branches; the rest is inference, however reasonable, not the evidence of fossils."[12] Gould went on to identify two features in the history of most fossil species:

1. **Stasis**—most species exhibit no directional change during their tenure on earth. They appear in the fossil record looking much the same as when they disappear; morphological change is usually limited and directionless;

2. **Sudden appearance**—in any local area, a species does not arise gradually by the steady transformation of its ancestor; it appears all at once and fully formed.[13]

Another prominent paleontologist, Steven Stanley of Johns Hopkins University, has described the lack of fossil evidence for evolutionary transitions at the genus level: "Despite the detailed study of the Pleistocene mammals of Europe, not a single valid example is known of phyletic [evolutionary] transition from one genus to another."[14]

Thus far we've focused on the fossil record for animals. Yet the fossil record for plants is just as incomplete. Biologist Harold Bold, for instance, has urged scientists interpreting the fossil record to "rigorously distinguish between evidence and speculation." He then added: "At this time there are no known living or fossil forms which unequivocally link any two of the proposed divisions [in the plant kingdom]."[15] Even though that statement was written over thirty years ago, it remains true today.

The rarity of fossil transitional forms for both plants and animals is a vexing problem for Darwinian theory. What do these gaps signify? How should they be interpreted? Four explanations have been offered to resolve this problem:

1. **Imperfect Record.** The gaps result from imperfection of the fossil record. Only a small fraction of the organisms that lived in the past have been preserved as fossils. It is therefore unlikely that future research will fill in the gaps. Support for the theory of evolution must look beyond the fossil record.

2. **Insufficient Search.** The gaps result from a failure to examine the fossil-bearing strata thoroughly enough. The gaps will close as a more complete sampling of fossils is taken.

3. **Punctuated Equilibrium.** The gaps result from evolution being a saltational or "jerky" process. Darwin was wrong about evolution being smooth and gradual. Evolution happens rapidly in small, isolated populations, punctuating an otherwise stagnant equilibrium. Although transitional forms must have existed, few, if any, get preserved.

4. **Abrupt Emergence.** The gaps are real and reflect fundamental discontinuities in nature. Basic groups of organisms emerged suddenly. Transitional forms connecting these basic groups never existed.

Let's examine each of these explanations in turn.

## 3.4 IMPERFECT RECORD

The imperfection of the fossil record is the most popular explanation for the rarity of fossil transitional forms. This was Darwin's preferred explanation. As he put it: "The explanation [of the gaps] lies, as I believe, in the extreme imperfection of the geological record."[16] But how imperfect is this record really? Certainly, many organisms in the history of life died and decayed without ever being fossilized. Yet, in attributing imperfection to the fossil record, our concern is not with the sheer numbers of individual organisms that didn't make it into the record, but with the degree to which the fossil record is representative of the different types of organisms.

Given different types of organisms, the important thing for the fossil record is that each type be represented in it. If it's a question of fossilizing numerous instances of only a single type of organism or fossilizing only one instance for each of many different types of organisms, paleontologists prefer the latter. Seen in this light, the fossil record becomes less imperfect to the degree that it preserves increasingly many different types of organisms (and not to the degree that it repeats instances of the same organism).

How imperfect is the fossil record? Looked at not in terms of the sheer numbers of organisms that failed to be preserved, but in terms of the different types of organisms in fact preserved, and how representative they are, the fossil record actually looks quite good, at least in drawing the broad contours for the history of life. How do we know this? It's a simple matter to check how many known living types of organisms are represented in the fossil record. The percentage of living types recovered as fossils then gives a good indication of the degree to which different types of organisms have been preserved in the fossil record generally.

For instance, among 43 known living orders of terrestrial vertebrates (the level of classification just below classes and phyla), 42 have been found as fossils. Thus, 98 percent of extant terrestrial vertebrates at that level of classification were fossilized. It is therefore a good bet that if there were other orders of terrestrial vertebrates, they too would have been fossilized.

As one moves from broader to more specific levels of classification, the percentage of organism-types preserved in the fossil record gets smaller but is still significant. For instance, among 329 known living families of terrestrial vertebrates (the level of classification just below orders), 261 have been found as fossils—that's a fossilization percentage of almost 80 percent. If one removes birds (which tend to be poorly fossilized), then among 178 known living families of terrestrial vertebrates that are not birds, 156 of them have been found as fossils—which raises the fossilization percentage to 88 percent. A fossilization percentage of 66 percent at the level of genera is not uncommon.[17]

Similar analyses can be done with invertebrates as well as with plants. By looking to the percentage of living organisms at a given level of classification preserved as fossils, paleontologists get a benchmark for the percentage of organisms (living or extinct) preserved as fossils generally. The evidence here is clear and points overwhelmingly to the fossil record as being a faithful preserver of organisms at higher levels of classification.

To hold to a gradual form of evolution in the face of this evidence requires some additional assumption. Thus, one might assume that in times past the process of fossilization was less representative in preserving extinct rather than extant organisms. Alternatively, one might assume that geological processes have systematically wiped out the record of extinct organisms while preserving the record of extant organisms. But all such assumptions are ad hoc and unsupported by evidence. Our best evidence suggests that the process of fossilization has remained largely unchanged throughout geological history and that fossils themselves are extremely durable.

The absence from the fossil record of transitional forms connecting organisms at higher levels of classification is therefore evidence that no such transitional forms ever existed in the first place. If, for instance, two distinct animal phyla share a common ancestor (as they must if Darwin was right about life being a great tree in which all living forms are descended from a universal common ancestor), then there must be numerous distinct orders and families serving as transitional forms to connect the two. Moreover, given the high fossilization percentages of extant organisms at the level of orders and families, most of those transitional forms should have representatives in the fossil record. Yet, for the phyla of the Cambrian Explosion, the transitional forms are nowhere to be found. But is it that they are nowhere to be found in the sense of never having existed in the first place, or is it that we just haven't looked hard enough to find them?

### 3.5 INSUFFICIENT SEARCH

This brings us to the next explanation for the scarcity of fossil transitional forms—*insufficient search*. To claim that insufficient search explains the scarcity of fossil transitional forms is to accept that the fossils preserved throughout the earth are reasonably complete (at least at higher levels of classification), but that scientists simply have not expended sufficient time and effort to find them. In other words, the fossils are there all right, waiting to be found; scientists just need to get busy.

This explanation may have seemed plausible in Darwin's day, but today it has worn exceedingly thin. As already noted, all the main invertebrate types appear abruptly in the early Cambrian. Despite an enormous effort by paleontologists over the last century and a half to connect these organisms by transitional forms, the links have not been found. This lack of progress should give us pause. Darwin knew about the Cambrian rocks. A relatively small investment of paleontological effort was enough to discover the Cambrian Explosion. And yet, a huge subsequent investment of paleontological effort has yielded no insight into the supposed evolutionary precursors to the Cambrian Explosion. Michael Denton elaborates:

> In Darwin's time no fossils of any sort were known from rocks dated before six hundred million years ago, but since then fossils of unicellular and bacterial species have been found in rocks dating back thousands of millions of years before the Cambrian era. Also, several new types of organisms which were not known one hundred years ago have been discovered in the Burgess Shale and at Ediacara, in rocks of Cambrian and late pre-Cambrian age: however, none of these discoveries have thrown any light on the origin or relationships of major animal phyla. . . . Newly discovered hitherto unknown groups, whether living or fossilized, invariably prove to be distinct and isolated and can in no way be construed as connecting links in the sense required by evolution theory.[18]

Insufficient search is therefore an increasingly implausible explanation for the scarcity of transitional forms among the known fossils (especially at higher levels of classification such as the animal phyla). To be sure, one can always argue that it is impossible to conduct an exhaustive search of the earth's rock strata, and thus one can hope that the missing links will eventually turn up. But exhaustive searches confined to individual locales have been conducted in sediments where such transitional forms were thought to be most likely. Yet, even in these most favorable circumstances, paleontologists have consistently failed to discover the missing links that would connect organisms from higher levels of classification. It is fair to say that paleontologists have engaged in heroic efforts to find missing links, searching through thousands of feet of sediment and sorting through tons of hard rock (not just sandstone or shale but quartzite needing to be split into thin slabs).

So neither imperfection of the fossil record nor insufficient search provides a plausible explanation for the scarcity of transitional forms among known fossils. Michael Denton puts his finger on the problem:

> The fundamental problem in explaining the gaps in terms of an insufficient search or in terms of the imperfection of the record is their systematic character—the fact that there are fewer transitional species between the major divisions than between the minor. . . . And this rule applies universally throughout the living kingdom to all types of organisms, both those that are poor candidates for fossilization such as insects and those which are ideal, like molluscs. But this is the exact reverse of what is required by [Darwinian] evolution. Discontinuities we might be able to explain away in terms of some sort of sampling error, but their systematic character defies all explanation.[19]

It appears, then, that the fossil record is not nearly as imperfect as Darwinian evolution suggests it would have to be to account for the missing evidence. Moreover, the search of the fossil record has now been extensive, so there is little reason to expect that some new batch of fossils will suddenly turn up and overthrow the longstanding pattern of gaps between major divisions of fossilized organisms. To be sure, filling gaps between minor divisions continues to show some progress, as recent fossil finds for whales, turtles, and elephants make evident. But far from confirming Darwin's view of gradual evolution, such finds actually undercut it, because for every link connecting minor divisions there should be hundreds connecting major divisions.

## 3.6 PUNCTUATED EQUILIBRIUM

It appears, then, that the only way to salvage Darwin's theory is to concede that Darwin was mistaken when he taught that the evolutionary process is smooth and gradual. Instead of proceeding by creeps and crawls, evolution can also proceed by jerks and jumps, especially when bridging major divisions between organisms. Accordingly, we should expect gaps in the fossil record because living things evolve so quickly that few if any transitional forms get preserved.

Darwin represented evolution as a gradually branching tree of life (recall figure 3.2 earlier in this chapter). An alternative view, which sees evolution as a saltational process, is called punctuated equilibrium. Originally proposed by Niles Eldredge and

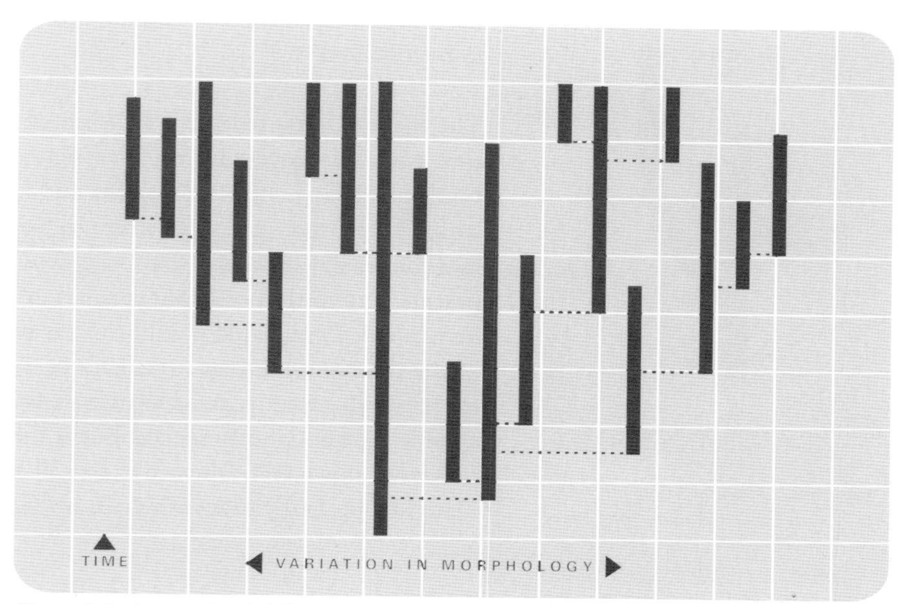


***Figure 3.8*** *A generalized phylogenetic tree, representing a punctuated pattern of evolution. The dotted lines represent unobserved speciation events.*

Stephen Jay Gould, the theory of *punctuated equilibrium* represents evolution as a tree whose branches proceed mostly straight up but at key moments suddenly jut out to the side (see figure 3.8). The vertical columns in this figure represent stasis—the fact that species are relatively stable during long periods of time (this is the "equilibrium" in "punctuated equilibrium"). By contrast, the dotted lines connecting major groups of organisms represent evolutionary changes too quick to be recorded in the fossil record (this is the "punctuation" in "punctuated equilibrium").

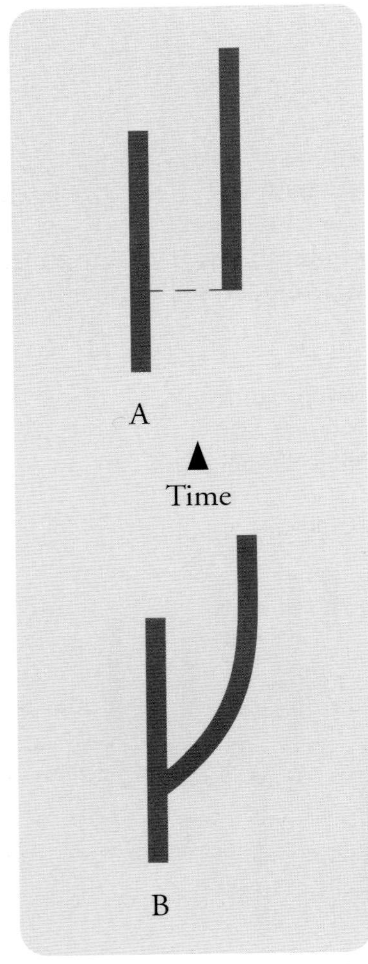


***Figure 3.9*** *The branching process of the tree of life as viewed by traditional evolutionists (A) and the branching process of the tree of life as viewed by proponents of punctuated equilibrium (A) and by traditional evolutionists (B).*

According to the theory of punctuated equilibrium, even though stasis and stability are the norm, in small populations this equilibrium is occasionally punctuated by rapid evolutionary change. New species are thus supposed to evolve in isolated locations, separated from the main population. There they supposedly originate in a geological blink of an eye lasting just a few thousands or tens of thousands of years (as opposed to the millions of years posited by traditional Darwinism). Punctuated equilibrium describes a more haphazard form of evolution than traditional Darwinism (with its gradualism in which organisms steadily accumulate advantageous traits). In the theory of punctuated equilibrium, a small population will evolve through several transitional "trial" species (represented by the short horizontal dotted lines in figure 3.9) before establishing a new species capable of developing into a stable population likely to be fossilized.

The theory asserts that such transitional species did at one time exist. Moreover, during those periods of rapid evolution, no drastic change between parents and immediate offspring would be evident (a cat, for instance, would not give birth to a mouse or vice versa). Offspring would still resemble their parents, and differences from generation to generation would be gradual. Punctuated equilibrium is non-Darwinian only in the limited sense that organisms at the start and end of (geologically) brief periods of intense evolutionary activity exhibit substantial differences. But if the fossil record preserved not merely all species but all individual organisms, Darwin's gradually branching tree of life could be recovered.

In the theory of punctuated equilibrium, transitional species that arise during bursts of evolutionary activity leave no fossil trace because of their small numbers and because of the short time during which they were transitioning. Punctuated equilibrium therefore predicts that the fossil record is imperfect and offers a rationalization for why it should be so: because transitional species were rapidly evolving to fill an environmental niche, most of them were "trial" species that didn't stick around very long. Other species were constantly evolving from them and outperforming them. As a consequence, most of these "trial" species quickly went extinct. In the end, only one or two new species, significantly different from the original species, could successfully fill an environmental niche. Only these flourished, endured, and were fossilized.

Whereas Darwinian gradualism views the pace of evolution as slow and steady, punctuated equilibrium sees the pace of evolution as speeding up on infrequent occasions but then settling back to a slow tempo for the majority of the time. Punctuated equilibrium is Darwinian evolution walking to the beat of a different drummer, but it is still Darwinian evolution. Punctuated equilibrium wants the best of both worlds. It wants the rapid changes that Darwin thought could not be reconciled with his theory, but at the same time it wants a mechanistic account of evolution consistent with Darwinism. It does this by cramming what amounts to a gradual form of evolution (when considered on a generation by generation basis) into periods so brief geologically that no record of such evolutionary bursts is left.

Yet, without an empirically confirmed material mechanism capable of accounting for these bursts in evolutionary activity, the theory of punctuated equilibrium finds its support not in any positive evidence but simply in the silence of the fossil record. Indeed, there is a deep irony that punctuated equilibrium finds its main evidential support in predicting the absence of transitional fossil forms. Punctuated equilibrium therefore does nothing to change the fact that no material mechanism proposed to date adequately accounts for the origin of the functional information present in all organisms. Punctuated equilibrium rightly notes that sudden appearance and stasis characterize the fossil record. But in interpreting the history of life as a history of short evolutionary bursts followed by long periods of evolutionary stagnation, the theory has exactly the same mechanisms at its disposal that neo-Darwinism does. Indeed, punctuated equilibrium is a variant of neo-Darwinism and not a fundamentally new theory in its own right.

## Gould's Final Capitulation to Darwin

Stephen Jay Gould, the co-inventor with Niles Eldredge of punctuated equilibrium, was throughout his varied career prone to make remarks critical of Darwinism, especially in the way it was uncritically received and taken over by Darwin's contemporary disciples. In fact, Gould intended the theory of punctuated equilibrium as a correction to certain faults that he perceived in the traditional understanding of Darwinism. Nevertheless, prior to his death in 2002, Gould conceded that he thought Darwin, in identifying natural selection as the prime mechanism of evolution, had hit the nail on the head.

The last of Gould's many books was a mammoth 1400-page volume titled *The Structure of Evolutionary Theory.* Serving as a definitive statement of Gould's contribution to evolutionary theory, it stated plain-as-day that when it comes to explaining the origin of "organized adaptive complexity," or what he also called "the intricate and excellent design of organisms," there is no alternative to natural selection.[20] Gould elaborated:

> I do not deny either the wonder, or the powerful importance, of organized adaptive complexity. I recognize that we know no mechanism for the origin of such organismal features other than conventional natural selection at the organismic level—for the sheer intricacy and elaboration of good biomechanical design surely preclude either random production, or incidental origin as a side consequence of active processes at other levels.[21]

Gould repeated the same point later in the book:

> [W]e do not challenge either the efficacy or the cardinal importance of organismal selection. As previously discussed, I fully agree with Dawkins (1986) and others that one cannot invoke a higher-level force like species selection to explain "things that organisms do"—in particular, the stunning panoply of organismic adaptations that has always motivated our sense of wonder about the natural world, and that Darwin described, in one of his most famous lines (1859, p. 3) as "that perfection of structure and coadaptation which most justly excites our admiration."[22]

Interestingly, in the two decades prior to his death, Gould had been at odds with Oxford evolutionist Richard Dawkins over natural selection's primacy in evolution. Their debate was so sharp that Kim Sterelny, the editor of the journal *Biology and Philosophy*, even wrote an entire book on it titled *Dawkins vs. Gould: Survival of the Fittest.*[23] Dawkins would be right to regard Gould's remarks in his last book as *full surrender*. Gould admits that anything Dawkins really cares about regarding biological structures—their origin, function, complexity, adaptive significance—is the product of natural selection. Gould was as much a Darwinist as Dawkins.

## 3.7 ABRUPT EMERGENCE

There is a fourth option for explaining the gaps in the fossil record. Besides imperfection of the record, insufficient search, and punctuated equilibrium, there is also abrupt emergence. To explain the gaps in the fossil record by means of abrupt emergence is to say that the gaps are real—that the discontinuities in the fossil record represent discontinuities in the history of life. Abrupt emergence isn't just saying that transitional links connecting major groups of organisms are absent from the fossil record. It's saying that the transitional links never existed.

Abrupt emergence is the face-value interpretation of the fossil record. It provides a straightforward and parsimonious explanation for the absence of fossil transitional forms. Even so, it immediately raises a further question: How did novel biological forms emerge in the first place (whether abruptly or otherwise)? Materialistic theories of biological evolution propose material mechanisms that supposedly can bring about novel biological forms. Invariably, the problem with such mechanisms (notably, the Darwinian mechanism of random variation and natural selection) is that on theoretical and empirical grounds they do not appear causally adequate to bring about the functional information present in biological systems.

The scarcity of transitional forms in the fossil record (and especially their overwhelming absence in linking the major taxonomic groups such as phyla) bears out this deficiency in current evolutionary theorizing. Indeed, if such mechanisms were causally adequate to drive full-scale biological evolution, then transitional forms would have existed and been candidates for fossilization. This is true even for the punctuated equilibrium model: it may explain why species- and genera-level transitions are absent from the fossil record, but it does not explain why higher-level transitions, such as at the level of phyla, are absent (since even a punctuated equilibrium form of evolution would preserve higher-level transitions in the fossil record).

Previously, we attempted to explain the absence of transitional forms in the fossil record under the assumption that they once existed. Now let's turn the question around. Suppose that abrupt emergence is a fact not just about the fossil record but about the history of life itself—in other words, transitional forms between major groups of organisms never existed. In that case, what exactly happened to bring about biological forms in all their vast complexity and diversity? In asking this question, we are not asking for the underlying cause or causes of biological complexity and diversity. Rather, we are asking what a video camera would see if it were scouring the past and recording key events in life's history. There are four possibilities:

1. **Nonbiogenic formation.** Organisms form without the direct causal agency of other organisms. In place of life begetting life, here we have nonlife begetting life.

2. **Symbiogenic reorganization.** Organisms emerge when different organisms from different species come together and reorganize themselves into a new organism.

3. **Biogenic reinvention.** Organisms reinvent themselves in midstream. At one moment they have certain morphological and genetic features, at the next they have a vastly different set of such features.

4. **Generative transmutation.** Organisms, in reproducing, produce offspring that are vastly different from themselves.

None of these possibilities is as bizarre as it first seems. Nonbiogenic formation has happened at least once, namely, at the origin of life. Symbiogenic reorganization has been the focus of Lynn Margulis's work, and there is some evidence for it. Many "organisms" are really a combination of organisms. For instance, lichens that we see growing on rocks and trees actually constitute a merger of a fungus and a green alga (which is photosynthetic). Keep the lichen in darkness, and only the fungal part will survive. Place the lichen in water, and the fungus will drown and the green alga will grow and flourish.

Biogenic reinvention (organisms changing in midstream) is also not that farfetched when one considers the life cycles of certain organisms that from one stage to the next are completely unrecognizable. Consider, for instance, the metamorphosis of a butterfly. If you didn't know that the caterpillar constituted a stage in the butterfly's lifecycle, you would think it was a completely different organism. The caterpillar, as we might say, reinvents itself. This sort of reinvention, however, happens within the lifecycle of a single organism. But what if, instead, it happened over a lineage of organisms, so that up to a certain point in the lineage the organisms have certain features, and then after that point they take on vastly different features? For now, there is little evidence for this possibility, but it is also one that remains largely unexplored.

Finally, there is generative transmutation. Generative transmutation is essentially a type of biogenic reinvention. In this case, the organism reinvents itself at the time of reproduction by producing offspring very different from itself. Thus, instead of reinventing itself by changing within its own lifecycle (as in biogenic reinvention), the organism reinvents its offspring. Berkeley geneticist Richard Goldschmidt considered this possibility in the 1940s, calling the transmuted offspring a "hopeful monster."

What, then, are we to make of these four possibilities—nonbiogenic formation, symbiogenic reorganization, biogenic reinvention, and generative transmutation? All four suggest a programmed view of abrupt emergence, where, like a computer program that kicks in at a certain time (recall the Michelangelo computer virus that kicked in on March 6, 1993), organisms change suddenly in one generation. French paleontologist Anne Dambricourt Malassé, for instance, has argued that generative transmutation (she calls it "dynamic ontogenetic determinism") gave rise to *Homo sapiens* and that it was programmed to occur cross-species at a given time.[24] Needless to say, programming points to a programmer and therefore to design.

How, then, are we to make sense of these four possibilities in light of intelligent design? Certainly, these possibilities are not mutually exclusive; they could all operate in natural history. What's more, if any of these possibilities did occur, they would explain the absence of transitional forms between major groups of organisms in the fossil record. In this case, the fossils would be missing because the transitional forms never existed in the first place. Nevertheless, these four possibilities are extremely difficult to square with a conventional evolutionary account of life's origin and subsequent development. We know of no material mechanism that could account for the drastic changes in organisms that three of the possibilities (symbiogenic reorganization being the exception) require. Apart from intelligent design, these three possibilities all seem to require vast doses of exceedingly good luck. Indeed, one might equally well hope that a tornado whirling through a junkyard will assemble a Boeing 747, to use an analogy by astronomer Fred Hoyle.

Let's consider these three possibilities more closely. Nonbiogenic formation is asking for a full organism or biological structure to form from nonliving matter. Here the tornado in the junkyard analogy could not be more apt. The functional complexity associated with living systems is unlike anything seen in the nonliving world, and so to ask for a sudden transition from one to the other without intelligent design is to embrace a "chance of the gaps." It is to abjure the hard work of science and simply invoke chance.

Biogenic reinvention and generative transmutation do not fare any better here, and for the same reasons that the biological community rejected the views of Schindewolf and Goldschmidt. Apart from some intelligent programming of organisms to change at given times and in given ways, there's no reason to think that organisms can undergo massive change within a single generation (like a reptile transforming into a bird).

That leaves symbiogenic reorganization. Several eminent biologists have recently advertised this view as the key to abrupt emergence. If one organism could, for instance, acquire the genomic complement of another, it could undergo massive change instantly. Such symbiogenic reorganization or hybridization has been the focus of Lynn Margulis's work now for many years.[25] Although she provides evidence for many interesting cases of symbiogenic reorganization, she has offered little justification for why this process should constitute a general solution to the problem of abrupt emergence. Symbiogenic reorganization can at best mix existing traits but it cannot create new ones. Genuine novelty is beyond its reach. And yet, genuine novelty is precisely what requires explaining when it comes to abrupt emergence.

Although abrupt emergence is often viewed as a form of miraculous intervention, it is important to see that abrupt emergence is also compatible with a nonmiraculous view of life's origin and subsequent development. As novel organisms and novel biological structures emerge (whether abruptly or not), they are fully capable of having coherent causal histories in space and time. In particular, intelligent design is not committed to the history of life being a magic show in which at key moments—*presto-chango!*—novel biological structures and organisms magically materialize without any causal antecedents. Intelligent design is open to the possibility that organisms and their structures have a fully traceable natural history. At the same time, it is open to the possibility that intelligence was integrally involved in that history and played an indispensable role in guiding it. Moreover, it will regard such guidance as detectable in its effects even if not detectable in the process of emergence.

To understand this last point, imagine Scrabble pieces moving and arranging themselves to spell a coherent English sentence. If we look at the motion of each Scrabble piece individually, we may see no sign of intelligence. If we look at the joint motion of the Scrabble pieces before they stop and spell a coherent English sentence, we may again see no sign of intelligence. Yet, once the pieces stop and spell a coherent English sentence, from the resulting pattern of linguistic information we have no choice but to infer that intelligence was guiding them. Note that in spelling a coherent English sentence, the Scrabble pieces and their motion had a fully traceable history—the pieces did not materialize out of thin air and their motions need not have violated any physical laws. Yet, their final configuration could not be explained without recourse to intelligence. Likewise, in the history of life, novel organisms and biological structures need not materialize out of thin air. They can have a fully traceable natural history. The task of intelligent design is to determine the effects of intelligence within natural history and to distinguish them from the material forces that also operate there.

## 3.8 USING FOSSILS TO TRACE EVOLUTIONARY LINEAGES

Evolution is about connecting dots. The dots are organisms and the connections are lines of genealogical descent. Yet, for the dots to be plausibly connected, they need to be reasonably close together. That's why gaps and missing links in the fossil record constitute a problem for evolution. To be sure, Darwinists do not regard such discontinuities as setbacks that threaten to overturn their theory but as opportunities for vindicating the theory once the missing intermediates are found. Consequently, whenever a presumed intermediate is found, it is regarded as a triumph for evolutionary theory.

According to some Darwinists, the greatest triumph in this regard is the mammal-like reptiles. According to Douglas J. Futuyma's standard textbook on evolutionary biology,

> The origin of mammals, via mammal-like reptiles, from the earliest amniotes [a broad class of vertebrates comprising reptiles, birds, and mammals] is doubtless the most fully, beautifully documented example of the evolution of a major taxon. In no other case are as many steps preserved in the fossil record. The origin of mammals definitely refutes creationists' claims that fossils fail to document evolution. It also shows how characteristics can become highly modified to serve new functions.[26]

Futuyma then describes a sequence of supposed transitional forms that, over the course of 150 million years, exhibit gradual changes in skeleton, anatomy, locomotion, digestion, feeding habits, and general way of life. He claims that these changes are readily traceable in the fossil record and occur at a fine level of detail. Notable here are bones in the reptilian jaw that are said to transform gradually into the delicate bones of the mammalian ear.

But things are not as clear-cut as Futuyma claims. According to Darwinists, mammal-like reptiles called therapsids played a key role in the class-level transition from reptiles to mammals. Among the several therapsid lineages were the dominant land-dwelling vertebrates from the middle of the Permian period to the middle Triassic (see figure 3.10). Evolutionary biologist James Hopson describes a series of eight therapsid skulls that made up a sequence of intermediate types, supposedly leading to a ninth, an early mammal named *Morganucodon*.[27]

Hopson detailed several characters in the series that appear to progress together toward the mammalian body plan. These include: (1) change in the way the limbs are connected; (2) increased mobility of the head; (3) fusing of the palate; (4) improved musculature of the jaw; and (5) migration of the articular and the quadrate bones from the back of the reptile's jaws toward the middle ear (where in the mammal they would be transformed into auditory ossicles). The simultaneous transformation of these several traits is supposed to demonstrate that the therapsids constitute a continuous lineage to the mammal. (Of course, the fossils don't record the potentially vast differences in systems such as the reproductive and circulatory systems, or the organs, glands, and other soft tissues they entail.)

As with all such arguments, however, what Hopson actually presents is a structural series—not a lineage. He conjectures that "the series of mammal-like reptiles ordered on the basis of morphology will also form a series in geologic time."[28] But there are several inconsistencies. The first three of Hopson's therapsids are contemporaries from two separate orders. What's more, evolutionary biologists dispute whether some of the therapsids Hopson lists are in fact mammalian ancestors. Also, there's the problem of getting structural similarities to match up with temporal ordering. Indeed, lining up fossils structurally so that transitions are as smooth as possible tends to throw off temporal ordering. For instance, the fourth of Hopson's therapsids is more recent than the fifth, and the final therapsid is more recent than the mammal (Morganucodon) that is supposed to be its descendant. One way around this is to assume that structural predecessors are in fact temporal predecessors, but that the fossil record (because of incompleteness or insufficient search) fails to indicate when the organisms in question first emerged. Such an assumption is, of course, entirely ad hoc.

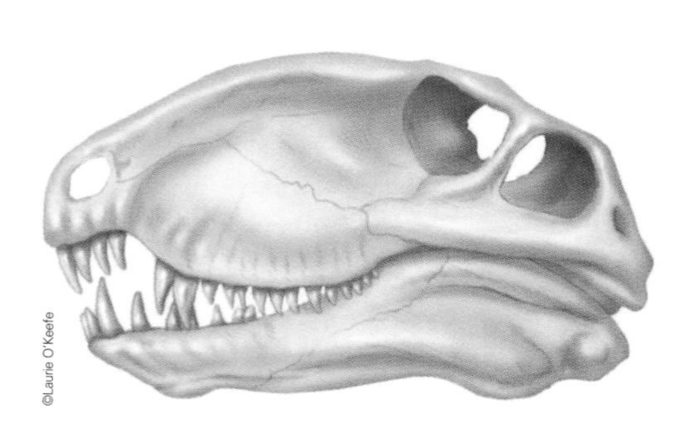

***Figure 3.10*** *Reconstructions of Mammal-like reptiles.*

It is legitimate to assemble a morphological series to assess how skulls, for instance, are structurally similar to other skulls (though note that with fossils, morphological similarity need tell us nothing about genetic relatedness). But it is illegitimate simply to take a series of structural similarities and declare that they constitute a single path of descent—in other words, an actual evolutionary lineage. There are numerous therapsid species in the fossil record. In fact, according to Douglas Futuyma, "The gradual transition from therapsid reptiles to mammals is so abundantly documented by scores of species in every stage of transition that it is impossible to tell which therapsid species were the actual ancestors of modern mammals."[29] With so many fossils to choose from, therapsid fossils can be arranged in not just one but multiple sequences that suggest Darwinian lineages. But the sheer number of ways of arranging therapsid fossils raises more problems than it solves.

For instance, if mammals arose from just one of these lineages, then the others are not ancestral to them. But if several unrelated species have the same mammal-like features as the "actual ancestor," how compelling are these features as evidence of ancestry? If the same mammal-like features of therapsids keep emerging without being related by common ancestry, then why should mammals be regarded as related to the therapsids by common ancestry? Do shared features require a Darwinian interpretation? Do they merely suggest it? Are they simply consistent with it? Or are they carefully chosen to illustrate it? Consider the fact that, according to some researchers, mammals arose several times from several different therapsid species. Yet it is extremely unlikely that accidental mutations capable of crafting the extraordinary, precisely integrated parts of the mammalian ear occurred independently time after time.

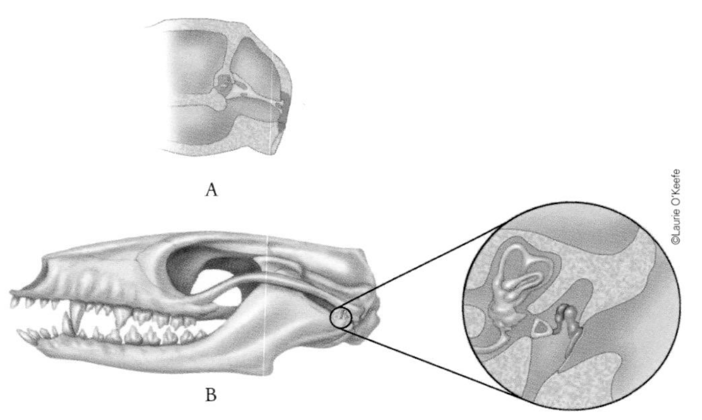


***Figure 3.11*** *The sectioned reptilian skull (A) shows the quadrate bone (red) and the articular bone (blue) where the upper and lower jaws meet. (B) shows their relocation according to Darwinian theory as Malleus (blue) and Incus (red) in the mammalian middle ear (after Romer).*

Let's examine the proposed evolution of the mammalian ear more closely. The skull and mandibles (lower jaws) of the therapsids are said to have bones similar (homologous) to those of the first mammals. The upper and lower jaws of reptiles articulate (fit together) with two bones (one each located at the back of the upper and lower jaws) not found in mammals. According to Darwinian theory, these two bones relocated in the middle ear of the mammals in the course of descent with modification (see figure 3.11). Darwinists describe the reptilian jaw bones as "migrating" to their new locations in the mammalian ear. Nevertheless, there is no fossil record of such an amazing process. Nor is it clear how the neo-Darwinian mechanism of natural selection acting on random genetic changes can cause bones to move and relocate. Consider that to make this change, one of these bones had to cross the hinge from the lower jaw into the middle ear region of the skull. Thereafter the neo-Darwinian mechanism would have had to reshape and refine these bones into a highly specialized, delicate instrument of sound transmission. Such an occurrence would be extraordinary enough by itself, but Darwinists propose that this happened more than once and without the need for any intelligent guidance!

The evidence for the evolution of the mammalian ear from the reptilian jaw therefore comes down to balancing a bone count. In every other respect, the bones are very different and there is no evidence of how evolution might have caused the reptilian jaw bones to "migrate" to the mammalian ear. Why, then, should we think that the identical bone count serves as reliable evidence for the macroevolution of reptiles into mammals? If a mere match of a bone count can provide conclusive evidence of macroevolution, then how can we dismiss the far more detailed similarities between species that evolutionists agree do not indicate evolutionary relationships?

For instance, in Chapter 5 we will examine the giant and red pandas. Both share an enlarged wrist bone, known technically as a "radial sesamoid" and more popularly as the "panda's thumb." This bone is used for stripping bamboo, the preferred food for both pandas. Yet, evolutionary theorists now agree that the two pandas evolved from separate ancestors, neither of which possessed such a structure. The "panda's thumb" therefore had to evolve twice, once in the giant panda and once in the red panda. But if sharing a highly specified structure such as the panda's thumb does not indicate an evolutionary relationship, then why should a matched bone count between the reptilian jaw and the mammalian ear indicate an evolutionary relationship? We can't have it both ways. How can we be confident of evolutionary relationships based on remote similarities (e.g., identical bone count) if striking similarities (e.g., the panda's thumb) do not indicate kinship?[30] We will return to such questions in detail in Chapter 5.

## Whale Evolution

Some Darwinists regard fossil evidence for the evolution of whales as a success story second only to fossil evidence for the evolution of mammals from mammal-like reptiles. In fact, the evidence for neither is compelling. Whales are thought to have evolved from land-dwelling mammals, but for many decades the expected transitional forms seemed to be missing from the fossil record. Even before Darwin published *The Origin of Species* in 1859, fossils of dolphin-like dorudons and serpent-like basilosaurs were known, yet these presumed whale ancestors appear to have been fully aquatic.

In 1983, a mammalian fossil skull was discovered in Pakistan (*Pakicetus*) that had some whale-like features, but the animal appears to have been a land-dweller.[31] Given the large number of changes in anatomy and physiology that would be required to turn a land mammal into an aquatic mammal, critics of Darwinian evolution argued that the absence of transitional forms between terrestrial and aquatic mammals posed a serious difficulty for whale evolution.

In 1994, however, paleontologist Hans Thewissen and his colleagues reported the discovery of a fossil with characteristics intermediate between a land mammal and a whale. They called their find *Ambulocetus natans* (literally, the swimming walking-whale)[32] (see figure 3.12). A few months later, Philip Gingerich and colleagues reconstructed a slightly younger fossil that had some features intermediate between Thewissen's find and modern whales. They called it *Rodhocetus kasrani*[33] (see figure 3.13). Other possible intermediates have also been found.

***Figure 3.12*** Ambulocetus natans. *Claimed by its discoverers to exhibit aquatic locomotion using appendages developed on land.*

***Figure 3.13*** Rodhocetus kasrani. *Claimed by its discoverers to be the first swimming Cetacean, or member of the order in which whales are classified.*

Stephen Jay Gould described these findings as "the sweetest series of transitional fossils an evolutionist could ever hope to find."[34] But does this structural series in fact represent a series of ancestors and descendants? According to Berkeley paleontologist Kevin Padian, all of the fossils in the whale series have "distinguishing characteristics, which they would have to lose in order to be considered direct ancestors of other known forms."[35] At best, therefore, each fossil represents a terminal side branch of the whales' hypothetical lineage.

Since the oldest specimens consist of skulls and teeth that are similar to an extinct group of hyena-like mammals called mesonychians, University of Chicago evolutionary biologist Leigh Van Valen proposed in the 1960s that modern whales are descended from mesonychians. In the 1990s, however, molecular studies suggested that the closest living relatives of whales are hippopotamuses.[36]

*Continued on next page*

Some evolutionary biologists remain skeptical of the whale-from-hippo, or "whippo," hypothesis. As John E. Heyning wrote in *Science* in 1999:

> "Previous experience suggests we should be cautious about wholeheartedly embracing such provocative hypotheses of relationships. More often than not, such controversial claims are found to be weakly supported or contradicted when scrutinized in more-detailed analyses."[37]

The "whippo" hypothesis sets at odds fossil and molecular evidence. Fossil similarities suggest that hippos are close evolutionary relatives of other even-toed hoofed mammals such as pigs and camels, but far removed from whales. On the other hand, molecular similarities suggest that hippos are close evolutionary relatives of whales, but far removed from pigs and camels. But if the original fossil similarities are not evidence for common ancestry, then by the same logic molecular similarities need not be, either. There's no compelling reason to trust either hypothesis. In fact, there is good reason to distrust both hypotheses (see section 3.9).

## 3.9 ABUSING FOSSILS TO TRACE EVOLUTIONARY LINEAGES

We have just examined what many evolutionary theorists regard as the best example of an evolutionary lineage in the fossil record, namely, the transition from reptiles into mammals via mammal-like reptiles. There are three fundamental problems with this and all other examples of inferring Darwinian evolution on the basis of fossil evidence. The first is that any specific hypothesis must use the fossil data selectively; the second is that similarities in fossil or living organisms may not be due to common ancestry; and the third is that fossils cannot, in principle, establish biological relationships.

1. **Using the fossil evidence selectively.** In the case of therapsids we saw that fossils more mammal-like can occur earlier in the fossil record than fossils that are less mammal-like. Yet to trace an evolutionary lineage on the basis of the fossil record requires that therapsids structurally more similar to mammals enter the history of life later than those that are structurally less similar. Evolution, after all, needs to follow time's arrow and cannot have offspring giving birth to parents.

A similar problem arises with geographical mismatches, in which fossil organisms that are supposedly next to each other in a structural progression are widely separated geographically. If the geographical separation is too great, how can one organism be ancestral to the other? Reproduction, after all, requires proximity—parents do not give birth to offspring at the other side of the globe.

The problem of temporal and geographical mismatches is widespread. The Darwinist's way around this problem is to assume that organisms that appear to enter the fossil record too late or too far away actually entered earlier or closer together. But such assumptions are entirely ad hoc and ignore the actual fossil evidence.

This illustrates a larger problem—what scientists call "cherry-picking." Given a sufficiently large data set, it's possible to find salient patterns simply by trying out enough different ways of combining items of data. Many structural progressions found in the fossil record are nothing more than "cherries"—in other words, they are statistical artifacts that result from trying out enough different ways of combining fossil data. The sheer quantity of fossil data is immense. Simply by combining and recombining these data in enough different ways and by attending to sufficiently many distinct features of structural similarity, it is possible to generate reasonably long fossil progressions arranged by structural similarity.

Two well-known results from statistics give rise to the cherry picking fallacy. One is the *birthday paradox*. Although there are 365 days in the year, it only takes 23 people (chosen at random) for there to be a better than even chance that at least two of them share a birthday.[38] That's because in calculating the probability of a shared birthday, we must factor in all possible ways of pairing these 23 people. As it turns out, there are 253 pairings and thus 253 ways that any two of them might share a birthday (it's not coincidental that 253 is over half of 365; that's why 23 people are more likely than not to share a birthday). Because of the birthday paradox, the fossil record readily yields fossils that match up on a given feature of similarity quite apart from any underlying cause.

The other result from statistics that gives rise to the cherry picking fallacy is the *file-drawer effect*. Suppose you claim that a coin you are flipping is biased because you just now flipped it ten times and each time it came up heads. The degree to which you are justified in claiming that the coin is biased will depend on the unreported number of times you flipped the coin before actually reporting ten heads in a row. The file-drawer effect refers to the unsuccessful studies that go unreported and languish in a researcher's file-drawer.[39] The bigger the file-drawer, the greater the number of unsuccessful studies that went unreported and, consequently, the less compelling is any eventual report of success. Even with a fair coin, after a few thousand coin flips, one is virtually assured of flipping ten heads in a row. Thus, if your file-drawer contains thousands of unreported coin flips, the ten heads in a row you report can't confirm that the coin is biased.

Likewise, for every "successful" structural progression in the fossil record (like the reptile-to-mammal progression), there are all too many "unsuccessful" ones, conveniently unreported and languishing in evolutionary biology's "file-drawer." Evolutionary biology's file-drawer of failed attempts at finding such fossil progressions is huge. For instance, where are the progressions based on structural similarity that connect the different animal phyla—progressions that should be there if evolutionary theory is correct? Despite a massive search of the fossil record by paleontologists and evolutionary biologists, no such progressions are known. In consequence, there is every reason to be suspicious of using "successful" fossil progressions to infer evolutionary lineages.

2. **Similarity may not be due to common ancestry.** In evolutionary theory, convergence refers to the origination of identical or highly similar structures through independent evolutionary pathways rather than inheritance from a common ancestor. We've already mentioned the "thumb" of the giant and red pandas, and we'll see several more examples of convergence in Chapter 5. Darwinian theory attributes convergence to similar environments that apply similar selection pressures and thereby produce similar structures.

   This explanation is on its face implausible because there is no reason to think that Darwin's opportunistic mechanism has the fine discrimination to produce virtually identical complex structures in widely separated environments that do not influence one another. Yet organisms possess many similar features not thought to arise from a common ancestor. Convergence is a widespread fact. As a result, even if Darwinian theory were true, one could never be sure whether similar features shared by two fossils resulted from convergence or from common ancestry. *If similar structures can evolve and re-evolve repeatedly, then fossils cannot distinguish convergence from common ancestry, and tracing evolutionary lineages in the fossil record becomes impossible.* In fact, similarities may not be due to Darwinian evolution at all. In a 1990 book intended to refute critics of Darwinian evolution, biologist Tim Berra used pictures of various models of Corvette automobiles to illustrate how the fossil record provides evidence for descent with modification. "If you compare a 1953 and a 1954 Corvette, side by side," he wrote, "then a 1954 and a 1955 model, and so on, the descent with modification is overwhelmingly obvious."[40] But automobiles are designed, not descended from other automobiles. Berra actually proved the opposite of what he intended, namely, that a series of similarities could be a product of intelligent design rather than Darwinian evolution.

The case for Darwinian evolution would be greatly strengthened if scientists could demonstrate (rather than merely gesture at) a plausible mechanism for producing macroevolution. But they have been unable to do so. Even if we assume that a structural progression such as the therapsid-to-mammal sequence is an evolutionary lineage, the fact remains that we know of no material mechanism capable of producing it. To be sure, one can tell a story about how a Darwinian mechanism might have caused the progression, but that's all it would be—a fanciful story.

Take the evolution of the mammalian ear from the reptilian jaw. How exactly did those two bones from the reptilian jaw "migrate" to the mammalian ear? The word "migrate" in this context is empty of scientific content. What genetic changes and selection pressures were in fact operating, and how, specifically, did they bring about the evolutionary pathway in question? No such details are known. Yet, without such details, there is no way to assess whether the Darwinian mechanism was even capable of, much less responsible for, evolving the mammalian ear.

Perhaps a sufficiently adept designing intelligence could change the reptilian jaw into the mammalian ear. But an intelligently guided process would not be Darwinian.

3. **Fossils cannot, in principle, establish ancestor-descendant relationships.** Imagine finding two human skeletons in the same location, one apparently about thirty years older than the other. Was the older individual the parent of the younger? Simply by looking at the skeletons, one can't say. Without independent evidence (e.g., genealogical, dental, or molecular), it is impossible to answer the question. Yet in this case we're dealing with two skeletons from the same species that are only a generation apart. It follows that even if we had a fossil representing every generation and every imaginable intermediate between, say, reptiles and mammals—if there were *no missing links whatsoever*—it would still be impossible, in principle, to establish ancestor-descendant relationships.

   In 1978, fossil expert Gareth Nelson, of the American Museum of Natural History in New York, wrote: "The idea that one can go to the fossil record and expect to empirically recover an ancestor-descendant sequence, be it of species, genera, families, or whatever, has been, and continues to be, a pernicious illusion."[41]

   Henry Gee, a science writer for *Nature*, doesn't doubt Darwinian evolution, but he likewise admits that we can't infer descent with modification from fossils. "No fossil is buried with its birth certificate," he wrote in 1999. "That, and the scarcity of fossils, means that it is effectively impossible to link fossils into chains of cause and effect in any valid way." According to Gee, we call new fossil discoveries missing links "as if the chain of ancestry and descent were a real object for our contemplation,

> and not what it really is: a completely human invention created after the fact, shaped to accord with human prejudices." He concluded: "To take a line of fossils and claim that they represent a lineage is not a scientific hypothesis that can be tested, but an assertion that carries the same validity as a bedtime story—amusing, perhaps even instructive, but not scientific."[42]

In short, fossils cannot demonstrate Darwinian evolution. But what about living organisms? Can they demonstrate Darwinian evolution? We turn to them in the next chapter.

*The general notes on the CD included with this book expand on the material in this chapter and throw light on the following discussion questions.*

## 3.10 DISCUSSION QUESTIONS

1. What are fossils? How are they relevant to the study of biological origins? Does the fossil record provide direct or only circumstantial evidence for common descent? Does that evidence more readily support a Darwinian mechanism or a designing intelligence as the cause of biological innovation?

2. What is the "gravest objection" to Darwin's theory? Does the fossil record fit Darwin's "great tree of life" in which organisms gradually diversify from prior forms? Why is Darwinism committed to a gradual form of evolution? Is the fossil record full of the transitional forms predicted by Darwin's theory? Is *Archaeopteryx* a good example of a transitional form? Why or why not?

3. What are the three major features of the fossil record? What is the Cambrian Explosion? Why is it so significant for evaluating the truth of macroevolution? Are there clear evolutionary precursors to the phyla of the Cambrian period? What does it mean to say that the fossil record exhibits stasis? What does it mean to say that it exhibits gaps?

4. What are four ways of interpreting the fossil record? Why does appealing to an imperfect fossil record or to an insufficient search of fossil remains not adequately explain the mismatch between Darwin's theory and the fossil record? How well are extant living forms represented in the fossil record? What does this indicate about how good the fossil record is at preserving life's past? How diligently has the fossil record been searched? How does appealing to an abrupt emergence of biological forms address the mismatch between Darwin's theory and the fossil record?

5. Describe the evolutionary view known as *punctuated equilibrium*. How does it account for both stasis and gaps in the fossil record? According to the theory of punctuated equilibrium, did transitional forms once exist? If so, why, according to the theory, did they fail to fossilize?

6. What does it mean to explain the gaps in the fossil record as the result of abrupt emergence? Is abrupt emergence the face-value interpretation of the fossil record? What are the four possibilities in the history of life that could account for abrupt emergence? How does nonbiogenic formation differ from the other three possibilities? Why can't symbiogenic reorganization be the whole story in accounting for abrupt emergence? How does biogenic reinvention differ from generative transmutation?

7. Describe the work of Otto Schindewolf and Richard Goldschmidt. Does their thinking about macroevolution fall under abrupt emergence? Was their theory one of generative transmutation? Explain. What is a "hopeful monster"? What is a "hopeless monster"? How were Goldschmidt's ideas received by the biological community? How did he justify his position against the dominant neo-Darwinism of his day? (See general notes.)

8. How good is the fossil evidence that reptiles gradually evolved into mammals? What are some of the difficulties with this view? How do evolutionary biologists explain the formation of the mammalian ear from the jaw-bones of reptiles? Do you find this explanation compelling? Why or why not?

9. Imagine you are making a documentary about whether whales evolved from land-dwelling mammals. Search the world wide web and develop the best case either for or against whale evolution (argue for it if you think the evidence confirms it, argue against it if you think the evidence negates it).

10. What are the three fundamental problems with inferring Darwinian evolution on the basis of fossil evidence? Discuss how (i) similarity through convergence, (ii) temporal and geographical mismatches, (iii) the absence of a plausible material mechanism, and (iv) the statistical fallacy of cherry picking undercut attempts to reconstruct macroevolutionary sequences from the fossil record?

# CHAPTER FOUR The Origin of Species

## 4.1 EVOLUTION'S SMOKING GUN

No one doubts that species change. People have been observing them change for centuries—not only in their own families, but also in domestic plants and animals. Descent with modification within a species is utterly uncontroversial, and if Darwin had written a book titled *How Existing Species Change Over Time*, nobody would have paid much attention. But he wrote a book titled *On the Origin of Species by Means of Natural Selection.* Darwin's whole theory attempts to explain how new species originate from existing species—what biologists call "speciation."

Speciation is thus the starting point for everything else in evolutionary theory. "Darwin called his great work *On the Origin of Species*," wrote Harvard evolutionary biologist Ernst Mayr, "for he was fully conscious of the fact that the change from one species into another was the most fundamental problem of evolution."[1] Yet Darwin never solved this "mystery of mysteries," as he called it.[2] According to Mayr, "Darwin failed to solve the problem indicated by the title of his work. Although he demonstrated the modification of species in the time dimension, he never seriously attempted a rigorous analysis of the problem of the multiplication of species, of the splitting of one species into two."[3]

In 1997, evolutionary biologist Keith Stewart Thomson wrote: "A matter of unfinished business for biologists is the identification of evolution's smoking gun," and "the smoking gun of evolution is speciation, not local adaptation and differentiation of populations." Before Darwin, centuries of artificial selection had seemingly demonstrated that species can vary only within certain limits. "Darwin had to show that the limits could be broken," wrote Thomson, and "so do we."[4] As a purely scientific matter, to justify Darwinism's sweeping claims about the evolution of living things requires establishing the first step: how species originate by unguided material causes.

This task is not as straightforward as it might seem because biologists have never been able to agree on a single definition of "species." In their 2004 book *Speciation*, evolutionary biologists Jerry A. Coyne and H. Allen Orr pointed out that there are more than two dozen definitions of "species." How are we to choose among them? "Biologists want species concepts to be useful for some purpose," wrote Coyne and Orr, "but differ in what that purpose should be. We can think of at least five such goals." A species concept is useful, according to them, if it (1) helps biologists to classify organisms; (2) corresponds to entities actually in nature; (3) helps us understand how those entities originate; (4) represents evolutionary history; and (5) applies to as many organisms as possible. Coyne and Orr acknowledged that "no species concept will accomplish even most of these purposes," but they felt that, "when deciding on a species concept, one should first identify the nature of one's 'species problem,' and then choose the concept best at solving that problem."[5]

Like most other Darwinists searching for evolution's smoking gun, Coyne and Orr prefer Ernst Mayr's Biological Species Concept (abbreviated BSC), which defines species as "groups of interbreeding natural populations that are reproductively isolated from other such groups." Why do Coyne and Orr prefer this over other definitions of species? For them "the most important advantage of the BSC is that it immediately suggests a research program to explain the existence of the entities it defines." Coyne and Orr "feel that it is less important to worry about species status than to recognize that the *process* of speciation involves acquiring reproductive barriers."[6]

Unfortunately, the BSC applies only to living, sexually reproducing organisms. It does not apply to asexual organisms such as bacteria, among which species are distinguished on morphological and biochemical grounds. Nor does it apply to dead organisms, such as fossils, among which species are distinguished solely on morphological grounds. So the BSC ignores many organisms of interest to evolutionary biologists—including *all* of the organisms discussed in the last chapter—and limits itself to just one aspect of living things: the one that seems most likely to provide evolution's smoking gun.

Beyond these immediate deficiencies of the BSC, it is troubling that the primary motivation for Darwinists to define species the way they do is that the BSC leads to a line of research they find congenial. Many failed scientific projects of the past, including Ptolemaic astronomy and the caloric theory of heat, facilitated a program of observational and experimental research, but ultimately proved to be *dead wrong*. The ability of a scientific concept to facilitate research by itself says nothing about whether that research in the end provides true insights into nature. Coyne and Orr are like the proverbial man searching for something under a streetlamp because the rest of the street is dark. If what the man is looking for doesn't happen to be under the streetlamp, all his searching will be in vain. By choosing the BSC to define species, Coyne and Orr commit the same error.

## 4.2 SPECIES AS REPRODUCTIVELY ISOLATED POPULATIONS

If, for the sake of argument, species are defined as populations of interbreeding organisms that are reproductively isolated from other such populations, then new species presumably originate through the acquisition of barriers to reproduction. One way this could happen is that a geographic barrier (such as a mountain range or a body of water) separates two populations of a single species so that they are physically prevented from interbreeding (see figure 4.1). Over time, the two populations might diverge genetically until they are unable to interbreed even if the geographic barrier is removed. This is the theory of *allopatric speciation* ("allopatric" comes from Greek and means "different fatherland").

***Figure 4.1*** *Geographical barrier. A small subpopulation may be cut off from its parental population resulting in the elimination of much of the original gene pool and major changes in gene frequencies.*

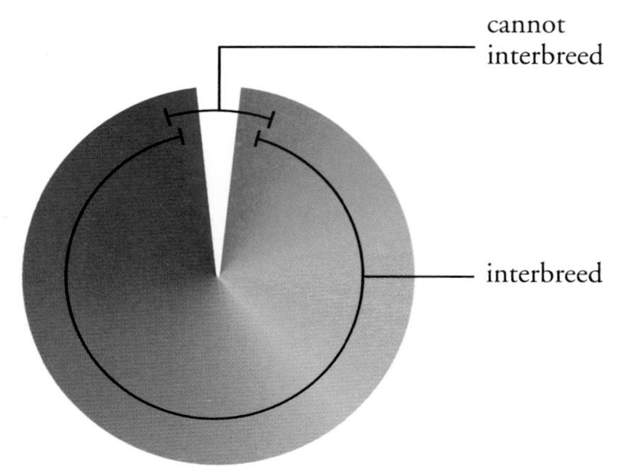


*Figure 4.2* *As subpopulations successively develop, subpopulations at opposite ends of the breeding chain can lose their ability to interbreed.*

There is circumstantial evidence for the theory of allopatric speciation. It includes the existence of breeding chains, in which one population may be able to interbreed with a second population, the second population may be able to interbreed with a third, and the third may be able to interbreed with a fourth, but the fourth population may be unlikely or unable to interbreed with the first (see figures 4.2 and 4.3). In these cases, the second and third populations might represent intermediate stages in the process of allopatric speciation between the first and fourth populations.

Allopatric speciation is currently the most popular theory among evolutionary biologists, but it is not the only one. According to the theory of *sympatric speciation* ("sympatric" comes from Greek and means "same fatherland"), two populations can become reproductively isolated without being geographically separated. There is circumstantial evidence for this as well. For example, two centuries ago the North American fruit fly *Rhagoletis pomonella* apparently consisted only of individuals that lived on hawthorn trees. The females laid their eggs in August on the hawthorn trees and at the end of September the larvae fed on the red hawthorn fruit. When apple trees were introduced years later, some flies laid their eggs on them and the larvae fed on apples. Since apples ripen earlier than hawthorn fruit, the apple-based flies now reproduce a month earlier than their hawthorn-based counterparts. Because of the one-month difference in their mating schedules, the two flies do not mate in the wild. Unlike allopatric speciation, the barrier to interbreeding in this case is temporal and behavioral, not geographic.[7]

*Figure 4.3* *A breeding chain. A species inhabiting an extended area may develop distinct subpopulations, each interbreeding among themselves or with adjacent subpopulations.*

Of course, it takes more than a physical or temporal barrier to turn one species into two. In theory, two populations that are reproductively isolated would undergo different genetic changes, possibly to the point where they would become genetically incapable of interbreeding. Several mechanisms have been proposed to explain how reproductively isolated populations might diverge genetically. Natural selection could non-randomly favor some genes and eliminate others as populations adapt to different environments. Or gene frequencies could simply change by chance, especially if one of the isolated populations is very small.

Small break-away populations tend not only to have a different ratio of genes from the larger main population (this is known as the *founder effect*), but they also tend more readily to "drift away" from whatever frequency distribution of genetic traits they started with (such chance deviations from expected gene frequencies is known as *genetic drift*). This tendency to depart from the gene frequencies of the ancestral population follows from the laws of genetics, which are statistical and therefore obey the laws of probability. When, for instance, a black guinea pig mates with a white one, the laws of genetics may tell us to expect a certain percentage of white offspring, and a certain percentage of black ones. But in particular matings the offspring may be all white or all black. Why? According to the laws of probability, the smaller the number of cases, the larger the chance that there will be a significant deviation from the expected outcome.

Coin tossing illustrates this principle. When you toss a coin, the laws of probability tell you that the chance of getting heads or tails is even. Thus, when you toss a coin, you expect to see the same number of heads and tails. But what happens if you toss a coin only four times? Will you see the expected outcome—that is, exactly two heads and two tails? Try it and see. With a small sample like this there is a good chance that the actual outcome will diverge considerably from the expected outcome. You may even find that you flipped four heads in a row or four tails in a row. The reason you can get such a large deviation from expected results is the extremely small number of trials. Yet, if you take a large sample by tossing the coin ten thousand times, you would find the ratio of heads to tails very close to the expected 50:50 ratio. The larger the number of events or trials, the more closely the results will conform to the expected probability. On the other hand, the smaller the number of events or trials, the more likely the results will deviate from the expected probability.

This same principle of deviation from expected outcomes could occur with small isolated subpopulations that have broken away from a large ancestral population. Members of these small subpopulations are forced to interbreed very closely. Even a neutral mutation (one that is neither harmful nor beneficial) could therefore become very widespread in such a population. Alternatively, some gene already present but not very prevalent in the ancestral population might come to completely dominate a small, breakaway population.

Consider, for instance, a subpopulation consisting of two gray guinea pigs of opposite sex, each of which has one albino parent. From genetics we know that these animals are heterozygous, that is, Gg (assuming that G stands for the dominant gray trait and g for the recessive albino trait). Now, suppose these gray guinea pigs mate and produce a litter of four offspring. How many of the offspring will be gray and how many will be white? Each parent guinea pig (having a Gg genotype) will produce two kinds of gametes, one containing G and another g. On average, the random combination of these gametes (see figure 4.4) should produce three gray offspring (GG, Gg, and gG) and one albino (gg). Theoretically, this is what should happen. But in the real world it might not. All of the offspring might turn out to be albino (gg) and thereafter only give birth to albino guinea pigs because the dominant G allele would have been lost.

The process of establishing a gene in this way by the loss of its allele is called *fixation.* Fixation is likely to occur in small populations because the effects of random events get magnified in small populations. In the guinea pig example, fixation occurred via the founder effect, in which a pair or handful of some species become isolated from its parent population and thereby establish an entirely new population. This small sample, which constitutes the founder group, acquires simply by chance new gene frequencies at variance with those of the larger group from which it comes. As an example of the founder effect, consider the Amish people of Pennsylvania. Because they are descendants of only about 200 settlers and have tended to marry among themselves, the Pennsylvania Amish have a greater percentage than the American average of genes for short fingers, short stature, a sixth finger, and certain diseases.

***Figure 4.4*** *Genetic drift. The expected ratio of offspring can be predicted if the genes of the parents are known and if the number of offspring is large. Small counts of offspring, however, often don't reflect the predicted ratios.*

Similar to the founder effect is the *bottleneck effect.* The bottleneck effect occurs when some severe environmental event greatly reduces the size of a population. Such events could include glaciation (as perhaps in the case of Neanderthals), drought, or even a severe winter. The survivors then become the founders of a new population. Because the founders are few in number, genetic drift can alter gene frequencies quickly. Further, the gene frequencies of the survivors may vary greatly from those in the original population (some genetic factor rare in the general population might have been responsible for their survival). The gene frequencies of a small surviving population might therefore vary greatly from those of the original population. Thus, as the survivors increase and establish themselves, they might be quite different from the population that preceded them. In theory, a new population might eventually emerge that would be unable to interbreed with the original population.

The origin of species through the joint action of reproductive isolation and the mechanisms of genetic change remains for now a speculative possibility. Theories of speciation based on such Darwinian processes do not, at present, adequately account for the origin of species. Such theories remain controversial because the evidence is simply not there to confirm that any one of them is true. To be sure, these theories sketch scenarios for speciation that have a certain plausibility. But they are not confirmed by known facts. In particular, there are no credible reports of anyone witnessing the origin of a new species through the joint action of reproductive isolation and the mechanisms of genetic change.

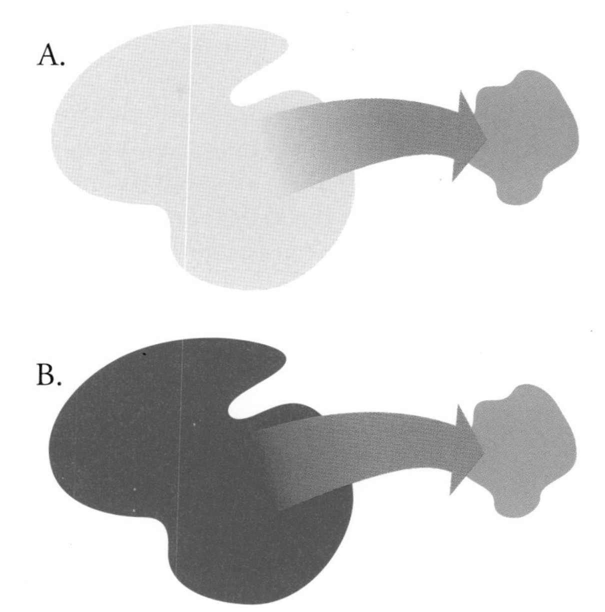


***Figure 4.5*** *Bottleneck and founder effects. The bottleneck effect (A) occurs when some natural event destroys most of a population, thereby reducing the variation within the species' gene pool. The founder effect (B) occurs when two or a few founders migrate to a new habitat, thus reducing the variation within the new gene pool of the daughter population.*

## 4.3 ALLEGED INSTANCES OF OBSERVED SPECIATION

Despite the absence of evidence for the ability of reproductive isolation to harness the mechanisms of genetic change and thereby to produce new species, some Darwinists still claim that there are many instances of observed speciation.[8] But most of these alleged instances are in fact analyses of existing species that are used to defend one or another theory of how they might have originated—such as the theories of allopatric and sympatric speciation, or the bottleneck and founder effects, discussed in the previous section. Analyzing existing species to support one or another theory of speciation, however, is not the same as observing speciation in action.

There actually *are* some confirmed cases of observed speciation, but these are due to an increase in the number of chromosomes, or "polyploidy." Such cases, however, are limited to flowering plants and result from hybridizing two species to form a new one.[9] Furthermore, according to evolutionary biologist Douglas Futuyma, speciation that results from polyploidy (also called "secondary speciation") "does not confer major new morphological characteristics . . . [and] does not cause the evolution of new genera" or higher taxonomic levels.[10] Darwinian evolution, by contrast, depends on taking a single existing species and splitting off new species from it (called "primary speciation"), which then in turn diverge and split, diverge and split, over and over again. Only primary speciation, and not secondary speciation, could produce the branching-tree pattern required by Darwinian evolution.

Of the many instances of observed speciation alleged by Darwinists, only five come close to claiming observed *primary* speciation. First, in 1962, from a single lab population of *Drosophila* (fruit flies), J. M. Thoday and J. B. Gibson bred only those flies with the highest and lowest number of bristles (the insect equivalent of hair). After twelve generations, the experiment produced two populations that not only differed in bristle number but also showed "strong though partial isolation." Yet Thoday and Gibson not did claim to have produced a new species. Furthermore, other laboratories were unable to reproduce their results.[11]

Second, in 1958 Theodosius Dobzhansky and Olga Pavlovsky started a laboratory population of fruit flies using a single female of a strain from Colombia. Crosses between that fly and several other strains produced fertile hybrids in the laboratory. In 1963, however, similar crosses yielded sterile hybrids. In 1966, Dobzhansky and Pavlovsky concluded that the strain they had introduced in 1958 had become "a new race or incipient species . . . in the laboratory at some time between 1958 and 1963."[12] But Coyne and Orr, writing in 2004, suspect their results were "due to contamination of cultures by other subspecies."[13] In any case, Dobzhansky and Pavlovsky reported only a "new race or incipient species," not a new species.

Third, in 1964 biologists collected some marine worms in Los Angeles Harbor and used them to start a lab colony. When they went back to the same location twelve years later, the original population had disappeared, so they collected worms from two other locations several miles away, and these were used to start two new lab colonies. In 1989, researchers found that the two new colonies could interbreed with each other but not with the Los Angeles Harbor colony that had been started twenty-five years earlier. In 1992, James Weinberg and his colleagues called this an observed instance of "rapid speciation," based on the assumption that the original colony had "speciated in the laboratory, rather than before 1964."[14] A few years later, however, tests performed by Weinberg and two others showed that the original population was "already a species different from" the two new colonies "at the time when it was originally sampled in 1964."[15] No speciation had occurred.

Fourth, in 1969 E. Paterniani reported an experiment on maize in which breeding was permitted only between individuals possessing two extremes of a particular trait. Paterniani noted the development of "an almost complete reproductive isolation between two maize populations" but did not claim that a new species had been produced.[16]

Fifth and last, in the 1980s William R. Rice and George W. Salt subjected a population of fruit flies to eight different environments. They then took the flies that preferred the two most extreme environments and allowed only them to breed. Within thirty generations the flies had sorted themselves into two populations that did not interbreed. Even so, Rice and Salt did not claim to have produced two new species. More modestly, they believed only that "incipient speciation" had occurred.[17]

So, of the five alleged instances of observed primary speciation, only one (Weinberg's) claimed to have observed actual speciation—and it was later retracted. The other four (one of which could not be reproduced by other scientists and one of which was not controlled for contamination) claimed only some degree of reproductive isolation or "incipient speciation."

What is "incipient speciation"? Darwin wrote: "According to my view, varieties are species in the process of formation, or are, as I have called them, incipient species."[18] But how can we possibly know whether two varieties (or races) are in the process of becoming separate species? Saint Bernards and Chihuahuas are two varieties of dog that cannot interbreed naturally, but they are members of the same species. Maybe they are on their way to becoming separate species, or maybe not. The two varieties of *Rhagoletis pomonella* described in the previous section do not interbreed in the wild, but they look exactly alike and are still capable of mating in the laboratory. Like different breeds of dogs, they are still members of the same species. Calling them "incipient species" amounts to no more than a prediction that they will eventually become separate species. But maybe they won't. Short of waiting to see whether the prediction comes true, we can't really know. And given our limited lifespans, we don't have time to wait (at least not by conventional evolutionary timescales).

Darwinists therefore discount the lack of observed instances of primary speciation by saying that it takes too long to observe them. But if it takes too long for scientific investigators to observe primary speciation, then there will never be anything more than indirect evidence for the first and most important step in Darwinian evolution. Darwinists *claim* that all species have descended from a common ancestor through variation and selection. But until they can point to a single observed instance of primary speciation, their claim must remain an unverified assumption, not an observed scientific fact. University of Bristol bacteriologist Alan H. Linton made precisely this point when in 2001 he assessed the direct evidence of speciation:

> None exists in the literature claiming that one species has been shown to evolve into another. Bacteria, the simplest form of independent life, are ideal for this kind of study, with generation times of twenty to thirty minutes, and populations achieved after eighteen hours. But throughout 150 years of the science of bacteriology, there is no evidence that one species of bacteria has changed into another. . . . Since there is no evidence for species changes between the simplest forms of unicellular life, it is not surprising that there is no evidence for evolution from prokaryotic [e.g., bacterial] to eukaryotic [e.g., plant and animal] cells, let alone throughout the whole array of higher multicellular organisms.[19]

So except for secondary speciation, which is not what Darwin's theory needs, there are no observed instances of the origin of species. As evolutionary biologists Lynn Margulis and Dorion Sagan wrote in 2002: "Speciation, whether in the remote Galápagos, in the laboratory cages of the drosophilosophers, or in the crowded sediments of the paleontologists, still has never been directly traced."[20] Evolution's smoking gun is still missing.

## 4.4 MICROEVOLUTION, MACROEVOLUTION, AND EVO-DEVO

According to the 2005 edition of Douglas Futuyma's college textbook *Evolution*, primary speciation "is the sine qua non of diversity" required to produce Darwin's branching tree of life. It "stands at the border between microevolution—the genetic changes within and among populations—and macroevolution"—the evolution of biological categories above the level of species.[21] So another way of stating the importance of speciation is by distinguishing between "microevolution," the uncontroversial changes within species that people observed long before Darwin, and "macroevolution," the production of new organs and body plans in the branching-tree pattern that is the essence of Darwinism.

Evolutionary biologists have long distinguished between microevolution and macroevolution. In 1937, Theodosius Dobzhansky noted that there was no hard evidence to connect small-scale changes within existing species (which he called "microevolution") to the origin of new species and the large-scale changes we see in the fossil record (which he called "macroevolution"). According to Dobzhansky, "there is no way toward an understanding of the mechanisms of macroevolutionary changes, which require time on a geological scale, other than through a full comprehension of the microevolutionary processes observable within the span of a human lifetime." He therefore concluded: "For this reason we are compelled at the present level of knowledge reluctantly to put a sign of equality between the mechanisms of macro- and microevolution, and proceeding on this assumption, to push our investigations as far ahead as this working hypothesis will permit."[22]

## Some Darwinists Deny the Distinction Between Microevolution and Macroevolution

When confronted with criticisms of evolutionary theory, some Darwinists try to soft-pedal the distinction between micro- and macroevolution, or they even deny that there is a distinction at all. For instance, when asked in 2005 to review proposed Kansas science standards requiring students to study evidence both for and against Darwinian evolution, Brown University biologist Kenneth R. Miller wrote that "the artificial distinction between micro- and macroevolution should be dropped," because "macroevolution has been observed repeatedly in nature."[23] To prove his point, Miller cited a 2004 article in the *Proceedings of the National Academy of Sciences USA* about experiments performed on two existing fly species. But the authors of the article didn't even claim that they had observed the formation of a new race, much less a new species.[24]

Another Darwinist who reviewed the proposed standards, Gary Hurd, claimed that the distinction between microevolution and macroevolution was just a creationist fabrication. Hurd wrote to the Kansas State Board of Education: "I am confident that there are other qualified commentators who will have pointed out the absurdity of differentiating 'macro' and 'micro' evolution—terms which have no meaning outside of creationist polemics."[25]

Yet it was a neo-Darwinist, Theodosius Dobzhansky, who many years ago introduced the terms "microevolution" and "macroevolution" into the English language. Moreover, he did so because there is a very real difference between the small-scale evolutionary changes that humans can directly observe and the large-scale evolutionary changes that humans cannot directly observe but which, according to Darwinian theory, must nonetheless have occurred over the history of life.

So Dobzhansky had to *assume* that microevolutionary processes are sufficient to account for macroevolution, and his assumption has been scientifically controversial ever since. In 1940, Berkeley geneticist Richard Goldschmidt published a book arguing that "the facts of microevolution do not suffice for an understanding of macroevolution."[26] Goldschmidt concluded: "Microevolution does not lead beyond the confines of the species, and the typical products of microevolution, the geographic races, are not incipient species."[27]

Likewise, in 1996, biologists Scott Gilbert, John Opitz and Rudolf Raff wrote in the journal *Developmental Biology*: "Genetics might be adequate for explaining microevolution, but microevolutionary changes in gene frequency were not seen as able to turn a reptile into a mammal or to convert a fish into an amphibian. Microevolution looks at adaptations that concern the survival of the fittest, not the arrival of the fittest." They concluded: "The origin of species—Darwin's problem—remains unsolved."[28] And in 2001, biologist Sean B. Carroll wrote in *Nature*: "A long-standing issue in evolutionary biology is whether the processes observable in extant populations and species (microevolution) are sufficient to account for the larger-scale changes evident over longer periods of life's history (macroevolution)."[29]

Darwin thought that species were infinitely plastic, that is, capable of unlimited evolutionary change. Centuries of accumulated evidence, however, seriously challenge Darwin's view. For example, both natural and artificial selection produce changes only within existing species. The birds studied by Hermon Bumpus (see Chapter 2) varied only within narrowly fixed limits. The textbook example of natural selection in action used to be the increase of dark-colored peppered moths during the industrial revolution, but dark-colored moths are just a variety of the same species as light-colored moths. Selective breeding has resulted in better beef, tastier chickens, and increased protein content in corn, but cattle remain cattle, chickens remain chickens, and corn remains corn. Domestic plant and animal production has increased substantially through exploiting genetic variation, but in all such cases the variation is eventually exhausted, and further change does not occur. It bears repeating that nobody has ever shown that selection can produce a new species, much less the new organs and body plans needed for macroevolution.

Neo-Darwinism assumes that the limits of selection can be overcome by genetic mutations, but the evidence does not support this. For many years Hermann J. Muller conducted mutation experiments with the fruit fly *Drosophila*. His goal was to demonstrate unlimited change, but his fruit flies remained fruit flies. In 1980, German geneticists Christiane Nüsslein-Volhard and Eric Wieschaus reported using a technique called "saturation mutagenesis" to search for every possible mutation involved in fruit fly development. They discovered dozens of mutations that affect development at various stages and produce a variety of malformations, but they did not turn up a single morphological mutation that would benefit a fly in the wild or turn a fruit fly into another species.[30]

As we saw in section 2.7, evolutionary developmental biology, or "evo-devo," attempts unsuccessfully to show how genes involved in development can provide the raw materials for macroevolution. In that section we focused mainly on the conceptual difficulties inherent in evo-devo. Here we focus briefly on the evidential problems in employing developmental genetics to underwrite macroevolution. Consider the announcement, in 2002, by biologists at the University of California (San Diego) that they had discovered mutations enabling shrimp-like animals, "with limbs on every segment of their bodies, to evolve 400 million years ago into a radically different body plan: the terrestrial six-legged insects."[31]

***Figure 4.6*** *The fruit fly* Drosophila *(under special lighting). Small size and rapid reproduction make these insects excellent subjects for experiments designed to investigate mutations.*

But the actual research, published in *Nature*, was far more modest.[32] A fruit fly embryo normally contains a protein that inhibits leg formation in its abdomen (from which no legs develop). At the same time, none of this protein is present in its thorax (from which legs do develop). By contrast, although a shrimp embryo contains a similar protein, it does not inhibit leg formation in shrimp. The shrimp embryo therefore develops legs in its abdomen as well as in its thorax. The San Diego biologists showed that when the protein from the abdomen of a fruit fly is inserted into the thorax of a fruit fly embryo, leg development is inhibited; but when the comparable protein from a shrimp abdomen is inserted into the thorax of a fruit fly embryo, normal fruit fly leg rudiments develop. They concluded that the gene for the protein in the abdomen of an ancient shrimp must have mutated into the fruit fly version that now suppresses leg development.[33]

Perhaps this is what in fact happened. But all the San Diego biologists actually did was to produce normal leg rudiments in a fruit fly embryo. They did not reduce the number of legs in a shrimp, which is what supposedly happened in the course of evolution. Furthermore, even if they had shown how ancient shrimp lost a few legs, their experiment would not have begun to explain how a water-dwelling shrimp could acquire the ability to breathe air and fly.

Evo-devo advocates have not given up, though. When a severe drought killed most of the finches on an island in the Galápagos in 1977, biologists observed that the survivors had, on average, slightly larger beaks. In 2004, a research team reported that Galápagos finches with larger beaks have more of a particular protein in their embryos.[34] When the researchers experimentally altered the amount of this protein in chicken embryos, they found changes in the shapes of the chicken embryos' beaks, though the protein had other effects as well. The researchers speculated that changes in the protein might have contributed to the beak changes in Galápagos finches. Yet they did not produce a breed of chickens or finches with modified beaks, much less a new species of finch. And neither did the 1977 drought—when the rains returned, the average beak size reverted to normal.[35]

So evo-devo has provided us with evidence that changes in proteins can affect embryo development. But the effects of these changes are always either trivial or harmful. Mutant fruit flies have taught us something about developmental genetics, but nothing about evolution. All of the evidence points to one conclusion: No matter what we do to a fruit fly embryo, there are only three possible outcomes—a normal fruit fly, a defective fruit fly, or a dead fruit fly. What we never see is primary speciation, much less macroevolution.

## 4.5 SPECIATION AND INTELLIGENT DESIGN

New species have originated many times in the history of life. About this there is no doubt. This fact, however, poses a problem for Darwinian evolution in a way that it does not for intelligent design. Speciation is the starting point for everything else in evolutionary theory. If Darwinism cannot account for speciation, it is dead in the water. But there is only one way for Darwinism to account for speciation, and that is by extrapolating from microevolution to macroevolution. Only if macroevolution is an extension of microevolution can the discontinuities we see among living things be reasonably attributed to the splitting of populations caused by primary speciation. This is why it is essential for Darwinism to propose not only a mechanism for primary speciation but also compelling evidence for it. Without such evidence there is no reason to think that the extrapolation from microevolution to macroevolution is valid.

The continuing controversy over species definitions and theories of speciation, along with the fact that primary speciation has never been observed, seriously undercuts Darwinian theory. But what if biologists could eventually demonstrate primary speciation empirically? Would that pose a problem for intelligent design? The answer is No, and here's why. If species are defined by the Biological Species Concept (i.e., as populations of interbreeding organisms that are reproductively isolated from other such populations), and biologists could produce (or observe the emergence of) permanent reproductive isolation between two populations, biologists would have demonstrated primary speciation. Nevertheless, the two populations would not differ in ways that could plausibly lead to macroevolution. Their distinctiveness would involve only variations on a common theme, not the introduction of entirely new themes.

Take, for instance, Hawaiian honeycreepers. These are among the most unusual and colorful birds found anywhere in the world. Their plumage reflects a beautiful range of colors, and the shapes of their bills vary widely, too. One variety (*Hemignathus munroi*) possesses a unique adaptation: the lower bill is straight and heavy and is used like a chisel, woodpecker-style, to bore into the wood to find insects; at the same time the upper bill is long and curved and is used as a probe to pry out insects. By contrast, honeycreepers on the North American mainland are drab birds. Why are the Hawaiian honeycreepers so different from their counterparts on the mainland?

According to Darwinian theory, a few birds (or a pregnant female) of a single variety somehow made it to Hawaii from the mainland in the distant past. The descendants then speciated in their new environment to produce the striking diversity we now see. But throughout such changes, the basic body type and structural features of the honeycreepers were preserved. In other words, there was diversification, but only within limited boundaries. Different forms of Hawaiian honeycreepers are (as far as we know) reproductively isolated from each other—and are thus separate species according to the Biological Species Concept. Nevertheless, they differ from each other no more than, say, different breeds of dogs, which are all members of the same species. In fact, the morphological differences between some dog breeds are greater than the morphological differences between species of Hawaiian honeycreepers.

***Figure 4.7*** *Hawaiian honeycreepers. There are distinct differences among the various species of Hawaiian honeycreepers, yet they presumably descended with modification from a single variety that originally arrived in Hawaii from the mainland.* Hemignathus munroi *is shown at lower right.*

So even if primary speciation does occur by Darwinian descent with modification (and this is a plausible hypothesis for the origin of different species of Hawaiian honeycreepers), the real question for Darwinian evolution remains unanswered: What is the origin of novelties such as new organs and body plans? Evo-devo biologists Marc W. Kirschner (at Harvard) and John C. Gerhart (at Berkeley) published a book in 2005 to address this very point. According to them, Darwinian theory suffers from a "major gap," namely, our ignorance about the origin of biological novelties such as new variations, physiologies, anatomies, and behaviors. "Ignorance of novelty" is "a major weakness" in evolutionary theory, they wrote, "casting all else in doubt."[36]

To remedy these problems with evolutionary theory, Kirschner and Gerhart proposed a theory of "facilitated variation" in which Darwinian random variations give way to variations biased to be useful in evolution. But whence this bias? Such a bias certainly sounds teleological and thus potentially compatible with a form of intelligent design. Yet that is precisely where Kirschner and Gerhart are not prepared to go (indeed, they are quite critical of intelligent design in their book). Instead they look to system constraints that limit the variations an organism can experience. But how does an organism acquire those system constraints in the first place (constraints that must be in place for facilitated variation to work at all)? This is the million dollar question they never answer. Their theory of facilitated variation therefore suffers from as big a gap as the Darwinian theory they sought to correct.[37]

The problem of the origin of biological novelty therefore remains unresolved in Darwinian or quasi-Darwinian terms. Kirschner and Gerhart's theory of facilitated variation fails because the constraints that supposedly facilitate the variations themselves remain unexplained (to say nothing that the authors provide no more evidence for macroevolution than we have already seen). Geographic or temporal separation of two previously interbreeding populations does not solve the problem either, since mere reproductive isolation does not introduce any biological novelty. Nor is the problem solved by the mechanisms invoked by Darwinists to explain how reproductively isolated populations could diverge genetically. Random changes in gene frequencies due to the founder effect or genetic drift are merely rearrangements of existing variation. They generate no new genes. And finally, natural selection merely shuffles existing variations, preserving some and eliminating others. As the Dutch botanist Hugo de Vries put it a century ago, "Natural selection may explain the survival of the fittest, but it cannot explain the arrival of the fittest."[38]

If anything, the mechanisms invoked by Darwinists to explain speciation would lead to a net *loss* of genetic information. The elimination of disadvantageous variations by natural selection, or the loss of certain genes through the founder effect or genetic drift, would actually *diminish* the genetic variability present in an ancestral population. It follows that the Darwinian mechanisms thought to cause speciation cannot be the source of the increases in functional information responsible for the spectacular pageant of life on Earth. Evolutionary change that occurs through the loss of information must eventually come to an end. If organisms lose enough genetic information in the course of speciation, their very survival will be threatened, and the result will be extinction rather than evolution.

Intelligent design, by contrast, does not view reproductive isolation as the first step to major evolutionary change. Proponents of intelligent design argue that material mechanisms at best account for relatively minor changes in a population. In particular, known material mechanisms capable of causing genetic modification cannot account for the vast *increases* in functional information required for macroevolution. The striking thing about the diversity of organisms, whether in living or fossil forms, is the significant amount of new functional information that characterizes each form as one moves up the scale of biological complexity. According to intelligent design theorists, the source of that functional information is a designing intelligence.[39]

The theory of intelligent design (ID) neither requires nor excludes speciation—even speciation by Darwinian mechanisms. ID is sometimes confused with a static view of species, as though species were designed to be immutable. That is a conceptual possibility within ID, but it is not the only possibility. ID precludes neither significant variation within species nor the evolution of new species from earlier forms. Rather, it maintains that there are strict limits to the amount and quality of variations that material mechanisms such as natural selection and random genetic change can alone produce. At the same time, it holds that intelligence is fully capable of supplementing such mechanisms, interacting with and influencing the material world, and thereby guiding it into certain physical states to the exclusion of others. To effect such guidance, intelligence must bring novel information to expression inside living forms. Exactly how that happens remains for now an open question, to be answered on the basis of scientific evidence. The point to note, however, is that intelligence can itself be a source of biological novelties that lead to macroevolutionary changes. In this way, intelligent design is compatible with speciation.

*The general notes on the CD included with this book expand on the material in this chapter and throw light on the following discussion questions.*

## 4.6 DISCUSSION QUESTIONS

1. Did Darwin solve the problem of the origin of species? Darwin frequently appealed to domestic breeding. How far does domestic breeding go toward explaining the origin of species? In attempting to explain the origin of species, does Darwin's theory give an adequate account of how biological information increases?

2. What is the Biological Species Concept? Is it completely adequate for defining species? If not, what are some problems with it? Can you think of a better definition?

3. What is reproductive isolation? What are some possible causes of reproductive isolation? Why is reproductive isolation thought to be a necessary first step toward speciation? Does reproductive isolation add or subtract genetic information? Explain.

4. Define allopatric speciation. How does it work? Why does it work better when the daughter population is small rather than large? Describe how probability theory is relevant here. Does allopatric speciation result in a gain or loss of genetic information? Compare allopatric speciation to sympatric speciation.

5. Four mechanisms thought to be involved in speciation are genetic drift, fixation, the founder effect, and the bottleneck effect. Describe each of these briefly and assess their efficacy in propelling macroevolutionary change. Do these mechanisms add or subtract genetic information? Explain.

6. What is polyploidy? Is speciation by polyploidy enough to account for Darwinian evolution?

7. What is the difference between primary and secondary speciation? Which one does Darwin's theory need? Has primary speciation ever been observed?

8. Why does Darwinian theory regard primary speciation as the first step in macroevolution? What does macroevolution require that speciation does not provide?

9. If someone eventually observes primary speciation, would it undercut the theory of intelligent design? Would intelligent design still be needed to explain speciation? What is it about macroevolution that Darwinism fails to explain? Can intelligent design do any better?

10. What do you imagine were the original species at the origin of life? What species do you see emerging subsequently? Was the information in these later emerging species already present at the origin of life (though unexpressed), or did it have to be introduced from elsewhere? Is the Darwinian mechanism an adequate source for that information?

# CHAPTER FIVE Similar Features

## 5.1 CLASSIFICATION AND INTERPRETATION

Biologists since the time of Aristotle have observed that different organisms are never so different that they don't share some similar features. Such similarities form the basis of taxonomy, the science that classifies living things. The taxonomist's goal is to group organisms by their similarities and to distinguish them by their differences. But taxonomy is not merely about classification; it is also about interpretation and explanation. Once similar features have been described, the taxonomist seeks to interpret them and to explain what they mean.

It is remarkable that we can classify organisms at all. Why can we classify living things into distinct categories as representatives of species, genera, families, orders, classes, phyla, and kingdoms? Why are all vertebrates (animals with backbones) constructed on essentially the same body plan despite the many obvious differences that separate them? It's conceivable that features of distinct organisms could have varied so randomly or clustered so strangely as to preclude any coherent classification scheme. Yet, in fact, most similar features associated with organisms fall into neat group-within-group arrangements.

The more similarities that are shared and the greater the degree to which they are held in common, the closer we classify organisms that share them. A dog is more like a wolf than a fox; as a result, the dog and the wolf are classified in the same genus (*Canis*) and the fox is classified in a different genus. Yet a dog is more like a fox than a cat; so the first two are classified in the same family (Canidae) and the cat is classified in a different family. But a dog is more like a cat than a horse; so the first two are placed in the same order (Carnivora), and the horse is placed in a different order. Still, a dog is more like a horse than a fish; therefore, the first two are in the same class (Mammalia) and the fish is in a different class. But a dog is more like a fish than a worm; both dog and fish belong to a single phylum (Chordata) and the worm belongs to an entirely different phylum. The dog has more in common with a worm, however, than it has with an oak tree; therefore, the dog and worm are in the same kingdom (Animalia) and the tree is in a different kingdom (Plantae).

For Darwin, similarity was a result of common ancestry. He interpreted similarity as "family resemblance": two organisms are similar because they are descendants of a common ancestor. Imagine a photograph of a large extended family. The family features are obvious; brothers and sisters resemble one another most closely, cousins somewhat less, and so on. Likewise, say Darwinists, degrees of similarity reveal how closely organisms are related to a common ancestor. For instance, all mammals are built on a common body plan (see figure 5.1). Darwinists interpret this to mean that mammals descended from a common ancestor that originally had that body plan. Differences among mammals are thus said to reveal how the basic plan has been adapted in each species under the pressure of natural selection. But for Darwin, similarities are generally a consequence of kinship.

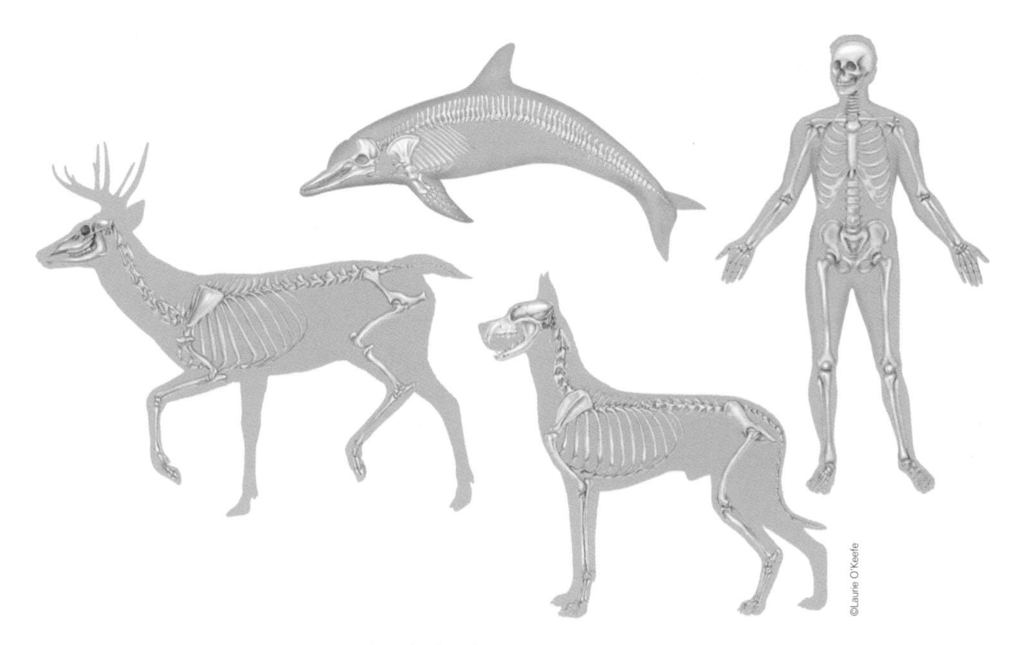


***Figure 5.1*** *Examples of the mammalian body plan.*

Because fossils are not living organisms, we cannot use them to establish relationships of ancestry and descent. Imagine digging up two recent human skeletons. Without identifying marks and written records, we cannot tell how the two are related to each other. (The only exception would be if we extracted identical DNA from them, in which case we could establish that they were twins.) If we cannot tell how two recent skeletons from the same species are related, we certainly cannot tell how ancient fossils of different species are related—or if they're related at all.

Consequently, paleontologists must rely on similarities to construct hypotheses about evolutionary relationships. According to Darwinian theory, the greater the number of similarities between two organisms, the closer their evolutionary relationship is likely to be. Yet to discern and interpret similarities is not as simple as it may sound. Once we proceed beyond the rather obvious similarities (e.g., birds have feathers and fish have scales), it is not always easy to decide which organisms should be classified together. Similarities appear in a patchwork pattern that makes classification difficult.

Consider the marsupials—mammals that complete their embryo development in an outer pouch on the mother's belly (in contrast to placental mammals, such as humans, that complete their embryo development inside the mother's womb). Marsupials and placental mammals are sometimes strikingly similar (see figure 5.2). For instance, in skeletal structure, the North American wolf and the now-extinct Tasmanian wolf are very close (Tasmania is a large island adjacent to Australia that, like Australia, contains a large variety of marsupials—Tasmanian wolves ate the settlers' livestock and were therefore hunted to extinction). In some features, such as their jaws and dentition, these wolves are nearly indistinguishable (see figure 5.3). The behavior and life-style of the Tasmanian wolf was likewise quite similar to that of the North American wolf.

***Figure 5.2*** *Examples of the many dramatic parallels in placental (above) and marsupial (below) mammals.*

Yet the two animals differ fundamentally in their early development. Despite the striking similarities in the adults, taxonomists focus on this difference and therefore classify the two in widely different categories. Thus, they group the North American wolf with the dog and the Tasmanian wolf with the kangaroo. Darwinists in turn interpret this anatomical difference to indicate that the two types of wolves are only remotely related, and that each had a long and separate evolutionary history dating back to the time when Australia became a separate continent.

(A)

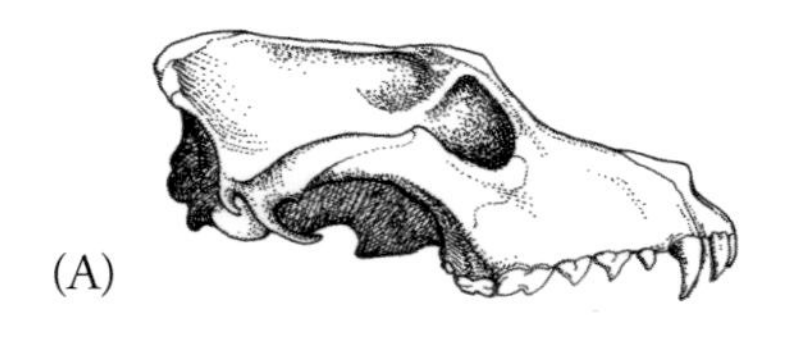

(B)

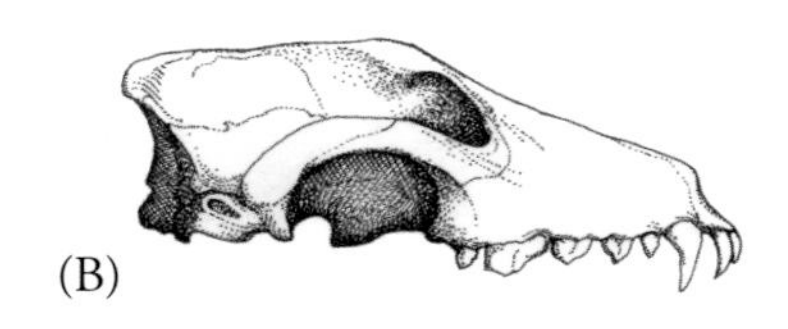

(C)

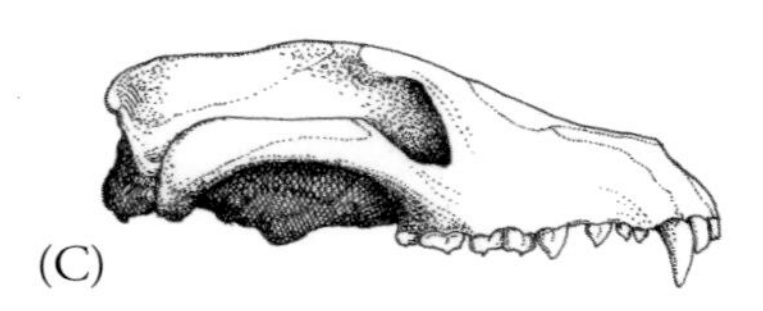

***Figure 5.3*** *The skulls of a dog (A), a North American wolf (B), and a Tasmanian wolf (C). Notice that the skull of the North American wolf is similar to the dog's, which is said to be closely related to it by common ancestry, but nearly identical to the Tasmanian wolf's, which is alleged to be only distantly related to it by common ancestry.*

According to Darwinists, both types of wolves evolved independently into wolf-like forms, a phenomenon known as *convergent evolution.* Thus, separate evolutionary paths supposedly gave rise to similar features that were independently adapted to meet similar environmental demands. The Darwinian assumption is that the selective regime and environmental niches that produced the North American wolf closely approximated those in Australia, so that in adapting to similar environments the two widely separated wolves increasingly resembled one another until they became superficially almost identical.

Yet there are two problems with this line of reasoning: (1) the evidence doesn't support the assumption that environmental demands in the evolutionary history of both wolves were similar; and (2) there's no reason to think that even with similar environmental demands, two separately evolving organisms should evolve not just a single similar feature, but rather a whole suite of similar features that match up point for point. But the coincidences don't stop there. Besides marsupial wolves, Australia is also home to a host of other marsupial look-alikes: marsupial cats, squirrels, ground hogs, anteaters, moles, and mice. Thus, not only were individual organisms on different continents independently evolving whole suites of similar features, but the two entire subclasses of marsupial and placental mammals on two different continents were independently evolving the same morphological types, each sharing whole suites of similar features. Without some form of design or teleological guidance, convergent evolution requires a piling of coincidences on coincidences that strains credulity.

## 5.2 ANALOGY AND HOMOLOGY

Marsupials raise an interesting question for taxonomy: If similarity is the basis for classification, what should we do when one set of similarities clashes with another? The Tasmanian wolf is strikingly similar to the North American wolf in most features. Yet the Tasmanian wolf is a marsupial and hence similar in this one significant feature to the kangaroo. Upon which similar features do we therefore build our classification scheme?

In biology, similar features come in two basic forms: functional and structural. For instance, bird wings and insect wings are used for flying. Both function in the same way: air currents push against the surface of the wings to provide lift, and flapping the wings provides forward thrust. Yet the internal structure of a bird wing is very different from that of an insect wing. The bird wing consists of flesh, supplied with nourishment and oxygen by a network of blood vessels. Its structural stability comes from bones on the inside. The insect wing, on the other hand, has no bones or blood vessels. It consists of a thin membrane stretched tightly around a network of wiry structures, similar to a kite.

Bird wings and insect wings serve the same function but are very different in structure. Here the similarity is functional rather than structural. But that's not the only possibility. Similarity of structure and divergence of function is common. For instance, the pattern of bones in a bat's wing is similar to that in a porpoise's flipper, though the wing is used for flying and the flipper is used for swimming.

Which type of similarity is more relevant for biological classification—similarity of function or similarity of structure? The founder of modern taxonomy, Carolus Linnaeus, faced this problem in the late 1700s and chose to classify organisms according to structural rather than functional similarity. Thus, he classified the flying insects with other insects because of their similarity in structure, and not with birds. In the 1840s, British anatomist Richard Owen called functional similarity "analogy" and structural similarity "homology." At the time, Owen intended the distinction simply as an aid to Linnaeus's biological classification scheme: analogy suggests independent adaptations to external conditions, while homology suggests deeper structural affinities. Both Linnaeus and Owen regarded structural similarities as a more reliable guide for classifying organisms.

The classic examples of homologous structures are the forelimbs of vertebrates. Although a bat has wings for flying, a porpoise has flippers for swimming, a horse has legs for running, and a human has hands for grasping, the bone patterns in their forelimbs are similar (see figure 5.4). Such skeletal similarities, along with other internal affinities such as warm-bloodedness and milk production, justify classifying all these creatures as mammals despite their external differences.

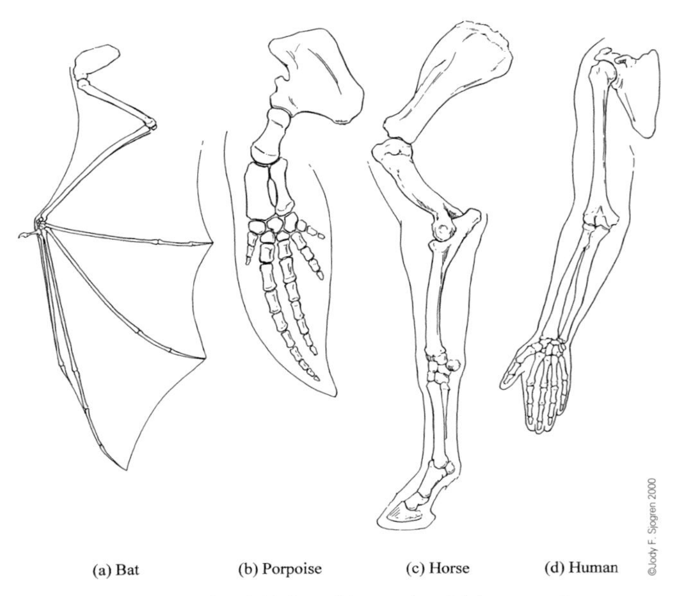


***Figure 5.4*** *Forelimbs of (a) bat, (b) porpoise, (c) horse, and (d) human, showing bones considered to be homologous.*

Like other pre-Darwinian biologists, Owen considered homologous features to be derived from a common pattern or "archetype." An archetype, however, could be understood in various ways: a disembodied Platonic idea, a plan in the mind of the Creator, an Aristotelian form inherent in the structure of nature, or a prototypical organism, among others. Both Owen and Darwin regarded the archetype as a prototypical organism. Owen, however, was not an evolutionist. Owen regarded organisms as constructed on a common plan (in the same way that a building is constructed from architectural plans). Darwin regarded them as descended from a common ancestor.

According to Darwin, homologies provide evidence of evolutionary descent. Thus, Darwinian theorists interpret the similarity in skeletal structure among vertebrates as evidence that all descended from a common ancestral form. Take the human hand and the dog's forepaw. Both contain the very same types of bones, and the bones of both have even been assigned the same names (see figure 5.5). Although the bone pattern is similar, the individual bones are quite different in shape and function. The dog cannot grasp objects with its digits or oppose its thumb to the rest of its toes, either individually or collectively. Human hands, of course, are not made for walking or standing for any length of time. Their similarity to dogs' paws is therefore independent of function. The only alternative, according to Darwinists, is that their similarity results from inheritance from a common ancestor that possessed this basic arrangement of bones.

What was this common ancestor? According to Darwinists, it was some type of insectivore (the order of mammals that includes moles and shrews). This insectivore, in turn, is assumed to have evolved from an ancestral reptile, which in turn received its forelimbs from a semiaquatic fish that supposedly used its bony fins to drag itself from mudhole to mudhole during seasonal droughts. A fin might not seem the best structure to use for a horse's leg or a dog's paw, but it was the only raw material available (and evolution can only work with available materials). Accordingly, both a human's hand and a dog's paw are homologous to the fin of a fish. Darwinian evolution is opportunistic. Lacking novel ideal parts, the Darwinian mechanism must fashion and remold existing parts. In this way, the bones in a fin are said to have evolved into the bones in a human hand and the bones in a dog's paw.

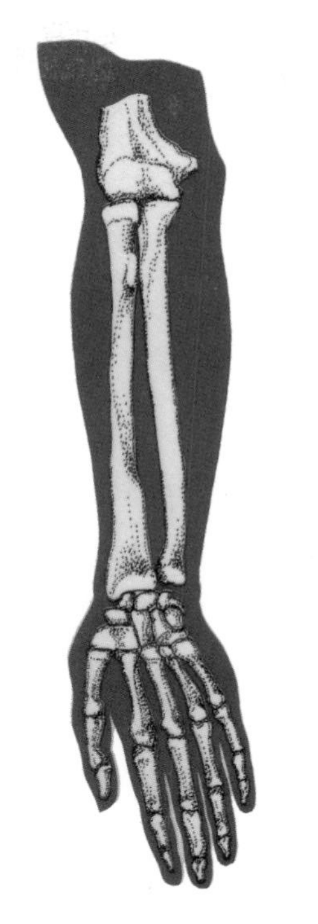

***Figure 5.5*** *The front appendages of a human and a dog. Observe how the bones in each correspond to the bones in the other.*

Darwinian theorists are committed to fitting the history of life within an evolutionary scheme. While they have offered many rationalizations for why gaps should exist in the fossil record, they continue to look for intermediate fossils as evidence for their theory. Yet unbroken series of fossils showing the descent of one species from another are never found. The common ancestors needed by Darwin's theory are missing from the fossil record, creating large gaps between the major groups of organisms. Therefore, homology becomes the only real means to establish evolutionary relationships. But this means that Darwinian theorists need to interpret analogies (i.e., functional similarities) that are not also homologies (i.e., structural similarities) as the result of convergent evolution (i.e., the evolution of similar adaptations—presumably in response to similar environments). Accordingly, the evolution of flight in insects and birds occurred by independent processes that converged in response to similar environmental pressures.

Yet the distinction between homology and analogy is not always easy to draw. Homologous structures may be quite different in both appearance and function (compare a bat's wing and a horse's foreleg in figure 5.4). On the other hand, very similar-looking structures that perform similar functions may be merely analogous and thus irrelevant both for classification and for drawing evolutionary relationships. Think of the body shape of the fish and the whale (see figure 5.6). Early in his career, Linnaeus classed the Cetaceans (whales) as fish, not realizing that their fish-like shape was not a homologous but merely an analogous resemblance. Repeatedly in the study of taxonomy, structures of astonishing similarity at first regarded as homologous were later determined to be merely analogous. Many structures are in fact mixed, being both homologous and analogous. We turn next to one such structure.

***Figure 5.6*** *The fish-like shape of the whales is an example of similarity that does not come from close kinship. It is therefore held to be an analogous resemblance.*

## 5.3 THE PUZZLING PANDAS

The giant panda and the lesser, or red, panda illustrate the problem of distinguishing homologous and analogous structures. Both pandas are native to the bamboo forests of southwest China. For over a century, scientists studying the two pandas were

*Figure 5.7 Giant panda and lesser, or red, panda.*

unable to agree whether they are members of the bear family or of the raccoon family. Since the first serious attempt to classify these animals in 1869, more than forty major scientific studies have been published on the topic. The astonishing thing is that these studies have split almost down the middle on the bear/raccoon question, half concluding that they are bears, half concluding that they are raccoons. One scientist described this failure to resolve the issue as a "taxonomic game of ping pong."

Then in 1964, Dwight Davis, Curator of Vertebrate Anatomy at the Field Museum of Natural History in Chicago, published what soon became widely accepted as the definitive discussion on the matter, which finally settled the argument to the satisfaction of most biologists.[1] Davis concluded that the giant panda was not a raccoon but a bear, and that the red panda was not a bear but a raccoon! More recently, accumulating biochemical data have extended the list of similarities between the giant panda and other bears.

But why, then, were biologists convinced for so long that the two pandas were close relatives, both in the same family? One reason is geographical. If the red panda is a raccoon and the giant panda is a bear, then the red panda is the only raccoon outside the Western hemisphere. But having a lone raccoon stranded in China struck many biologists as implausible. More plausible was that neither the giant nor the red panda was a raccoon or that both were.

Besides residing in the same geographical area, the giant and the red panda share an impressive number of behavioral and physical traits. For instance, the muzzle or snout of each is similar in shape, as are their upper jaws—see figure 5.8. Notice how short the muzzles of both are in comparison to the polar bear and how sharply the jaw bones widen toward the back of the head. Both pandas also have massive premolar teeth and enlarged chewing muscles that work in coordination with these traits.

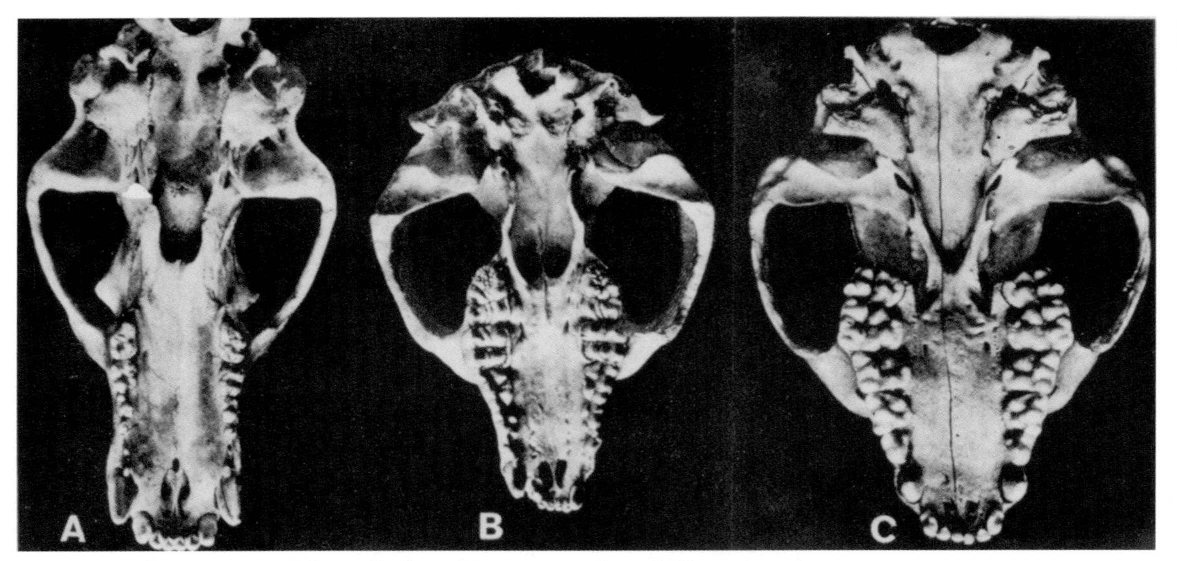


*Figure 5.8* *The upper jaws of (A) a polar bear, (B) a giant panda, and (C) a red panda.*

These similarities are obvious to even casual observers, but others are not. The two pandas differ from other bears in sharing several subtle characteristics. For instance, their stomachs, alimentary tracts, and livers are similar and differ significantly from those of other bears. Evolutionists regard the shared bamboo diet of the two pandas as at least in part responsible for this convergence. But that seems not to be the whole story. Genetically, the giant panda has much in common with the other bears; yet it has only forty-two chromosomes, far closer to the red panda's thirty-six than to the seventy-four chromosomes of most bears.

Among biologists, the giant panda is best known for its "thumb," which gives it a dexterity not found among other bears. This structure operates like an opposable thumb, although it is not a true thumb and is only partially opposable (see figures 5.9 and 5.10). Actually, it is an enlarged bone of the panda's wrist, known as the radial sesamoid. The cluster of bones in the panda's wrist function together smoothly, many of them operating surface to surface, through all of the paw's manipulations—opening, closing, swiveling, swatting, etc. The giant panda's facility at handling and stripping bamboo is truly remarkable. This activity consumes a major part of its day, and the enlarged radial sesamoid is the key to its success at this activity.

Like the giant panda, the red panda also has an enlarged radial sesamoid that it uses to handle and strip bamboo, though the red panda's thumb is not as prominent as the giant panda's. Not only are these structures similar in both pandas, but so are their support structures. For instance, the special way the tendon from the abductor muscle fits into the radial sesamoid is the same in both pandas, as is the working surface among the cluster of wrist bones. In addition, the red and the giant panda have many similar behavioral characteristics. For instance, unlike most other bears, neither one hibernates.

The two pandas share an impressive list of characteristics that, on the face of it, provide persuasive evidence for their common ancestry. Indeed, biologists initially thought of these detailed similarities as homologies, interpreted them as the result of common ancestry, and classified the pandas together in one family. But suppose we accept the view that currently predominates among biologists, namely, that the two pandas belong to different families and that the unique features they share are not homologous but rather analogous—examples of convergent evolution. This means that the two pandas did not receive these features by descent from their respective ancestral families, and that a structure as peculiar as the panda's thumb (the enlarged radial sesamoid) evolved twice from scratch. Coincidences like this (and there are many in the living world) raise the question whether similar features ever provide reliable information about evolutionary relationships.

***Figure 5.9*** *The panda's "thumb." Actually an enlarged bone in the wrist, located where the panda can use it with his claws to grip objects.*

***Figure 5.10*** *Giant panda exhibiting paw and "thumb."*

## 5.4 DARWINISM'S REDEFINITION OF HOMOLOGY

In The Origin of Species Darwin argued that the best explanation for homology is descent with modification: "If we suppose that an early progenitor—the archetype as it may be called—of all mammals, birds and reptiles, had its limbs constructed on the existing pattern," then "the similar framework of bones in the hand of a man, wing of a bat, fin of the porpoise, and leg of the horse . . . at once explain themselves on the theory of descent with slow and slight modifications."[2] Darwin considered homology important evidence for evolution, listing it among the facts that "proclaim so plainly, that the innumerable species, genera and families, with which this world is peopled, are all descended, each within its own class or group, from common parents."[3]

But some similar structures are *not* acquired through common ancestry, as we saw in the case of marsupial and placental mammals. We've also just seen this with the panda's thumb—biologists no longer regard the giant panda's thumb as related by common ancestry to the red panda's thumb. Likewise the structure of an octopus eye is remarkably similar to the structure of a human eye, yet biologists do not think that the common ancestor of octopuses and humans possessed such an eye. Darwinists now regard similar structures such as this to be the result of convergent evolution.

To ensure that only structures inherited from a common ancestor would be called homologous, Darwin's followers therefore *redefined* homology to mean similarity due to common ancestry. According to Ernst Mayr, one of the principal architects of neo-Darwinism, "After 1859 there has been only one definition of *homologous* that makes biological sense. . . . Attributes of two organisms are homologous when they are derived from an equivalent characteristic of the common ancestor."[4]

Even after homology was redefined, however, the Darwinian account remained incomplete without a mechanism to explain why homologous features were so similar in such different organisms. When neo-Darwinism arose in the 1930s and 1940s, it promised to resolve this problem: Homologous features were the result of similar genes inherited from a common ancestor. But biologists have known for decades that homologous features can arise from dissimilar genes, and that similar genes can underlie non-homologous features. So the mechanism that produces homology remains unknown (see the general notes on the CD included with this book).

Furthermore, if homology is *defined* as similarity due to common ancestry, it cannot be used as *evidence* for common ancestry except by reasoning in a circle. Recall the example of similar bone patterns in vertebrate forelimbs (figure 5.4), which Darwin regarded as evidence for the common ancestry of the vertebrates. A neo-Darwinist who wants to determine whether vertebrate forelimbs are homologous must first determine whether they are derived from a common ancestor. In other words, there must be evidence for common ancestry before limbs can be called homologous. But then to turn around and argue that homologous limbs point to common ancestry creates a vicious circle: Common ancestry establishes homology, which in turn establishes common ancestry.

Various biologists and philosophers have noticed and criticized this circularity. In 1945 J. H. Woodger wrote that the new definition was "putting the cart before the horse."[5] Alan Boyden pointed out in 1947 that neo-Darwinian homology requires "that we first know the ancestry and then decide that the corresponding organs or parts" are homologous. *"As though we could know the ancestry without the essential similarities to guide us!"*[6] When neo-Darwinian paleontologist George Gaylord Simpson tried to use homology-as-common-ancestry to infer evolutionary relationships, biologists Robert Sokal and Peter Sneath criticized him for his "circularity of reasoning."[7]

Even so, many neo-Darwinists try to defend their use of homology against the charge of circularity. In 1966 Michael Ghiselin pointed out that the neo-Darwinian definition is not circular because homology is not defined in terms of itself.[8] But this did not solve the problem, because although the definition itself is not circular, the reasoning based on it is. The following year, David Hull argued that the reasoning is not circular, but merely an example of the scientific "method of successive approximation."[9] According to Hull, evolutionary biologists start by assuming a particular hypothesis of descent and then use similarities to refine the hypothesis. But the method—which critics at the time derided as "groping"—works, if it works at all, only by assuming the truth of common ancestry. If the question is whether Darwin's theory is true in the first place, then Hull's method of successive approximation is just another circular argument.[10]

The controversy has raged ever since. Neo-Darwinists continue to defend their conjunction of homology with common ancestry, whereas critics object that it confuses definition with explanation and leads to circular reasoning. Thus philosopher of biology Ronald Brady, one of the more outspoken critics of neo-Darwinism in the last generation, observed, "By making our explanation into the definition of the condition to be explained, we express not scientific hypothesis but belief. We are so convinced that our explanation is true that we no longer see any need to distinguish it from the situation we were trying to explain. Dogmatic endeavors of this kind must eventually leave the realm of science."[11]

There are only three ways to avoid the circular reasoning brought on by simultaneously defining and explaining homology in terms of common ancestry:

1. To embrace the neo-Darwinian definition of homology but give up trying to infer common ancestry from it—in other words, to acknowledge that homology results from common ancestry but no longer provides evidence for it. "Common ancestry is all there is to homology," wrote evolutionary biologist David Wake in 1999; thus "homology is the anticipated and expected consequence of evolution. Homology is not evidence of evolution."[12]

2. To retain the pre-Darwinian definition of homology as structural similarity but acknowledge that this reopens the question of whether common ancestry is the best explanation for it. Intelligent design proponents are currently at the forefront in asking this question. Unlike the last option, this one does not give common ancestry the benefit of the doubt as the best explanation of homology.

3. To define homology in terms of common ancestry but then seek evidence for common ancestry that is independent of structures in living organisms—for example, patterns in DNA sequences and the fossil record. Alternatively, evidence may come from processes such as developmental pathways and developmental genetics.

Of these three options, the third is currently the most popular. It correctly distinguishes between defining homology as a consequence of common ancestry and employing homology as evidence for it. Also, unlike the first option, it does not merely presuppose common ancestry but looks for independent evidence of it. We examined the fossil evidence for common ancestry in Chapter 3, and we examine the developmental evidence for common ancestry in the general notes (note that neither of these convincingly supports common ancestry). In the next section (5.5), we look at the molecular evidence.

## 5.5 MOLECULAR PHYLOGENY

A phylogeny is a presumed evolutionary history of a group of organisms. Until recently, phylogenies were inferred from anatomical and physiological features (such as bone structures or warm-bloodedness). With the advent of modern molecular biology, however, many phylogenies are now based on DNA and protein comparisons. All living organisms, from bacteria to humans, contain DNA, RNA and proteins. A DNA molecule is a long chain consisting of various combinations of four subunits, abbreviated A, T, C and G. The order of these subunits specifies the sequences of messenger RNAs, that in turn specify the sequences of amino acids in an organism's proteins. During reproduction, the sequence of subunits is copied from one DNA molecule to another, but molecular accidents, or mutations, sometimes make the copy slightly different from the parent molecule. Therefore, organisms may have DNA molecules (and thus proteins) that differ somewhat from the DNA and proteins of their ancestors.

In 1962 Emile Zuckerkandl and Linus Pauling suggested that comparisons of DNA sequences and their protein products could be used to determine how closely organisms are related.[13] Organisms whose DNA or proteins differ by only a few subunits are presumably more closely related in evolutionary terms than those that differ by more (see figure 5.11). If mutations have accumulated steadily over time, the number of differences between organisms might even serve as a "molecular clock" that indicates how many years have passed since their DNA or protein was identical—that is, how long ago they shared a common ancestor.

| | DNA Sequence |
|---|---|
| **Organism 1** | C A T C G A |
| **Organism 2** | C A T C T A |
| **Organism 3** | C A T G T A |

***Figure 5.11*** *All DNA molecules consist of linear sequences of four subunits, abbreviated A, T, C and G. In the short sequence shown here, Organism 2 differs from Organism 1 in one position, while Organism 3 differs from it in two positions. If this were the only sequence being compared, Organisms 1 and 2 would be considered to have a more recent common ancestor (i.e., to be more closely related) than Organisms 1 and 3.*

Much of the early work in molecular phylogeny relied on proteins, but determining protein sequences is slow work. With the development of faster techniques for determining DNA sequences, it became more common to analyze the DNA or RNA coding for proteins rather than the proteins themselves. The conversion of the information in DNA sequences into proteins depends on complex molecular structures in the cell called *ribosomes*, which consist partly of ribosomal RNA, or "rRNA." Since 1980 the DNA sequences that code for rRNA have provided many of the data for molecular phylogeny.

Comparing DNA or RNA sequences is simple in theory but complex in practice. Since an actual segment of DNA may contain thousands of subunits, lining them up to start a comparison is itself a tricky task, and different alignments can give very different results. Nevertheless, conclusions drawn from molecular comparisons have been brought to bear on key events in the history of life. For example, molecular analyses have been used in an attempt to show that the Cambrian explosion (see Chapter 3) was not as abrupt as the fossil record indicates. But the results of those analyses have been inconsistent and fail to solve the problem (see sidebar).

## How Accurate Is the Molecular Clock?

Did the animal phyla emerge suddenly in the Cambrian, as the fossils seem to indicate, or did they slowly diverge from a common ancestor millions of years before, as Darwin's theory implies? It's not possible to analyze DNA, RNA or proteins from Cambrian fossils, but molecular biologists are able to compare the sequences of these molecules in living species. Assuming that sequence differences among the major animal phyla are due to mutations and that mutations accumulate at the same rate in various organisms over long periods of time, biologists use sequence differences as a "molecular clock" to estimate how long ago the phyla shared a common ancestor.

Dates obtained by this method vary widely. In 1982, Bruce Runnegar offered an initial estimate of 900-1000 million years for the initial divergence of the animal phyla.[14] In 1996, Russell Doolittle and his colleagues proposed a date of 670 million years,[15] while Gregory Wray and his colleagues proposed 1200 million.[16] Obviously, 670 million years comes closer to fitting the fossil record than 1200 million. In 1997, Richard Fortey and his colleagues endorsed the older date, and in 1998 Francisco Ayala and his colleagues endorsed the younger one.[17] But these two dates represent a spread of 530 million years, or as much time as has elapsed between the Cambrian explosion and the present.

This "range of divergence estimates," in the opinion of American geneticist Kenneth Halanych, testifies "against the ability to date such ancient events" using molecular methods.[18] According to paleontologists James Valentine, David Jablonski, and Douglas Erwin, "the accuracy of molecular clocks is still problematical, at least for phylum divergences," since the estimates vary by hundreds of millions of years "depending on the techniques or molecules used." Valentine and his colleagues consider the fossil record to be the primary evidence, and maintain that the molecular data "do not muffle the [Cambrian] explosion, which continues to stand out as a major feature" in animal evolution.[19]

Not only have molecular analyses failed to resolve the timing of the Cambrian explosion, but they have also failed to resolve the relationships of the animal phyla that appeared in it. Anatomically, vertebrates (animals with backbones, such as humans) are considered more closely related to arthropods (crustaceans and insects) than to nematodes (tiny roundworms found in soil and marine sediments). In 1997, however, Anne Marie Aguinaldo and her colleagues proposed a radical revision of animal relationships based on comparisons of their rRNA. According to the revised phylogeny, we are no more closely related to insects than we are to roundworms.[20]

This didn't resolve the question of human relatedness to other organisms, however, because molecular phylogenies of the animal phyla differ depending on which rRNA is analyzed, or even which laboratory does the analysis. In 1999, evolutionary biologist Michael Lynch wrote: "Clarification of the phylogenetic relationships of the major animal phyla has been an elusive problem, with analyses based on different genes and even different analyses based on the same genes yielding a diversity of phylogenetic trees."[21]

Lynch was optimistic that with improved methods molecular phylogeny would clarify relationships among the animal phyla. Despite the continuing efforts of many researchers, however, conflicts among molecular phylogenies of the major animal groups have not just persisted, but grown. In 2000, French molecular biologist André Adoutte and his colleagues affirmed their confidence in the new rRNA-based animal phylogeny.[22] In 2002, however, Pennsylvania State University biologist Jaime E. Blair, along with his colleagues, compared over 100 protein alignments and concluded: "The grouping of nematodes with arthropods is an artifact that arose from the analysis of a single gene, 18s rRNA. The results presented here suggest caution in revising animal phylogeny from analyses of one or a few genes . . . Our results indicate that insects (arthropods) are genetically and evolutionarily closer to humans (vertebrates) than to nematodes."[23]

In 2004, National Center for Biotechnology Information researcher Yuri I. Wolf and his colleagues analyzed over 500 proteins using three different phylogenetic methods. They concluded that "the majority of the methods . . . grouped the fly with humans to the exclusion of nematodes."[24] The following year, French biologist Hervé Philippe and colleagues analyzed 146 genes and 35 species representing 12 animal phyla. They concluded that their data grouped arthropods with nematodes to the exclusion of vertebrates.[25]

The debate continues. Round and round it goes, and where it stops—or even whether it will stop—nobody knows. In a commentary accompanying Philippe's report, University of Edinburgh evolutionary biologists Martin Jones and Mark Blaxter wrote: "Despite the comforting certainty of textbooks and 150 years of argument, the true relationships of the major groups (phyla) of animals remain contentious."[26] Although Jones and Blaxter favored Philippe's view, they predicted that the molecular tree of life "will sprout new shoots—and new controversies—very soon." Indeed, in December 2005, biologist Antonis Rokas and colleagues used two different methods to analyze fifty genes from seventeen animal groups. They noted that "different phylogenetic analyses can reach contradictory inferences with absolute support" and concluded that the evolutionary relationships among the phyla "remain unresolved."[27]

Clearly, molecular analyses have failed to resolve both the timing of the Cambrian explosion and the relationships of the phyla that emerged in it. But what about biologists using molecules to look for the common ancestor of all living things? Are molecular methods more successful in that case? No. In fact, the failure of molecular methods becomes even more pronounced. According to Darwin, "all the organic beings which have ever lived on this earth may be descended from some one primordial form."[28] In the 1980s, University of Illinois (Urbana) microbiologist Carl R. Woese pioneered the use of 18s rRNA to construct an all-encompassing tree of life and identify the "universal common ancestor."

But by the late 1990s, serious problems had emerged. "When scientists started analyzing a variety of genes from different organisms," wrote University of California (Los Angeles) molecular biologists James A. Lake, Ravi Jain and Maria A. Rivera, they "found that their relationships to each other contradicted the evolutionary tree of life derived from rRNA analysis alone."[29] According to Hervé Philippe and Patrick Forterre: "With more and more sequences available, it turned out that most protein phylogenies contradict each other as well as the rRNA tree."[30]

In 1998, Woese wrote: "No consistent organismal phylogeny has emerged from the many individual protein phylogenies so far produced." He concluded that primitive organisms acquired many of their genes and proteins, not by Darwinian descent with modification, but by "lateral gene transfer," in which organisms directly swap genes with other organisms. "The universal ancestor," he wrote, "is not an entity, a thing, but a community of complex molecules—a sort of primordial soup—from which different kinds of cells emerged independently."[31]

At about the same time, Dalhousie University evolutionary biologist W. Ford Doolittle concluded that lateral gene transfer among ancient organisms meant that molecular phylogeny might never be able to discover the "true tree" of life, not because it is using the wrong methods or the wrong genes, "but because the history of life cannot properly be represented as a tree." He concluded: "Perhaps it would be easier, and in the long run more productive, to abandon the attempt to force" the molecular data "into the mold provided by Darwin."[32] Instead of a tree, Doolittle proposed "a web- or net-like pattern."[33]

Philippe and Forterre proposed refinements of Doolittle's methods that enabled them to re-root the universal tree. The new root, however, was not what everyone else was looking for. In the standard Darwinian scenario, life began with simple cells (i.e., cells without nuclei) that then evolved into complex cells (i.e., cells with nuclei). Philippe and Forterre turned this scenario on its head, rooting their phylogeny in a complex cell with a nucleus from which simpler cells without nuclei would then have *de*-evolved.[34] To be sure, the Darwinian scenario is compatible with the evolution of simplicity from complexity via natural selection pruning away complexity. Thus it is a perfectly legitimate within Darwinism to explain the evolution of cells without nuclei from cells with nuclei. But that leaves unexplained how cells with nuclei first formed. The whole impetus of Darwinism is, after all, to explain the complex in terms of the simpler. Philippe and Forterre thus replaced the problem they set out to resolve with an even bigger one, namely, how to explain the origin of the first living cell—which, in their hypothesis, is much more complicated than anything previously imagined by origin-of-life researchers (see Chapter 8).

The controversy over the universal tree of life continues. In 2002, Woese suggested that biology should go beyond Darwin's doctrine of common descent.[35] In 2004, he wrote: "The root of the universal tree is an artifact resulting from forcing the evolutionary course into a tree representation when that representation is inappropriate."[36] In 2004, Doolittle and his colleagues proposed replacing the tree of life with a net-like "synthesis of life,"[37] and in 2005 they recommended that "representations other than a tree should be investigated."[38] Meanwhile, other scientists continue to defend the hypothesis that there was a universal ancestor but it was complex rather than simple.[39]

Whatever merits these hypotheses might have, one thing is clear: molecular phylogeny has failed, utterly and completely, to establish that universal common ancestry is true. It has failed to provide reliable dating for, much less chronicle, the history of life. It has failed to specify what evolutionary relationships, if any, exist among living forms. And it has failed, spectacularly, to give an unambiguous indication of what the universal common ancestor—if there even was one—might have looked like. Was it an organism, a group of organisms, or a batch of chemicals? The molecular evidence for common ancestry, like the fossil and embryo evidence, is plagued with inconsistencies and unanswered questions. In fact, Darwinism must be *assumed* in order to explain the evidence (or, as is increasingly the case, to explain it *away*).

## 5.6 VESTIGIALITY: THE BEST EVIDENCE FOR EVOLUTION?

In biology, a vestigial structure is one that performed a useful function in the past but now no longer performs it. The structure's continued existence therefore appears to be the result of what could be called "generative inertia." In other words, the structure continues to be reproduced from one generation to the next because it's easier just to keep reproducing it than to prune it away and thereby eliminate it once and for all. Of course, this assumes that the structure isn't harmful, in which case natural selection would work directly to eliminate it. Biology, however, seems to present us with vestigial structures that are neither harmful nor useful but are merely carried along for the ride.

Whether a structure is truly vestigial in the sense of having lost all use is often debatable. Many organs once regarded as vestigial were later found to have a function after all. For instance, evolutionary biology texts often cite the human coccyx as a "vestigial structure" that hearkens back to vertebrate ancestors with tails. The coccyx consists of several vertebrae at the base of the spine. In dogs, the coccyx plays a crucial role in the mobility of the tail. But humans have no tail. Is the coccyx in humans therefore truly a vestigial structure, one that serves no use or function? In anatomical and medical textbooks, one finds that the coccyx is an important point of contact with muscles that attach to the pelvic floor. In fact, textbooks such as *Gray's Anatomy* will describe this and other functions of the human coccyx quite matter-of-factly. The same is true for the human appendix. Formerly thought to be vestigial, the appendix is now known to be a functioning component of the immune system.

Although many supposedly vestigial structures have on closer examination been found to have a function after all, there also exist structures about which there can be no doubt that they are useless and therefore vestigial. Consider, for instance, the "eyes" of certain salamanders and fish. Where other salamanders and fish have fully formed and functioning eyes, "blind salamanders" and "cave fish" merely have nubs but no actual eyes. These organisms inhabit environments from which light is totally absent. In a classic case of use-it-or-lose-it, the ancestors of these organisms appear at some point to have lost their eyes (see figure 5.12).

***Figure 5.12*** *Blind salamander and cave fish*

Vestigial structures are entirely consistent with intelligent design, suggesting structures that were initially designed but then lost their function through accident or disuse. Nevertheless, vestigial structures also provide evidence for a limited form of evolution. From both a design-theoretic and an evolutionary perspective, a vestigial structure is one that started out functional but then lost its function. Yet, in the case of evolution, vestigiality explains only the loss of function and not its origination. Vestigiality at best documents a *degenerative form of evolution* in which preexisting functional structures change and lose their function. In consequence, vestigiality can provide no evidence for a complexity-increasing, function-adding form of evolution (as neo-Darwinism requires). Vestigiality can at best confirm limited common ancestry. It offers no insight into any mechanisms of evolution that might be responsible for creative innovations.

The case of vitamin C biosynthesis in mammals offers some interesting insights into vestigiality and the difficulty of establishing even limited common ancestry. Although most mammals can synthesize their own vitamin C (ascorbic acid), humans and guinea pigs cannot. (Primates more generally and fruitbats also lack this ability.) To manufacture their own vitamin C, mammals need an enzyme that's encoded by a gene known as GULO (L-gulono-gamma-lactone oxidase). What happened to that gene in guinea pigs and humans? They both have non-functional mutant copies of the GULO gene.[40] In a case of use-it-or-lose-it, both guinea pigs and humans appear to have started out with a gene for synthesizing vitamin C and then lost it (presumably because their diets were rich in vitamin C and there was no need to synthesize it).

Defective genes that no longer code for functional proteins are known as *pseudogenes*. Evolutionary biologists regard these apparently vestigial remnants of the GULO genes both as evidence contradicting intelligent design (why would any sensible designer install non-functional gene sequences in an organism?) and as evidence of common ancestry. Evolutionary biologists note that the genetic lesions found in the GULO pseudogenes of non-human primates, such as the chimpanzee, closely match those found in humans. But why would these "shared errors" exist if humans did not stem via material processes of descent with modification from a common ancestor shared with other primates?

Remarkably, however, a 2003 study by Inai *et al.* comparing the GULO pseudogene from guinea pigs to that of humans shows many of the same "shared errors" between the two species.[41] But guinea pigs and primates are not closely related, or so-called "sister" groups, among the mammals. Thus, the guinea pig GULO pseudogene and the human GULO pseudogene could not have originated from the same ancestral pseudogene—despite their "shared errors." The disabling substitutions to the GULO exons (i.e., coding sequences of DNA) must have occurred in parallel.

But if the same substitutions occurred in parallel, or independently, across unrelated mammalian groups (i.e., guinea pigs and humans), how do we know that the "shared errors" exhibited by other primates (e.g., chimpanzees) and humans could not also have occurred independently? Inai *et al.* therefore argue that the GULO sequences may be subject to "hot spots"—in other words, they may contain regions more likely to undergo substitutions.[42] But in that case what appear to be shared errors may reflect not common ancestry but an increased probability of independent substitutions along coding sequences. When we observe so-called "shared errors," therefore, we may be seeing changes that occurred in parallel among groups that do not necessarily stem from a common ancestor.

In such shared-error arguments, there are always two issues at play: (1) Are what appear to be errors really errors at all? and (2) If so, how and to what degree do shared errors confirm evolutionary relationships? We take up these questions in detail in the general notes. The basic idea behind a shared-error argument is that the same error is too unlikely to happen twice (or more often) independently, and thus must have a common cause, which in most biological contexts is taken to be descent from a common ancestor. Nonetheless, unlike human contexts (in which what counts as an error is often straightforward), in biological contexts unidentified functions are always a possibility. Unless all possible functions can be ruled out, shared errors arguments in living organisms may suggest evolutionary relationships, but they don't decisively settle the issue.

Another argument for common ancestry from molecular similarity—though strictly speaking it does not appeal to vestigiality—has the same flavor as the argument from shared errors. The molecular similarity in question is the coding of similar genes for similar proteins. The argument hinges on the fact that the genetic code is "degenerate" in the strict technical sense that more than one codon (i.e., triplet of DNA) can map to a given amino acid (see the general notes to section 2.4). This means that for a given amino acid sequence and any associated protein product, many different DNA sequences could, in principle, code for it.

Think of protein as the meaning of a sentence and DNA as the actual words of the sentence. The wording of the sentence can be changed without altering the meaning. Thus the sentences "give me your money" and "give your money to me" convey the same meaning even though the wording of the sentences differs. Now, it turns out that for similar proteins in different organisms, the "wording" of the genes that code for those proteins tends also to be similar. For instance, not just the protein products but the very genes coding for them are very similar in humans and chimpanzees (see section 1.3).

In contrast, degeneracy of the genetic code permits the codings to be drastically different while still coding for identical sequences of amino acids and therefore (presumably) identical proteins. This many-ways-to-code-for-the-same-protein property of the genetic code therefore raises the question why separately designed organisms should display similar codings and whether similar codings don't in fact confirm common ancestry. After all, if the whole point of genes is to encode proteins (a point that itself remains to be established), then the particular choice of coding is arbitrary—one coding is as good as another provided it leads to the same protein. Why, then, should the codings be similar or even identical in diverse organisms? Similar codings for similar proteins is certainly consistent with common ancestry. But it is hardly a knock-down argument for it. In fact, the argument here is considerably weaker than in the shared-error case.

Shared errors (if they truly are shared errors) seem inherently unlikely unless they are inherited from a common biological ancestor. But shared codings may simply reflect parsimony of design, in which the same coding is used repeatedly in widely different organisms because changing it would be a waste of energy. If, for instance, an engineering company has a perfectly good specification describing the function of a machine part, and if that part appears in different machines, there's no point in changing the wording of the specification in the different architectural blueprints and operating manuals for the different machines. Much easier is simply to cut and paste the same specification wherever it's needed. The same reasoning applies to shared codings for identical proteins. The argument for common ancestry and against design from similar or shared codings is therefore considerably weaker than the argument from shared errors.

Another problem with the argument from similar codings is that even though "silent mutations" can change the DNA without changing the amino acid sequence of the resulting protein, they may nevertheless affect the behavior of that protein.[43] DNA consists of four nucleotides, triplets of which make up each codon prescribing a particular amino acid. Since there are sixty-four possible codons but only twenty amino acids, the genetic code is "degenerate" in the sense that some amino acids are prescribed by more than one codon. So a mutation that changes a single nucleotide in a codon may not change the amino acid prescribed by that codon. Such mutations are called "silent" (or "synonymous") in the sense that they do not alter the amino acid sequence of the resulting protein.

Yet research by Chava Kimchi-Sarfaty at the National Cancer Institute in Bethesda, Maryland, has shown that proteins with identical amino acid sequences that have been synthesized from different DNA sequences can have different properties. Experiments show that "silent mutations" can make a marked difference in their associated protein products, changing the rate at which the proteins pump drugs from cells and thereby determining the drug resistance of certain cancers.[44] Such findings seriously undercut the shared-codings argument, which presupposes that when genes encode the same amino acid sequence they also produce the same protein. That is not the case. The genetic code is in fact not degenerate in the way degeneracy has traditionally been conceived. Yes, changes in "wording" (i.e., different DNA sequences) can leave amino acid sequences unchanged; nonetheless, "wording" can profoundly affect the "meaning" of those amino acid sequences (i.e., the final protein products).[45]

So vestigiality arguments do not decisively establish common ancestry. Of the arguments presented here, the most persuasive is the one based on shared errors involving pseudogenes and retroposons (see general notes). But even with these arguments, the possibility of undiscovered functions remains open. Also, all of these vestigiality arguments operate below the phylum level. Hence, they do not establish common descent in the full-blown sense of Darwin's great tree of life, in which all organisms trace their lineage back to a universal common ancestor.

Yet regardless, even if we were to grant that all the arguments presented in this section decisively establish common ancestry (albeit below the phylum level), they would tell us nothing about the *mechanism* of evolution, much less about how novel functional structures arose in the first place. Vestigial structures are able to confirm common ancestry only if they are truly functionless (or, in the case of shared codings, functionally equivalent). So vestigiality arguments depend on useless or selectively neutral features being carried along for the ride by generative inertia. They don't show evolution as a creative force. At best, they show the remnants of evolution after evolution has done all its interesting work. Most significantly, they are entirely consistent with evolution as a process directed by a designing intelligence. Bottom line: vestigiality does not make the design problem go away.

## 5.7 RECAPITULATION

Darwin held that neither the fossil record nor homologous structures supported his theory as conclusively as the evidence from embryology: "It seems to me," Darwin wrote in *The Origin of Species*, "the leading facts in embryology, which are second to none in importance, are explained on the principle of variations in the many descendants from some one ancient progenitor."[46]

Those leading facts, according to Darwin, were that "the embryos of the most distinct species belonging to the same class are closely similar, but become, when fully developed, widely dissimilar."[47] Reasoning that "community in embryonic structure reveals community of descent," Darwin concluded that early embryos "show us, more or less completely, the condition of the progenitor of the whole group in its adult state."[48] In other words, similarities in early embryos not only demonstrate that they are descended from a common ancestor, but they also reveal what that ancestor looked like. Darwin considered this "by far the strongest single class of facts in favor of" his theory.[49]

Because Darwin was not an embryologist, he relied for his evidence on the work of others. One of those was German biologist Ernst Haeckel (1834-1919). Haeckel coined the terms "ontogeny" to designate the embryonic development of the individual, and "phylogeny" to designate the evolutionary history of the species. He maintained that embryos "recapitulate" their evolutionary history by passing through the adult forms of their ancestors as they develop. When new features evolve, they are tacked on to the end of development, in a process Stephen Jay Gould called "terminal addition," making ancestral features appear earlier in development than more recently evolved features. Haeckel called this the "Biogenetic Law" and summarized it in the now-famous phrase, "ontogeny recapitulates phylogeny."

Haeckel made drawings of embryos from various classes of vertebrates to show that they are virtually identical in their earliest stages, and become noticeably different only as they develop (see figure 5.13). It was this supposed pattern of early similarity and later difference that Darwin found so convincing in *The Origin of Species:* "It is probable, from what we know of the embryos of mammals, birds, fishes and reptiles, that these animals are the modified descendants of some ancient progenitor."[50] In *The Descent of Man*, Darwin extended the inference to humans: "The [human] embryo itself at a very early period can hardly be distinguished from that of other members of the vertebrate kingdom."[51] Since humans and other vertebrates "pass through the same early stages of development, . . . we ought frankly to admit their community of descent."[52]

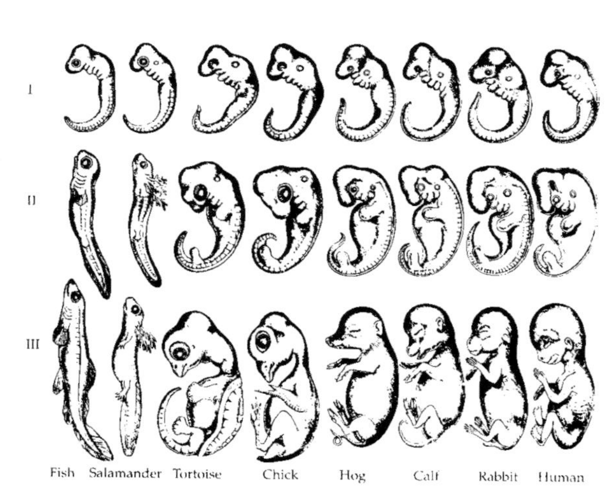


***Figure 5.13*** *Haeckel's Embryos. The embryos are (left to right) fish, salamander, tortoise, chick, hog, calf, rabbit, and human. Note that only five of the seven vertebrate classes are represented, and that half the embryos are mammals. This version of Haeckel's drawings is from George Romanes's 1892 book,* Darwinism Illustrated."

Haeckel's embryos seem like such a self-evident consequence of Darwin's theory that some version of them can be found in many modern textbooks dealing with evolution. Yet biologists have known for over a century that

Haeckel faked his drawings. Vertebrate embryos never look as similar as he made them out to be. Furthermore, the stage Haeckel labeled the "first" is actually midway through development. Indeed, stages exhibiting striking differences appear earlier in development, preceding those whose similarities Haeckel exaggerated. Darwin's "strongest single class of facts," therefore, constitutes a classic example of scientific evidence being skewed to fit a theory.

According to historian Jane Oppenheimer, Haeckel's "hand as an artist altered what he saw with what should have been the eye of a more accurate beholder. He was more than once, often justifiably, accused of scientific falsification, by Wilhelm His and many others."[53] In some cases, Haeckel used the same woodcut to print embryos that he claimed were from different classes. In others, he doctored his drawings to make the embryos appear more alike than they really were. Haeckel's contemporaries repeatedly criticized him for these misrepresentations, and charges of fraud abounded in his lifetime.

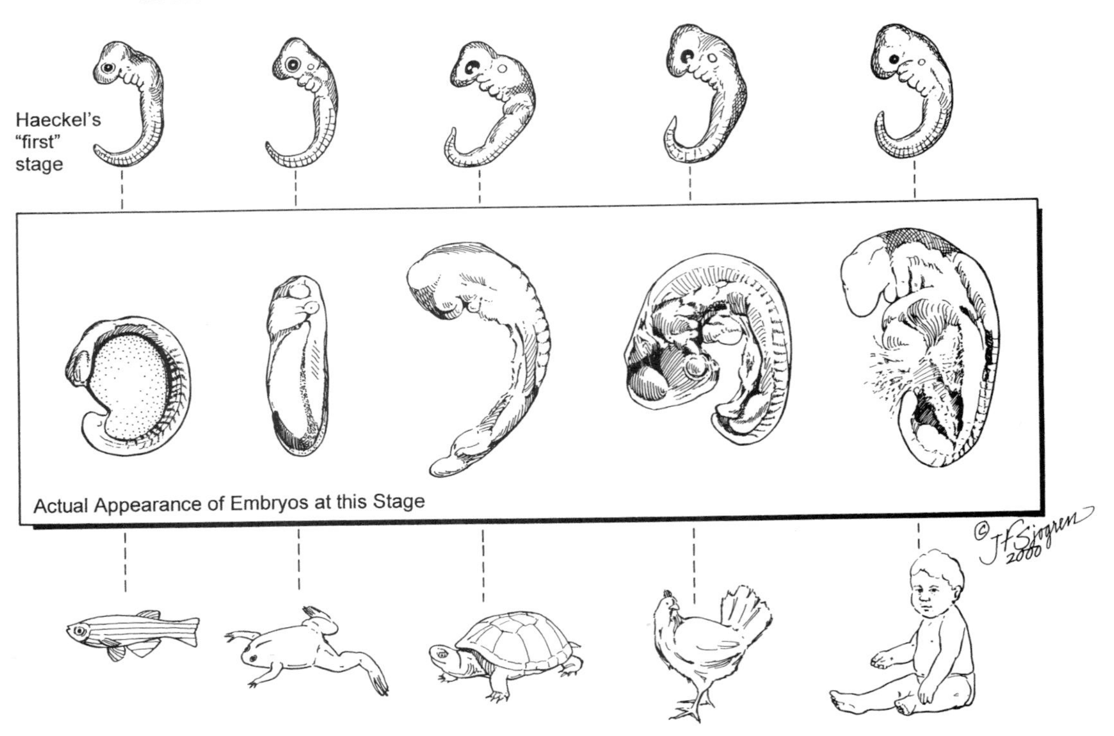


*Figure 5.14 A Comparison of Haeckel's Drawings with Actual Vertebrate Embryos. The top row is Haeckel's. The middle row consists of drawings of actual embryos at the stage Haeckel falsely claimed was the earliest. They are (left to right): a bony fish (zebrafish); an amphibian (frog); a reptile (turtle); a bird (chicken); and a placental mammal (human). To represent amphibians, Haeckel used a salamander, which fits his theory better than a frog; a frog is used here to highlight this fact. Other groups not included by Haeckel (such as jawlessstruct and cartilaginous fishes, and monotreme and marsupial mammals) are significantly different from the embryos shown here.*

When Haeckel's embryos are viewed side-by-side with actual embryos, there can be no doubt that his drawings were distorted to fit his theory (see figure 5.14). Writing in the March 2000 issue of *Natural History*, Stephen Jay Gould noted that "Haeckel had exaggerated the similarities by idealizations and omissions. He also, in some cases—in a procedure that can only be called fraudulent—simply copied the same figure over and over again."[54] Michael Richardson, interviewed by *Science* after he and his colleagues published in 1997 their now-famous comparisons between Haeckel's drawings and actual embryos, put it bluntly: "It looks like it's turning out to be one of the most famous fakes in biology."[55]

From its inception, Haeckel's biogenetic law was a theoretical deduction rather than an empirical inference. It exerted considerable influence in the late nineteenth and early twentieth centuries, but by the 1920s it began to lose favor. In 1922, for instance, British embryologist Walter Garstang criticized Haeckel's biogenetic law as "demonstrably unsound," because "ontogenetic stages afford not the slightest evidence of the specially adult features of the ancestry."[56] According to Garstang, Haeckel's theory that newly evolved features are simply tacked onto the end of development makes no sense: "A house is not a cottage with an extra story on the top. A house represents a higher grade in the evolution of a residence, but the whole building is altered—foundations, timbers, and roof—even if the bricks are the same."[57]

From 1940 to 1958 British embryologist Gavin de Beer published three editions of a book on embryology and evolution in which he criticized Haeckel's biogenetic law. "Recapitulation," wrote de Beer, "i.e., the pressing back of adult ancestral stages into early stages of development of descendants, does not take place."[58] But the problem was not merely the claim that *adult* forms are recapitulated. The problem went deeper. According to de Beer, "Variations of evolutionary significance can and do arise at the earliest stages of development."[59] In other words, the earliest stages of development show important differences, contrary to Darwin's belief that they are the most similar. De Beer concluded that recapitulation is "a mental straitjacket" that "has thwarted and delayed" embryological research.[60]

In summary, although biologists have known for over a century that recapitulation doesn't fit the evidence and although it was supposedly discarded in the 1920s, to this day recapitulation distorts our perceptions of embryos. Furthermore, although biologists have known for over a century that Haeckel's drawings are fakes and that the earliest stages in vertebrate development are not the most similar, textbooks continue to use those drawings (or carefully selected but misleading photos) to argue that Darwin's theory rests on embryological evidence. Recapitulation forces the evidence of embryology into the mold of Darwinian evolutionary theory. In so doing, recapitulation distorts our understanding of both embryology and evolution.

## 5.8 COMMON DESIGN, COMMON ANCESTRY, OR BOTH?

Organisms share many similar features. In fact, many of the similarities are so striking that they could only arise from a common cause. The key question, therefore, concerns the nature of that cause. Is it common design or common ancestry, or perhaps a combination of the two? In the absence of design, common ancestry becomes the default explanation for the similarities that pervade the biological world. Any materialistic theory of evolution, by repudiating design, therefore automatically commits itself to common ancestry. But once design is back in the picture, common ancestry can no longer be taken for granted. To be sure, common design and common ancestry could work together. But common design all by itself could as well be responsible for the similar features that pervade biology.

Many objects in our experience, despite sharing similarities, do not derive from an evolutionary process that traces back to a common ancestor. Consider human artifacts such as cars, paintings, or carpenter's tools. What makes all Corvettes look similar, or all Rembrandts, or all screwdrivers, is that they derive from a common design or pattern in the mind of a designing intelligence. We know from experience that when people design things (such as a car engine), they begin with a basic concept and adapt it to different ends. As much as possible, designers piggyback on existing patterns and concepts instead of starting from scratch. Our experience of how human intelligence works therefore provides insight into how a designing intelligence responsible for life might have worked.

Theories of intelligent design and materialistic evolution offer an explanation for why living things share common features. And since both theories are able to account for similarities, the sheer existence of similarities cannot count as evidence for or against either theory. Yet there is more to consider, and that is the erratic, patchwork pattern of similarities. Recall the puzzle of the marsupials. According to Darwinian theory, what appear to be wolves, cats, squirrels, ground hogs, anteaters, moles, and mice all evolved twice: once as placental mammals and again, totally independently, as marsupials. This amounts to the astonishing claim that an undirected process of random variation and natural selection somehow hit upon identical features many times in widely separated organisms. Or take the problem of flight. The capacity for powered flight requires a tremendously complex set of adaptations that affect virtually every organ of the body (see figure 5.15). Yet Darwinists claim that flight evolved independently—and without design—not once but four times: in birds, in insects, in mammals (bats), and in pterosaurs (extinct flying reptiles).

In biology, similarities do not form a simple branching pattern suggestive of evolutionary (genealogical) descent. Instead, they occur in a complex mosaic or modular pattern. We observe discrete, biologically significant blocks that can be assembled in various ways, not unlike subroutines in a computer program. The genetic programs of different organisms can be viewed as hierarchically organized collections of subroutines carefully chosen from a comprehensive library of subroutines. Or, to use another analogy, similarities among living things are like preassembled units that can be plugged into a complex electronic circuit board. They can be varied according to an organism's need to perform particular functions in air or water or on land. Organisms are mosaics made up of such units or modules. Accordingly, achieving the diversity of biological forms we see today is a matter of differentially combining such distinct "design modules."

***Figure 5.15*** *Not only the wings, but the skeleton, the heart, the respiratory system, and many other parts of the bird are attuned to its ability to fly.*

To say that organisms are hierarchically organized, integrated sets of design modules is not the whole story. The effects of design are, to be sure, present in biology. But so are the accidents of history. The accidents of history can rearrange, modify, or even break design modules. Once such changes occur in biological systems, they can be passed on in reproduction through successive generations of organisms provided selection doesn't weed them out. Similarities due to accidents of history result not from common design but from what in section 5.6 we called "generative inertia"—features that get carried along for the ride simply because the process of reproduction is not selective about what gets reproduced. This is why similar features that exhibit no function are so much more effective at arguing for common ancestry than similar features that do have a function. Similar features that are functional seem, on their face, to be design modules. But similar features that are nonfunctional seem, on their face, to be accidents of history carried along by generative inertia.

It's important here not to get caught in a false dilemma. A false dilemma presents a choice between two options, neither of which is entirely acceptable, but which together purport to be mutually exclusive and exhaustive. The false dilemma here is common design versus common ancestry. It's logically possible to have common design without common ancestry and common ancestry without common design. But common design and common ancestry are not mutually exclusive. The two can work together. We can think of life consisting of hierarchically arranged design modules that over the course of natural history have sustained substantial evolutionary change through the activity of natural forces as well as through the guidance of a supervising intelligence. Just how much evolutionary change has occurred in each way remains an open question.

And that brings us to a true either-or. If the choice between common design and common ancestry is a false either-or, the choice between intelligent design and materialistic evolution is a true either-or. Materialistic evolution does not merely embrace common ancestry; it also rejects any real design in the evolutionary process. Intelligent design, by contrast, contends that biological design is real and empirically detectable regardless of whether it occurs within an evolutionary process or in discrete independent stages. The verdict is not yet in, and proponents of intelligent design themselves hold differing views on the extent of the evolutionary interconnectedness of organisms, with some even accepting universal common ancestry (i.e., Darwin's great tree of life).[61]

Common ancestry in combination with common design can explain the similar features that arise in biology. The real question is whether common ancestry apart from common design—in other words, materialistic evolution—can do so. The evidence of biology increasingly demonstrates that it cannot.

*The general notes on the CD included with this book expand on the material in this chapter and throw light on the following discussion questions.*

## 5.9 DISCUSSION QUESTIONS

1. Why is it important to the study of biological classification that organisms share similar features? What sorts of similarities do they share? What is the difference between homology and analogy? Which of these is more important for the purposes of biological classification? Why?

2. Why is it important to the study of evolution that organisms share similar features? What sorts of shared similarities do Darwinists use to argue for common evolutionary ancestors? Which is more important here, homology or analogy? Why?

3. What is convergent evolution? Give some examples of supposed convergent evolution. How do you explain the marsupial look-alikes to placental wolves, cats, squirrels, ground hogs, anteaters, moles, and mice? Review the convergence of the panda's thumb in both the giant panda and the red panda. According to evolutionary theory, did this structure arise from a common ancestor or evolve independently? Does convergent evolution pose a challenge to Darwinian evolution? Why or why not?

4. Who came up with the term "homology"? Before Darwin, how was homology explained? How did Darwin define homology? How did Darwin explain homology? How do contemporary Darwinists define homology? Does homology provide evidence for evolution? If homology is defined as similarity due to common ancestry, can it legitimately be used as evidence for common ancestry? Explain your answers.

5. What is the difference between structural and molecular homologies? Give some examples of each. What is a phylogeny? How are homologies used to establish phylogenies? Are molecular phylogenies that are established on the basis of molecular homologies more secure than ordinary phylogenies that are established on the basis of anatomical morphology? Explain your answer. Is the following statement true or false: The more molecules that are factored in to formulate a molecular phylogeny, the more consistent is the picture of life's history that emerges. Explain your answer.

6. According to the text, it has been known for decades that homologous features can arise from dissimilar genes, and that similar genes can underlie non-homologous features. How does this fact pose a challenge to molecular phylogenies?

7. What is the "molecular clock"? Who first proposed it and how far back does this proposal go? How reliable is the molecular clock for establishing when key events in evolution happened? How useful is it for understanding the evolutionary history leading up to the Cambrian explosion? What range of dates for a common metazoan ancestor to the Cambrian explosion has the molecular clock hypothesis generated?

8. What is a vestigial structure? How could vestigial structures arise at the molecular level (in genes, say)? How do "shared errors" provide evidence for common evolutionary ancestors? Lay out the "shared error argument" in detail. How much evolution do shared errors support? Do shared errors indicate how evolution produced novel complex structures that are not errors but serve functions important to the organism? Explain your answer.

9. What is recapitulation? Does Darwin's theory predict recapitulation? What is the evidence for recapitulation? What does the statement "ontogeny recapitulates phylogeny" mean? Who first made this statement? Review Ernst Haeckel's embryo drawings. Were they accurate? What stage in embryonic development did they purport to show? What stages in fact precede it? Are the facts of embryology consistent with Darwin's theory?

10. If you had to explain similar features in organisms by (i) common design, (ii) common ancestry, or (iii) both, which would you choose? Explain your answer.

# CHAPTER SIX Irreducible Complexity

## 6.1 MOLECULAR MACHINES

Highly intricate molecular machines play an integral part in the life of the cell and are increasingly attracting the attention of the biological community. For instance, in February 1998 the premier biology journal *Cell* devoted a special issue to "macromolecular machines." All cells use complex molecular machines to process information, convert energy, metabolize nutrients, build proteins, and transport materials across membranes. Bruce Alberts, president of the National Academy of Sciences, introduced this issue with an article titled "The Cell as a Collection of Protein Machines." In it he remarked,

> We have always underestimated cells. . . . The entire cell can be viewed as a factory that contains an elaborate network of interlocking assembly lines, each of which is composed of a set of large protein machines. . . . Why do we call the large protein assemblies that underlie cell function protein *machines*? Precisely because, like machines invented by humans to deal efficiently with the macroscopic world, these protein assemblies contain highly coordinated moving parts.[1]

Almost six years later (December 2003), *BioEssays* published its own special issue on "molecular machines." In the introductory essay to that issue, Adam Wilkins, the editor of *BioEssays*, remarked,

> The articles included in this issue demonstrate some striking parallels between artifactual and biological/molecular machines. In the first place, molecular machines, like man-made machines, perform highly specific functions. Second, the macromolecular machine complexes feature multiple parts that interact in distinct and precise ways, with defined inputs and outputs. Third, many of these machines have parts that can be used in other molecular machines (at least, with slight modification), comparable to the interchangeable parts of artificial machines. Finally, and not least, they have the cardinal attribute of machines: they all convert energy into some form of 'work'.[2]

Alberts and Wilkins here draw attention to the strong resemblance between molecular machines and machines designed by human engineers. Nevertheless, as neo-Darwinists, they regard the cell's marvelous complexity as products of Darwinian evolution and thus as only apparently designed. In the 1990s, however, scientists began to challenge the neo-Darwinian view and argue that such molecular machines display evidence of design. Leading the way was Lehigh University biochemist Michael Behe.

## 6.2 MICHAEL BEHE'S DANGEROUS IDEA

In 1996, Michael Behe published a book titled *Darwin's Black Box*. In that book he detailed the failure of neo-Darwinian theory to explain the origin of complex molecular machines inside the cell. But he didn't stop there. He also argued that these molecular machines arose by actual design. Central to his argument was the idea of *irreducible complexity*. A functional system is *irreducibly complex* if it contains a multipart subsystem (i.e., a set of two or more interrelated parts) that cannot be simplified without destroying the system's basic function. We call this multipart subsystem the system's *irreducible core*.[3] This definition is more subtle than it might first appear, so let's consider it closely.

Irreducibly complex systems belong to the broader class of functionally integrated systems. A functionally integrated system consists of parts that are tightly adapted to each other and thus make the system's function sensitive to isolated changes of those parts. For an integrated system to continue to function given such changes typically requires multiple changes elsewhere in the system or else redundant backup. We therefore define the *core* of a functionally integrated system as those parts that are indispensable to the system's basic function: remove parts of the core, and you can't recover the system's basic function from the other remaining parts. To say that a core is *irreducible* is then to say that no other systems with substantially simpler cores can perform the system's basic function.

The *basic function* of a system consists of three things: (1) What the system does in its natural setting or proper context; this is known as the system's *primary function* (also *main function*). (2) The minimal level of function needed for the system to perform adequately in its natural setting or proper context; this is known as the system's *minimum function*. (3) The way or manner in which the system performs its primary function; this is known as the system's *mode of function*. Because the basic function of a system includes its mode of function, basic function is concerned not just with ends but also with means. Glue and nails, for instance, may perform the same primary function of fastening together pieces of wood and do so equally well in certain contexts, but the way in which they do it is completely different.

Pencil Sharpener

The Professor gets his think-tank working and evolves the simplified pencil sharpener.

Open window (**A**) and fly kite (**B**). String (**C**) lifts small door (**D**), allowing moths (**E**) to escape and eat red flannel shirt (**F**). As weight of shirt becomes less, shoe (**G**) steps on switch (**H**) which heats electric iron (**I**) and burns hole in pants (**J**). Smoke (**K**) enters hole in tree (**L**), smoking out opossum (**M**) which jumps into basket (**N**), pulling rope (**O**) and lifting cage (**P**), allowing woodpecker (**Q**) to chew wood from pencil (**R**), exposing lead. Emergency knife (**S**) is always handy in case opossum or the woodpecker gets sick and can't work.

***Figure 6.1*** *Which of these systems is not functionally integrated?*

So too, consider an outboard motor whose basic function is to propel a small fishing boat around a lake by means of a gasoline- or electric-powered engine that turns a propeller. The outboard motor is irreducibly complex and its irreducible core includes, among other things, a propeller, an engine, and a drive shaft connecting engine to propeller. Now, we can imagine simplifying this arrangement by replacing the engine and drive shaft with a rubber band that, when wound up, turns the propeller. But it's unlikely that the level of performance attainable from such an arrangement will propel a boat around a lake. In other words, minimum function is unlikely to be preserved with the rubber band. Yet even if it was, this new arrangement would not perform the primary function in the same way as the original outboard motor: the original outboard motor depended on the turning of rotors and not the torsion of an elastic medium.

As another example of an irreducibly complex system, consider an old-fashioned pocket watch. The basic function of the watch is to tell time by means of a winding mechanism. Several parts of the watch are indispensable to that basic function, for instance, the spring, the face, and the hour hand. These belong to the irreducible core.

But note that other parts of the watch are dispensable, for instance, the crystal, the metal case, and the chain. Because these parts are unnecessary or redundant to the system's basic function, they do not belong to the irreducible core. Whether other parts of the watch belong to the irreducible core depends on the minimum level of function demanded of the watch. The hour hand by itself is adequate for telling the hour and even certain ranges of minutes. But if it is important to know the exact minute, then the minute hand will also be required and belong to the irreducible core. Notice that many irreducibly complex systems are like the pocket watch in containing parts that are not crucial to the system's basic function—parts that therefore lie outside the system's irreducible core.

***Figure 6.2*** *Outboard motor and pocket watch.*

For an irreducibly complex system, each of the parts of the irreducible core plays an indispensable role in achieving the system's basic function. Thus, removing parts—even a single part—from the irreducible core results in loss of the system's basic function. Nevertheless, to determine whether a system is irreducibly complex, it is not enough merely to identify those parts whose removal renders the basic function unrecoverable from the remaining parts. To be sure, identifying such indispensable parts is an important step for determining irreducible complexity in practice. But it is not sufficient. Additionally, we need to establish that no systems with simpler cores achieve the same basic function.

Consider, for instance, a three-legged stool. Suppose the stool's basic function is to provide a seat by means of a raised platform. In that case each of the legs is indispensable for achieving this basic function (remove any leg and the basic function can't be recovered among the remaining parts). Nevertheless, because it's possible for a substantially simpler system to exhibit this basic function (for example, a solid block), the three-legged stool is not irreducibly complex.

To determine whether a system is irreducibly complex therefore requires an analysis of the system, and specifically of those parts whose removal renders the basic function unrecoverable. This analysis needs to demonstrate that no system with (substantially) fewer parts exhibits the basic function. For instance, from analyzing the outboard motor described earlier, one sees that no system performing the motor's basic function can omit a propeller, engine, or drive shaft. Consequently, these parts belong to the irreducible core, a fact that receives empirical confirmation by removing the parts experimentally and showing that the basic function is unrecoverable from the remaining parts.

## 6.3 THE BACTERIAL FLAGELLUM

Biology is chock-full of irreducibly complex protein machines. Indeed, it's been found that most proteins work together with multiple proteins in performing cellular functions. Thus, most proteins belong to functionally integrated systems of proteins. Of these, many are irreducibly complex. Michael Behe, in his book *Darwin's Black Box,* documents several cases of irreducible complexity, including the cilium, the blood-clotting cascade, and the bacterial flagellum. Biochemistry and cell biology textbooks are filled with such systems. Indeed, life as we know it would be impossible without them.[4]

The irreducibly complex protein machine that has especially captured the imagination of the biological community is the bacterial flagellum (see figure 6.3). In public lectures, Harvard biologist Howard Berg calls the bacterial flagellum "the most efficient machine in the universe." The flagellum is an acid-powered rotary motor with a whip-like tail whose rotating motion drives a bacterium through its watery environment. This whip-like tail acts as a propeller. It spins at tens of thousands of rpm and can change direction in a quarter turn. The intricate machinery of the flagellum includes a rotor, a stator, O-rings, bushings, mounting disks, a drive shaft, a propeller, a hook joint for the propeller, and an acid-powered motor. Many different bacteria have flagella. The flagellum belonging to *E. coli* has received the most attention in the biological literature.

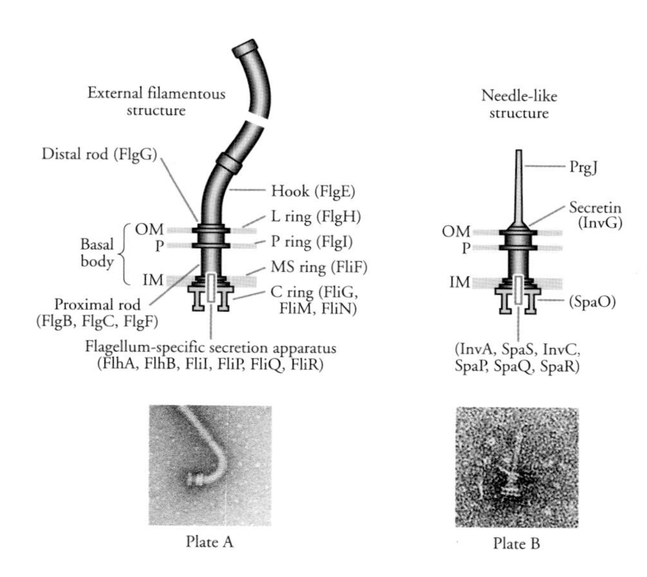


***Figure 6.3*** *Grainy electron micrographs and clean schematics of the bacterial flagellum and type III secretory system.*

The basic function of the bacterial flagellum is to propel the bacterium through its watery environment by means of a fast-spinning, bidirectional, whip-like tail (the propeller, also known as a filament). Note that a whip-like tail with these properties is not a luxury but rather a necessity if the flagellum is to be of any use as a motility structure for seeking food. In propelling a bacterium through its watery environment, the flagellum must overcome Brownian motion (the random motion of water molecules, which jostles small objects suspended in water). Flagella need to rotate bidirectionally because Brownian motion sets bacteria off their course as they wend their way up a nutrition gradient. Reversing direction of the rotating tail causes the bacterium to tumble, reset itself, and try again to get to the food it needs. The minimal functional requirements of a flagellum, if it is going to do a bacterium any good at all in propelling it through its watery environment up a nutrition gradient, is that the whip-like tail (or filament) rotate bidirectionally and extremely fast. Flagella of known bacteria spin at rates well above 10,000 rpm (actually, closer to 20,000 rpm, and even as high as 100,000 rpm). Anything substantially less than this will prevent a bacterium from overcoming the disorienting effects of Brownian motion and thus prevent it from finding the concentrations of nutrients it needs to survive, reproduce, and flourish.[5]

The flagellum's intricate machinery requires the coordinated interaction of about thirty proteins and another twenty or so proteins to assist in their assembly. Yet the absence of any one of these proteins would result in the complete loss of motor function.[6] These proteins form the irreducible core of the flagellum. How complex is this core? John Postgate describes some of the complexity:

> A typical bacterial flagellum, we now know, is a long, tubular filament of protein. It is indeed loosely coiled, like a pulled-out, left-handed spring, or perhaps a corkscrew, and it terminates, close to the cell wall, as a thickened, flexible zone, called a hook because it is usually bent. . . . One can imagine a bacterial cell as having a tough outer envelope within which is a softer more flexible one, and inside that the jelly-like protoplasm resides. The flagellum and its hook are attached to the cell just at, or just inside, these skins, and the remarkable feature is the way in which they are anchored. In a bacterium called *Bacillus subtilis* . . . the hook extends, as a rod, through the outer wall, and at the end of the rod, separated by its last few nanometers, are two discs. There is one at the very end which seems to be set in the inner membrane, the one which covers the cell's protoplasm, and the near-terminal disc is set just inside the cell wall. In effect, the long flagellum seems to be held in place by its hook, with two discs acting as a double bolt, or perhaps a bolt and washer. . . .[7]

This passage merely scratches the surface of the complexities involved with the bacterial flagellum. Here Postgate describes what amounts to a propeller and its attachment to the cell wall. Additionally, there needs to be a motor that runs the propeller. This motor needs to be mounted and stabilized. It must be capable of bidirectional rotation. Moreover, there needs to be signal transduction circuitry that takes information from the environment and tells the flagellum when and in which direction to spin. The complexities quickly mount, and a conceptual analysis reveals that the bacterial flagellum possesses an extremely complicated irreducible core.

So how did the bacterial flagellum originate? On a Darwinian view, a bacterium with a flagellum evolved via the Darwinian selection mechanism from a bacterium lacking not only a flagellum but also all the genes coding for flagellar proteins (including any genes homologous to the genes for the flagellum). For the Darwinian mechanism to produce a bacterial flagellum, random genetic changes therefore had to bring about the genes that code for flagellar proteins, and then selection had to preserve these proteins, gather them to the right location in the bacterium, and properly assemble them. How plausible is this? As we shall see in this and the next chapter, *Not at all.*

## 6.4 COEVOLUTION AND CO-OPTION

To explain irreducible complexity, Darwinists in the end always fall back on indirect Darwinian pathways. In an indirect Darwinian pathway, not only does a structure evolve but so does its associated function. By contrast, in a direct Darwinian pathway, at the same time that natural selection enhances or improves a structure that already serves a given function, the function itself does not change. Since the function of an irreducibly complex system is not attained until all the parts of the irreducible core are in place, a direct Darwinian pathway would therefore have to produce such a system in one fell swoop. But that's absurd. These systems are incredibly complicated and must, if they are to be produced apart from design, arise by, as Darwin put it, "numerous, successive, slight modifications."[8] Moreover, each of these small steps must, without exception, either enhance the organism's ability to survive and reproduce or be selectively neutral (in particular, no step along the way can be deleterious to the organism). Thus, the only way for Darwinism to explain irreducible complexity is by means of an indirect Darwinian pathway in which structures and functions coevolve.

One way this could happen is for parts previously targeted for other systems to break free and be co-opted into a novel system. It is as though pieces from a car, bicycle, motorboat, and train were suitably recombined to form an airplane. Evolutionary theorists sometimes denote such systems as *patchworks* or *bricolages*. Thus, any such airplane would be a patchwork or bricolage of preexisting materials originally targeted for different uses. Clearly, there is no logical impossibility that prevents such patchworks from forming irreducibly complex systems. But a patchwork, if sufficiently intricate and elegant, begs a precise causal account of how it arose. The bacterial flagellum, for instance, is an engineering marvel of miniaturization and performance. Simply to call such a system a patchwork of co-opted preexisting materials is therefore hardly illuminating. In fact, it misrepresents the system's true complexity and does nothing to answer how it originated.

The problem with trying to explain an irreducibly complex system (e.g., the bacterial flagellum) as a patchwork is that it requires multiple coordinated co-options. It is not just that one thing evolves for one function, and then, perhaps without any modification at all, gets used for some completely different function (imagine a rock first being used as a paperweight and then being co-opted for use as a doorstop). The problem is that multiple protein parts from different functional systems all have to break free and then all have to coalesce to form a newly integrated system (as with the airplane formed by taking parts from a car, bicycle, motorboat, and train).

Even if all the parts (i.e., proteins) for a bacterial flagellum are in place within a cell but serving other functions, there is no reason to think that those parts can spontaneously break free of the systems in which they are currently functioning to form a tightly integrated new system such as the flagellum. The problem here is that parts performing functions in separate systems are unlikely to be adapted to each other so that they can work together coherently within a novel system. Imagine a bolt that's part of one system and a nut that's part of another system. If these systems originated independently, as they would for separately evolved biological systems, it is unlikely that the bolt will be adapted to the nut so that the fit is mechanically useful (i.e., neither too tight, thereby preventing the bolt from screwing into the nut at all, nor too loose, thereby preventing the bolt from properly meshing with the nut).

This problem is magnified in the cell. Take the evolution of the bacterial flagellum. Besides those proteins that go into a flagellum, a cell evolving a flagellum will have many other proteins that play no conceivable role in a flagellum. The majority of proteins in the cell will be of this sort. How then can those, and only those, proteins that go into a functional flagellum be brought together and guided to their proper locations in the cell without interfering cross-reactions from the other proteins? It is like going through a giant grocery store blindfolded, taking items off the racks, and hoping that what ends up in the shopping cart are the precise ingredients for a cake. Such an outcome is highly unlikely. University of Rochester biologist Allen Orr, who is no fan of intelligent design, agrees:

> We might think that some of the parts of an irreducibly complex system evolved step by step for some other purpose and were then recruited

> wholesale to a new function. But this is also unlikely. You may as well hope that half your car's transmission will suddenly help out in the airbag department. Such things might happen very, very rarely, but they surely do not offer a general solution to irreducible complexity.[9]

Such co-option scenarios strain credulity because they require multiple coordinated co-options from multiple functional systems to bring about an irreducibly complex system in essentially one fell swoop. This requires a stacking up of coincidences that seems beyond the reach of purely material forces and points far more readily to design. In any case, empirical evidence does not support such co-option scenarios—they have never been shown to occur in nature.

But what if, instead, co-option occurred more gradually and incrementally? In the evolution of the bacterial flagellum, imagine natural selection gradually co-opting existing protein parts into a single evolving structure whose function co-evolves with the structure. In that case, an irreducibly complex system might arise by gradually co-opting parts that initially were dispensable but eventually become indispensable (as required of the parts that belong to core of an irreducibly complex system). Here is how Allen Orr sketches this possibility:

> An irreducibly complex system can be built gradually by adding parts that, while initially just advantageous, become—because of later changes—essential [i.e., indispensable]. The logic is very simple. Some part (A) initially does some job (and not very well, perhaps). Another part (B) later gets added because it helps A. This new part isn't essential, it merely improves things. But later on, A (or something else) may change in such a way that B now becomes indispensable. This process continues as further parts get folded into the system. And at the end of the day, many parts may all be required.[10]

Let's evaluate this argument. Orr posits a gradual increase in complexity in which novel parts that enhance function are added and alternately rendered indispensable. But which function (or "job," as Orr puts it) are we talking about? Obviously, functions along the way must be different from the final function because the final function is exhibited by an irreducibly complex system and hence cannot be exhibited by any system with a substantially simplified irreducible core. But then we run into the same problem as before: there is no empirical evidence that irreducibly complex biochemical systems such as the bacterial flagellum came about by this method of adding a component, making it indispensable, adding another component, making it indispensable, etc.

Indeed, Orr, along with the rest of the Darwinian community, never offers anything more than highly abstract scenarios for how irreducible complexity might arise. But clearly, something more is required. Minimally what's required are detailed, testable reconstructions or models that demonstrate how indirect Darwinian pathways might reasonably have produced actual irreducibly complex biochemical machines such as the bacterial flagellum. Orr, by contrast, merely gestures at unspecified abstract systems designated schematically by the letters "A" and "B." Evolutionary biologists have nothing resembling detailed evolutionary pathways leading to irreducibly complex systems such as the bacterial flagellum.

The closest thing that biologists have been able to find as a possible evolutionary precursor to the bacterial flagellum is what's known as a type III secretory system (TTSS). The TTSS is a type of pump that enables certain pathogenic bacteria to inject virulent proteins into host organisms. One bacterium possessing the TTSS is *Yersinia pestis,* the organism responsible for the black plague that during the fourteenth century killed a third of the population of Europe. The TTSS was the delivery system by which *Yersinia pestis* inflicted its massive destruction of human life. Now it turns out that the ten or so proteins that go into the construction of the TTSS are similar to proteins found in the bacterial flagellum. What's more, the TTSS corresponds roughly to the part of the flagellum used in the construction of its filament (i.e., the long whip-like tail). But note: it is not possible simply to substitute the TTSS for the corresponding part of the bacterial flagellum and have a functioning flagellum. Because the proteins in the TTSS are not adapted to the proteins of the bacterial flagellum, the resulting kludge would be nonfunctional

Despite such difficulties relating the TTSS to the bacterial flagellum, suppose we treat the TTSS as a subsystem of the flagellum. As such, it performs a function distinct from the flagellum. Notwithstanding, finding a subsystem of a functional system that performs some other function does not show that the subsystem evolved into the system. To see this, imagine arguing that because a motorcycle's motor can function as a heater, therefore the motor evolved into the motorcycle. Such an argument is transparently feeble. Indeed, multipart, tightly integrated functional systems almost invariably contain multipart subsystems that could serve some different function. At best, the TTSS represents one possible step in the indirect Darwinian evolution of the bacterial flagellum. But that still wouldn't constitute a solution to the evolution of the bacterial flagellum. What's needed is a complete evolutionary path and not merely a possible oasis along the way. To claim otherwise is like saying we can travel by foot from Los Angeles to Tokyo because we've discovered the Hawaiian Islands.

There's another problem here. The whole point of bringing up the TTSS was to posit it as an evolutionary precursor to the bacterial flagellum. Yet insofar as there is an evolutionary connection between the two, biologists now tend to think that the TTSS evolved from the flagellum rather than into it.[11] It's easy to see why. The bacterial flagellum is a motility structure for propelling a bacterium through its watery environment. Water has been around since the origin of life. Indeed, evolutionary biologists surmise that the bacterial flagellum has been around for billions of years. But the TTSS is a poison delivery system for plants and animals. Its function, therefore, depends on the existence of multicellular organisms. Accordingly, the TTSS could only have been around since the rise of multicellular organisms, which evolutionary biologists place around 600 million years ago.

It follows that the TTSS does not explain the evolution of the flagellum. At best, the bacterial flagellum could explain the evolution of the TTSS. But even that isn't quite right. The TTSS is, after all, much simpler than the flagellum. The TTSS contains ten or so proteins that are similar to proteins in the flagellum. The flagellum requires an additional thirty or forty proteins, which in turn need to be accounted for. Evolution needs to explain the emergence of complexity from simplicity. But if the TTSS (de-)evolved from the flagellum, then we've done the exact opposite (i.e., we've explained the simpler in terms of the more complex).

The scientific literature shows a complete absence of detailed, testable, step-by-step proposals for how coevolution and co-option could actually produce irreducibly complex biochemical systems. In place of such proposals, Darwinists simply observe that because subsystems of irreducibly complex systems might be functional, any such functions could be selected by natural selection. And from this unexceptional observation, Darwinists blithely conclude that selection works on those parts and thereby forms irreducibly complex systems.[12] But this conclusion is completely unfounded, and accounts for cell biologist Franklin Harold's frank admission that "there are presently no detailed Darwinian accounts of the evolution of any biochemical or cellular system, only a variety of wishful speculations."[13] Biologist Lynn Margulis is equally forthright: "Like a sugary snack that temporarily satisfies our appetite but deprives us of more nutritious foods, neo-Darwinism sates intellectual curiosity with abstractions bereft of actual details—whether metabolic, biochemical, ecological, or of natural history."[14]

To sum up, the Darwinian mechanism requires a selectable function if that mechanism is going to work at all. What's more, functional pieces pulled together from various systems via coevolution and co-option are selectable by the Darwinian mechanism. But what are selectable here are the individual functions of the individual pieces and not the function of the yet-to-be-produced system. The Darwinian mechanism selects for existing function. It cannot select for future function. Once that function is realized, the Darwinian mechanism can select for it as well. But making the transition from existing function to novel function is the hard part. How does one get from functional pieces that are selectable in terms of their individual functions to a system that makes use of those pieces and exhibits a novel function? In the case of irreducibly complex biochemical machines such as the bacterial flagellum, the Darwinian mechanism is no help whatsoever.

## 6.5 THE ARGUMENT FROM IRREDUCIBLE COMPLEXITY

In *Darwin's Black Box,* Behe argued that the irreducible complexity of protein machines provides convincing evidence of actual design in biology. Since its publication in 1996, Behe's book has been widely reviewed, both in the popular press and in scientific journals.[15] It has also been widely discussed over the Internet.[16] By and large, critics have conceded that Behe got his scientific facts straight. They have also conceded his claim that detailed neo-Darwinian accounts for how irreducibly complex protein machines could come about are absent from the biological literature. Nonetheless, they have objected to his argument on theoretical and methodological grounds. Behe presents what may be described as an *argument from irreducible complexity.* This argument purports to show that irreducibly complex biological systems are beyond the reach of the Darwinian evolutionary mechanism and that design provides a better explanation of them.

How does the argument from irreducible complexity reach this conclusion? The argument from irreducible complexity may be understood as making three key points: a logical, an empirical, and an explanatory point. Moreover, these points reinforce each other in showing that irreducibly complex systems are beyond the reach of conventional evolutionary mechanisms. The logical point is this: Irreducibly complex structures are provably inaccessible to *direct* Darwinian pathways. Thus, because certain biological structures are irreducibly complex, they, too, must be inaccessible to direct Darwinian pathways. A direct Darwinian pathway is one in which a system evolves by natural selection incrementally enhancing a given function. As the system evolves, the function does not evolve but stays put.

Thus we might imagine that in the evolution of the heart, its function from the start was to pump blood. In that case, a direct Darwinian pathway might account for it. On the other hand, we might imagine that in the evolution of the heart its function was initially to make loud thumping sounds to ward off predators, and only later did it take on the function of pumping blood. In that case, an *indirect* Darwinian pathway would be needed to account for it. Here the pathway is indirect because not only do the system components evolve but also the system function changes. Now, in making a logical point, the argument from irreducible complexity is only concerned with precluding direct Darwinian pathways. This is evident from the definition of irreducible complexity where the irreducible core is defined strictly in relation to a single function, namely, the basic function of the irreducibly complex system (a function that could not exist without all the parts of the irreducible core being in place).

In ruling out direct Darwinian pathways to irreducibly complex systems, the argument from irreducible complexity is saying that irreducibly complex biochemical systems are provably inaccessible to direct Darwinian pathways. How can we see that such systems are indeed provably inaccessible to direct Darwinian pathways? Consider what it would mean for an irreducibly complex system to evolve by a direct Darwinian pathway. In that case, the system must have originated via the evolution of simpler systems that performed the same basic function. But because the irreducible core of an irreducibly complex system can't be simplified without destroying the basic function, there can be no evolutionary precursors with simpler cores that perform the same function.

It follows that the only way for a direct Darwinian pathway to evolve an irreducibly complex system is to evolve it all at once and thus by some vastly improbable or fortuitous event. Accordingly, to attribute irreducible complexity to a direct Darwinian pathway is like attributing Mount Rushmore to wind and erosion. It is perhaps imaginable that wind and erosion could sculpt Mount Rushmore, but it's not a realistic possibility. The proof that irreducibly complex systems are inaccessible to direct Darwinian pathways is therefore probabilistic. Thus, in saying that irreducibly complex biochemical systems are provably inaccessible to direct Darwinian pathways, design proponents are saying that the Darwinian mechanism has no intrinsic capacity for generating such systems except as vastly improbable or fortuitous events.

In any case, critics of the argument from irreducible complexity look to save Darwinism not by enlisting direct Darwinian pathways to bring about irreducibly complex systems but by enlisting indirect Darwinian pathways to bring them about (see the previous section on coevolution and co-option). In indirect Darwinian pathways, a system evolves not by preserving and enhancing an existing function but rather by continually transforming its function. Whereas with direct Darwinian pathways structures evolve but functions stay put, with indirect Darwinian pathways both structures and functions (co)evolve.

How does the argument from irreducible complexity handle indirect Darwinian pathways? Here the point at issue is no longer logical but empirical. The fact is that for irreducibly complex biochemical systems, no indirect Darwinian pathways are known. At best, biologists have been able to isolate subsystems of such systems that perform other functions. But any reasonably complicated machine always includes subsystems that can perform functions distinct from the complete machine. So the mere occurrence or identification of subsystems that could perform some function on their own is no evidence for an indirect Darwinian pathway leading to the system. What's needed is a seamless Darwinian account that's both detailed and testable of how subsystems undergoing coevolution could gradually transform into an irreducibly complex system. No such accounts are available or have so far been forthcoming. Indeed, if such accounts were available, critics of intelligent design would merely need to cite them, and intelligent design would be refuted.

To recap, the argument from irreducible complexity makes a logical and an empirical point. The logical point is that irreducible complexity renders biological structures provably inaccessible to direct Darwinian pathways. The empirical point is that the failure of evolutionary biology to discover indirect Darwinian pathways leading to irreducibly complex biological structures is pervasive and systemic and therefore reason to doubt and even reject the claim that indirect Darwinian pathways are the answer to irreducible complexity. The logical and empirical points together constitute a devastating indictment of the Darwinian mechanism, which has routinely been touted as capable of solving all problems of biological complexity once an initial life-form is on the scene (cf. chapter 3). Even so, the logical and empirical points together don't answer how one gets from Darwinism's failure in accounting for irreducibly complex systems to the legitimacy of employing design in accounting for them.

This is where the argument from irreducible complexity needs to make a third key point, namely, an explanatory point. Scientific explanations come in many forms and guises, but the one thing they cannot afford to be without is *causal adequacy*. A scientific explanation needs to call upon causal powers sufficient to explain the effect in question. Otherwise, the effect remains unexplained. The effect in question is the irreducible complexity of certain biochemical machines. How did such systems come about? Not by direct Darwinian pathways—irreducible complexity rules them out on logical and mathematical grounds. And not by indirect Darwinian pathways either—the absence of scientific evidence here is as complete as it is for leprechauns. Nor does appealing to unknown material mechanisms help, for in that case not only is the evidence completely absent but also the very theory for which there's no evidence is absent as well.

Thus, when it comes to irreducibly complex biochemical systems, there's no evidence that material mechanisms are causally adequate to bring them about. But what about intelligence? Intelligence is well known to produce irreducibly complex systems (e.g., humans regularly produce machines that exhibit irreducible complexity). Intelligence is thus known to be causally adequate to bring about irreducible complexity. The argument from irreducible complexity's explanatory point, therefore, is that on the basis of causal adequacy, intelligent design is a better scientific explanation than the Darwinian mechanism for the irreducible complexity of biochemical systems.

In making its logical and empirical points, the argument from irreducible complexity assumes a negative or critical role, identifying limitations of the Darwinian mechanism. By contrast, in making its explanatory point, the argument from irreducible complexity assumes a positive or constructive role, providing positive grounds for thinking that irreducibly complex biochemical systems are in fact designed. One question about these points is now likely to remain. The logical point rules out direct Darwinian pathways to irreducible complexity and the empirical point rules out indirect Darwinian pathways to irreducible complexity. But the absence of empirical evidence for direct Darwinian pathways leading to irreducible complexity is as complete as it is for indirect Darwinian pathways. It might seem, then, that the logical point is superfluous inasmuch as the empirical point dispenses with both types of Darwinian pathways. But in fact the logical point strengthens the case against Darwinism in a way that the empirical point cannot.

If you look at the best-confirmed examples of evolution in the biological literature (from Darwin to the present), what you find is natural selection steadily improving a given feature performing a given function in a given way. Indeed, the very notion of "improvement" (which played such a central role in Darwin's *Origin of Species*) typically connotes that a given thing is getting better in a given respect. Improvement in this sense corresponds to a direct Darwinian pathway. By contrast, an indirect Darwinian pathway (where one function gives way to another function and thus can no longer improve because it no longer exists), though often inferred by evolutionary biologists from fossil or molecular data, is much more difficult to establish.

The reason is not hard to see: By definition natural selection selects for existing function—in other words, a function that is already in place and helping the organism in some way. On the other hand, natural selection cannot select for future function—functions that are not already present or not in some way currently helping the organism to survive and reproduce are invisible to natural selection. Once a novel function comes to exist, the Darwinian mechanism can select for it. But making the transition from old to new functions is not a task to which the Darwinian mechanism is suited. How does one evolve from a system exhibiting an existing selectable function to a new system exhibiting a novel selectable function? Because natural selection only selects for existing function, it is no help here, and all the weight is on random variation to come up with the right and needed modifications during the crucial transition time when functions are changing. (Or, as Darwin put it, "unless profitable variations do occur, natural selection can do nothing."[17]) Yet the actual evidence that random variation can produce the successive modifications needed to evolve irreducible complexity is nil.

The argument from irreducible complexity, in making the logical point that irreducible complexity rules out direct Darwinian pathways, therefore rules out the form of evolution that is best confirmed. Indirect Darwinian pathways, by contrast, are so open-ended that there is no way to test them scientifically unless they are carefully specified—and invariably, when it comes to irreducibly complex systems, they are left unspecified, thus rendering them neither falsifiable nor verifiable. In making its logical point, the argument from irreducible complexity, therefore takes logic as far as it can go in limiting the Darwinian mechanism and leaves empirical considerations to close off any remaining loopholes. And since logical inferences are inherently stronger than empirical inferences, the argument from irreducible complexity's refutation of the Darwinian mechanism is as strong and tight as possible. It's not just that certain biological systems are so complex that we can't imagine how they evolved by Darwinian pathways. Rather, we can show conclusively that direct Darwinian pathways are causally inadequate to bring them about and that indirect Darwinian pathways, which have always been more difficult to substantiate, are utterly without empirical support. Conversely, we do know what has the causal power to produce irreducible complexity—intelligent design.

## Working with Available Evidence

Science must work with available evidence and on that basis (and that basis alone) formulate the best explanations of natural phenomena. This means that science cannot explain phenomena by appealing to the promise, prospect, or possibility of future evidence. In particular, unknown material causes or undiscovered ways by which those causes operate cannot be invoked to explain a phenomenon or to hinder explanation of it. If known blind material causes fail to explain a phenomenon, then it is an open question whether any blind material causes whatsoever are capable of explaining it. If, further, there is good evidence for thinking that certain biological systems are designed, then design itself becomes a reasonable explanation in biology.

By contrast, Darwinists typically claim that no evidence whatsoever can support design in biology. "Even if there were no actual evidence in favour of the Darwinian theory," writes Richard Dawkins in *The Blind Watchmaker*, biologists "should still be justified in preferring it over all rival theories," and that, of course, includes preferring Darwinian theory over intelligent design.[18] But consider what follows. Darwinian theory is thereby rendered immune to disconfirmation in principle because the universe of unknown blind material causes can never be exhausted. The Darwinist thus refuses any burden of evidence. This is not right. Darwinists need to hold themselves to the same standard of evidence that they demand from design theorists. Why? Because they themselves admit that biological systems appear on their face to belong to that class of things known to be intelligently designed.

If a creature looks like a duck, smells like a duck, quacks like a duck, feels like a duck, and swims like a duck, the burden of evidence lies with those who insist the creature isn't a duck. The same goes for the incredibly intricate molecular machines and organ systems that we see throughout the living world—the burden of evidence is on those who want to deny their design. And yet one does not find Richard Dawkins rolling up his sleeves and trying to eliminate every imaginable and as yet unimagined intelligent design scenario, pleading for patience while he rules out an infinite set of design possibilities. Instead, the burden of evidence is shifted entirely to the critics of Darwinism, who are then called upon to establish a universal negative by an exhaustive search and elimination of all conceivable materialistic possibilities—however remote, however unfounded, however unsupported by evidence. That is not how science is supposed to work.

*The general notes on the CD included with this book expand on the material in this chapter and throw light on the following discussion questions.*

## 6.6 DISCUSSION QUESTIONS

1. How important are molecular machines in the life of the cell? What sorts of tasks do these machines perform inside the cell? Is it fair to call them "machines" at all? How do these biological/molecular machines compare with the artifactual machines produced by humans? Since engineering principles are required to understand the molecular machines inside cells, is it reasonable or unreasonable to infer that these systems are actually designed? Explain.

2. What is an irreducibly complex system? What is the core of an irreducibly complex system? What is the basic function of an irreducibly complex system? Why isn't a three-legged stool irreducibly complex? Give an example of a biological system that is irreducibly complex. Contrast irreducible complexity with cumulative complexity.

3. What is the bacterial flagellum? What is its basic function? What's the fastest it can spin? Is it irreducibly complex? Name at least one type of bacterium that has a flagellum.

4. What is a direct Darwinian pathway? What is an indirect Darwinian pathway? Give an example of each. Why can't an irreducibly complex biochemical system evolve via a direct Darwinian pathway? What is the evidence that irreducibly complex biochemical systems can evolve via indirect Darwinian pathways?

5. What is the argument from irreducible complexity? How does the logical point of this argument complement the empirical point? Does the logical point preclude both direct and indirect Darwinian pathways or only direct Darwinian pathways? Explain. What is the explanatory point of the argument from irreducible complexity? Explain why causal adequacy is a necessary feature of scientific explanations. Is the Darwinian mechanism of evolutionary change causally adequate to explain the origin of irreducibly complex biochemical structures?

6. Define co-option and coevolution. What is the co-option/coevolution objection to the argument from irreducible complexity? According to this objection, what is the most plausible evolutionary route to irreducibly complex systems? Why, according to Allen Orr, is the patchwork or bricolage approach to irreducible complexity unlikely to succeed?

7. What is the scaffolding or Roman arch objection to the argument from irreducible complexity? (See general notes.) Why isn't the elimination of functional redundancy a plausible route to irreducible complexity?

8. Is the type III secretory system (or some variant thereof) a likely evolutionary precursor to the bacterial flagellum? Justify your answer. The six-step model proposed in the general notes to section 6.4 describes how the type III secretory system supposedly evolved into the bacterial flagellum. Is this model sufficiently detailed to be testable? If not, how much more detail would be required to make it testable?

9. There are many irreducibly complex molecular systems inside the cell besides the bacterial flagellum. Consult a biology textbook and find at least five other examples. Unlike the flagellum (which bacteria can live without), cellular life could not exist without certain irreducibly complex molecular systems. Consult a biology textbook and show that the ribosome is one such machine. What are some other molecular machines that are required for life as we know it?

10. Darwinists criticize the argument from irreducible complexity, and especially its conclusion that irreducible complexity provides evidence for design, as a "God-of-the-gaps argument." Design theorists, pointing to the absence of fully articulated Darwinian pathways, criticize the claim that Darwinian pathways can explain irreducibly complex biological systems as a "Darwinism-of-the-gaps argument." Who, if anyone, is right? Explain what's at stake in this disagreement.

# CHAPTER SEVEN Specified Complexity

## 7.1 THE MARK OF INTELLIGENCE

Irreducible complexity, as we saw in the last chapter, provides not only negative evidence against Darwinian evolution but also positive evidence for intelligent design. Nevertheless, irreducible complexity needs to be supplemented with another form of complexity if it is to become a precise analytic tool for detecting design in biological systems. Often, when an intelligent agent acts, it leaves behind an identifying mark that clearly signals its intelligence. This mark of intelligence is known as specified complexity. Think of specified complexity as a fingerprint or signature that positively identifies the activity of an intelligence. Unlike irreducible complexity, which is a qualitative notion, specified complexity can be quantified and falls within the mathematical theory of probability and information.[1] The connection between these two forms of complexity is now this: irreducibly complex biological systems can, under certain circumstances, be shown to exhibit specified complexity.

But what exactly is specified complexity? An object, event, or structure exhibits specified complexity if it is both complex (i.e., not easily reproducible by chance) and specified (i.e., displays an independently given pattern). Neither complexity nor specification by themselves are enough to implicate intelligence. For instance, a sequence of randomly arranged Scrabble pieces is complex without being specified. Conversely, a sequence that keeps repeating the same short word is specified without being complex. In neither of these cases is an intelligence required to explain these sequences. On the other hand, a sequence of meaningful text is both complex and specified. Its explanation does require an intelligence.

NCHPA ZCXMBXRPTSLQAAVLKDSPOQK QWEOVCUSJKLZCXU MSSTE DKL
THETHETHETHETHETHETHETHETHETHETHETHETHETHETHETHETHETHETHET
THE MOST VALUABLE COMMODITY IN THE WORLD IS INFORMATION

***Figure 7.1*** *In the film* Wallstreet, *Michael Douglas (playing the inside trader Gordon Gekko) remarks to Charlie Sheen (playing his protégé Bud Fox): The most valuable commodity in the world is information.*

A memorable example of specified complexity comes from the 1985 novel by astronomer Carl Sagan titled *Contact* (which was subsequently made into a movie starring Jodie Foster). In that novel, radio astronomers engaged in the Search for Extraterrestrial Intelligence (SETI) discover a long sequence of prime numbers from outer space. Because the sequence is long, it is hard to reproduce by chance and therefore complex. Moreover, because the sequence is mathematically significant, it can be characterized independently of the physical processes that brought it about. As a consequence, it is also specified. Thus, when the SETI researchers in *Contact* observe specified complexity in this sequence of numbers, they have convincing evidence of extraterrestrial intelligence. Granted, real-life SETI researchers have thus far failed to detect designed signals from outer space. The point to note, however, is that Sagan based the methods of design detection that his fictional SETI researchers used on actual scientific practice.

## The Key Moment in the Movie *Contact*

In the movie *Contact*, researchers engaged in the Search for Extraterrestrial Intelligence (SETI) discover a signal that convinces them that they have made contact with an alien intelligence. In fact, at the key moment in the movie when this signal is discovered, one researcher turns to the other and says, "This isn't noise, this has structure." What was this signal? Though real-life SETI researchers have never found a signal from outer space that convincingly demonstrates alien intelligence, in the movie they did, namely, the following:

110111011111011111110111111111110111111111111101111111111111
111101111111111111111111011111111111111111111111011111111111
111111111111111111011111111111111111111111111111110111111111

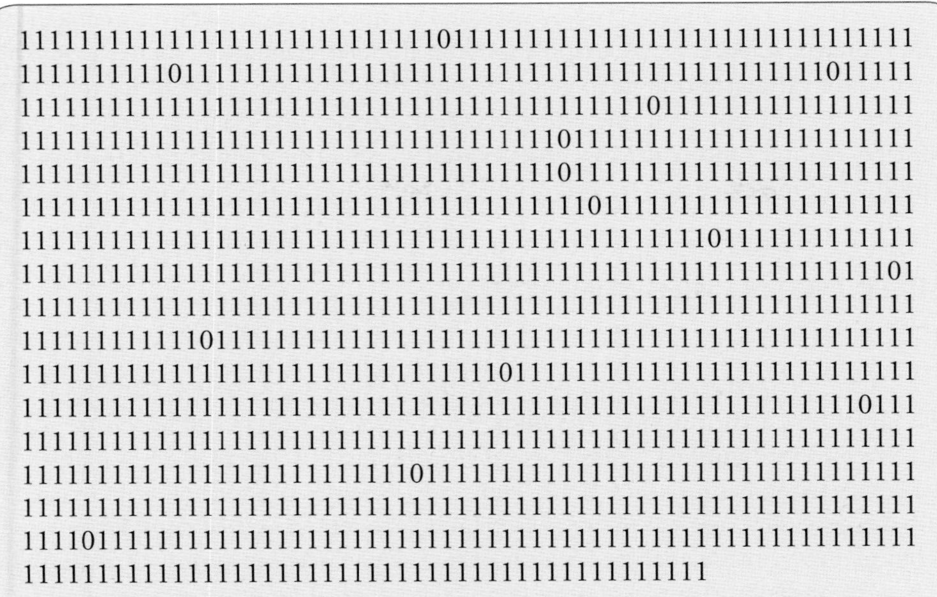
111111111111111111111111111101111111111111111111111111111111
111111111101111111111111111111111111111111111111111111011111
111111111111111111111111111111111111111111011111111111111111
111111111111111111111111111111111111011111111111111111111111
111111111111111111111111111111111111011111111111111111111111
111111111111111111111111111111111111110111111111111111111111
111111111111111111111111111111111111111111111101111111111111
111111111111111111111111111111111111111111111111111111111101
111111111111111111111111111111111111111111111111111111111111
111111111111011111111111111111111111111111111111111111111111
111111111111111111111111111111110111111111111111111111111111
111111111111111111111111111111111111111111111111111111110111
111111111111111111111111111111111111111111111111111111111111
111111111111111111111111110111111111111111111111111111111111
111111111111111111111111111111111111111111111111111111111111
111101111111111111111111111111111111111111111111111111111111
1111111111111111111111111111111111111111111111

***Figure 7.2*** *The signal discovered by SETI researchers in the movie* Contact, *comprising a long sequence of prime numbers.*

The SETI researchers in *Contact* received this signal as a sequence of 1,186 beats and pauses amplified over a speaker system. Here 1s correspond to beats and 0s to pauses. This sequence represents the prime numbers from 2 to 101 (prime numbers are numbers divisible only by themselves and one). In this example, a given prime number is represented by the corresponding number of beats (i.e., 1s), and individual prime numbers are separated by pauses (i.e., 0s). Thus, the sequence begins with 2 beats, then a pause, 3 beats, then a pause, 5 beats, then a pause, all the way up to 101 beats. The SETI researchers in *Contact* took this signal as decisive confirmation of extraterrestrial intelligence. This signal exhibits specified complexity.

***Figure 7.3*** *Jody Foster as a SETI researcher in the movie* Contact

***Figure 7.4*** *Arecibo observatory in Puerto Rico (the world's largest radio telescope)*

Many special sciences already employ specified complexity as a mark of intelligence—notably forensic science, cryptography, random number generation, archeology, and the search for extraterrestrial intelligence. Design theorists take this mark of intelligence and apply it to naturally occurring systems.[2] When they do, they claim to find that certain irreducibly complex molecular machines exhibit specified complexity and therefore are the product of intelligence. The aim of this chapter is to examine and justify this claim.

## 7.2 DEFINING SPECIFIED COMPLEXITY

The term *specified complexity* is now over thirty years old. Origin-of-life researcher Leslie Orgel first used it in his 1973 book *The Origins of Life*, where he wrote: "Living organisms are distinguished by their specified complexity. Crystals such as granite fail to qualify as living because they lack complexity; mixtures of random polymers fail to qualify because they lack specificity."[3] Orgel used the term specified complexity loosely, without giving a precise analytic account of it. Such an account is now in place thanks to the work of design theorists. Formulated as a statistical criterion for identifying the effects of intelligence, specified complexity incorporates three main elements: (1) a probabilistic version of complexity applicable to events; (2) a descriptive version of complexity applicable to patterns; and (3) a probabilistic gauge on the opportunities available for producing events by chance.

**Probabilistic complexity.** Probability can be viewed as a form of complexity. To see this, consider a combination lock. The more possible combinations the lock has, the more complex the mechanism and, correspondingly, the more improbable that the mechanism can be opened by chance. For instance, a combination lock whose dial is numbered from 0 to 39 and that must be turned in three alternating directions will have 64,000 (= 40 x 40 x 40) possible combinations. This number gives a measure of complexity for the combination lock but also corresponds to a 1/64,000 probability of the lock being opened by chance (assuming no prior knowledge of the lock combination). A more complicated combination lock whose dial is numbered from 0 to 99 and which must be turned in five alternating directions will have 10,000,000,000 (= 100 x 100 x 100 x 100 x 100) possible combinations and thus a 1/10,000,000,000 probability of being opened by chance. Complexity and probability therefore vary inversely: the greater the complexity, the smaller the probability. The "complexity" in "specified complexity" refers to improbability.

**Descriptive complexity.** As patterns, specifications exhibit varying degrees of complexity in terms of how easy or difficult it is to describe them. To see how patterns can vary in descriptive complexity, consider the following two sequences of ten coin tosses: HHHHHHHHHH and HHTHTTTHTH. Which of these is more readily attributed to chance? Both sequences have the same probability, approximately one in a thousand. Nevertheless, the pattern that specifies the first sequence is much simpler to describe than the second. For the first sequence, the pattern can be specified with the simple statement "ten heads in a row." For the second sequence, on the other hand, specifying the pattern requires a considerably longer statement, for instance, "two heads, then a tail, then a head, then three tails, then heads followed by tails and heads." A convenient way to think of descriptive complexity is therefore as the length of the minimum description that characterizes the pattern. There is a wide body of literature on this topic.[4]

For something to exhibit specified complexity it must have low descriptive complexity (as with the sequence HHHHHHHHHH consisting of ten heads in a row) but high probabilistic complexity (that is, its probability must be small). It's this combination of low descriptive complexity (a pattern or structure easy to describe in relatively short order) and high probabilistic complexity (something highly unlikely to happen by chance) that makes specified complexity such an effective "triangulator" of intelligence. Moreover, precisely because the descriptive complexity is low, it can be reconstructed independently of any physical events that give rise to it. That is why specifications are rightly regarded as "independently given patterns."

**Probabilistic resources.** Probabilistic resources refer to the number of opportunities for an event to occur and be specified. A seemingly improbable event can become quite probable once enough probabilistic resources are factored in. Alternatively, it may remain improbable even after all the available probabilistic resources have been factored in. Probabilistic resources come in two forms: replicational and specificational. Replicational resources refer to the number of opportunities for an event to occur. Specificational resources refer to the number of opportunities to specify an event. To see what's at stake with these two types of probabilistic resources, imagine that you are standing at a busy intersection and see ten brand new Pontiac Grand Ams each red with four doors drive past you in immediate succession. Did this happen by chance?[5] Sure, you witnessed a lucky coincidence, but what you need to determine is whether this coincidence was *so lucky* that no one could be expected to observe it if it happened by chance. That's the point. It's not just that you got lucky, but that no one else could have gotten that lucky.

To determine whether you were indeed too lucky, you need to know how many opportunities there are to observe this event. This requires knowing how many cars and how many makes of cars are on the road (if all cars in the world are brand new Pontiac Grand Ams each red with four doors, there is no coincidence to be explained). It also requires knowing how many people are standing on street corners around the world who might, in the course of a year, see the succession of cars you witnessed. The number of opportunities to observe this event constitutes the replicational resources. Replicational resources determine how likely it is for anyone, not just yourself, to observe the same succession of cars that you witnessed.

But that's not the only thing you need in determining whether the event happened by chance. No doubt, it would be striking to see ten brand new Pontiac Grand Ams each red with four doors drive past you in immediate succession. But what if the body type were altered to two doors? What if the color were changed from red to blue? And why just focus on Pontiac Grand Ams? What if the cars in succession had all been identical Honda Accords or Volkswagen Jettas? These questions point to different ways of specifying a succession of identical cars. Since any such succession would have been equally salient if you had witnessed it on the street corner, the number of such specifications must be factored into your judgment of whether the succession you observed happened by chance. This number constitutes the specificational resources. Specificational resources determine how likely it is to observe not just the succession of Grand Ams that you observed but any succession of identical cars.[6]

The full technical details for formulating specified complexity are complicated and best referred elsewhere.[7] Nevertheless, the main ideas underlying the concept are clear enough, and its three key elements are readily illustrated. We take up the task of elucidating these three elements in the next two sections.

## Real-Life SETI

Seth Shostak is the senior astronomer at the SETI Institute, a key think-tank devoted to the Search for Extraterrestrial Intelligence. The SETI Institute describes its mission as "to explore, understand and explain the origin, nature and prevalence of life in the universe."[8] With such a mission statement, it would appear that SETI and intelligent design have a natural affinity and can make common cause in detecting design in the universe. Nevertheless, in an online article,[9] Shostak criticized intelligent design, charging that its methods of design detection may not legitimately be compared with those of SETI. Why? Because intelligent design looks for complexity in biology to detect design whereas SETI looks for very simple signals to detect design, namely, radio signals with narrow bandwidth

*Seth Shostak*

transmissions. In particular, real-life SETI does not look for long, complex sequences of numbers as in Carl Sagan's novel *Contact.*

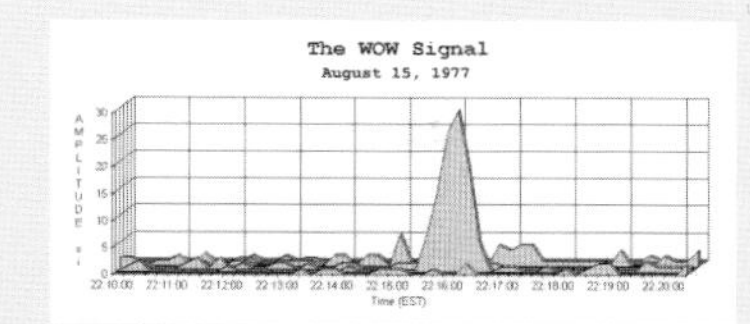


*Figure 7.5* *Image of narrow bandwidth transmission (above). On August 15, 1977 the Ohio State Radio Observatory found this narrow bandwidth signal that to date has been the most promising candidate for a transmission by an extraterrestrial intelligence. Jerry Ehrman wrote on the margin of the computer printout "WOW!" It has since been dubbed the "Wow! signal" (below).*

Shostak is in effect claiming that real-life SETI does not employ specified complexity when it attempts to identify alien intelligence from extraterrestrial radio signals. But he is mistaken. In fact, specified complexity is exhibited in precisely the signals that Shostak's SETI Institute is seeking, that is, narrow bandwidth transmissions. By definition, for a signal to exhibit specified complexity requires that it be *simple to describe but difficult to reproduce by chance.* That is the case here. Narrow bandwidth transmissions are easy to describe because humans, acting as designing agents, use such signals to communicate information. On the other hand, these signals are hard to bring about by chance because no undirected material process is known to produce them. In other words, these signals combine simplicity of description with complexity, in the sense of improbability, of outcome. That's specified complexity, the key marker for detecting design.

*Figure 7.6* *The monolith from* 2001: A Space Odyssey



Despite Shostak's protests, SETI researchers look for narrow bandwidth transmissions to detect intelligence from outer space precisely because such signals exhibit specified complexity. In fact, it appears that whenever people imagine what might constitute a sign of intelligence from outer space, such a sign invariably exhibits specified complexity. Take, for instance, the 1×4×9 monolith delivered by aliens to earth in Stanley Kubrick's film *2001, A Space Odyssey.* One detects design in this object because it exhibits specified complexity: it is easy to describe geometrically, but it is hard for undirected material processes to produce such hard homogeneous rectangular solids by purely undirected material forces.

## 7.3 THE REACH OF CHANCE

Specified complexity makes precise a fundamental human intuition, namely, that under the right circumstances highly improbable events are properly ascribed not to chance but to design.[10] In particular, specified complexity provides a rigorous way of determining what is within the reach of chance and what is outside the reach of chance. By chance here is meant not just brute randomness, as in flipping a coin, but any processes in which events characterized by probabilities play a role. Thus the neo-Darwinian mechanism, in which natural selection sifts gene mutations, is a chance process because mutations are genetic copying errors characterized by probabilities.

In determining whether an event is too improbable to occur by chance, one cannot simply appeal to brute improbability. For instance, one can take a fair coin, flip it a thousand times, and thereby participate in an incredibly improbable event. Whatever sequence of heads or tails happens to occur, its probability is approximately one chance in one followed by three hundred zeros:

1

1,000,000,000,000,000,000,000,000,000,000,000,000,000,000,000,000,000,000,000,
000,000,000,000,000,000,000,000,000,000,000,000,000,000,000,000,000,000,
000,000,000,000,000,000,000,000,000,000,000,000,000,000,000,000,000,000,
000,000,000,000,000,000,000,000,000,000,000,000,000,000,000,000,000,000,
000,000,000,000,000,000,000,000,000,000,000,000,000,000,000,000,000,000,
000,000,000,000,000,000,000,000,000,000

Though highly improbable, such an event would nonetheless be rightly ascribed to chance. Indeed, highly improbable events like this happen by chance all the time. In the definition of specified complexity given in the previous section, brute improbability corresponds to mere complexity.

Besides brute improbability, an event must also conform to a pattern if design rather than chance is be its proper explanation. Not just any pattern will do, however. Some patterns can legitimately be used to rule out chance and infer design whereas others cannot. To illustrate the difference, consider the case of an archer. Suppose an archer stands fifty paces from a large wall with bow and arrow in hand. The wall, let us say, is so large that the archer cannot help but hit it. Suppose each time the archer shoots an arrow at the wall, the archer paints a target around the arrow so that the arrow sits squarely in the bull's-eye. What can be concluded from this scenario? Absolutely nothing about the archer's ability as an archer. Yes, a pattern is being matched; but it is a pattern fixed only after the arrow has been shot. The pattern is thus purely *ad hoc*.

But suppose instead the archer paints a fixed target on the wall and then shoots at it. Suppose the archer shoots a hundred arrows, and each time hits a perfect bull's-eye. What can be concluded from this second scenario? Confronted with this occurrence, we are duty-bound to infer that here is a world-class archer, one whose shots cannot legitimately be attributed to luck but rather must be attributed to the archer's skill and mastery. Skill and mastery are, of course, instances of design.

The type of pattern in which an archer fixes a target first and then shoots at it is common to statistics, where it is known as setting a *rejection region* or *critical region* prior to an experiment.[11] In statistics, if the outcome of an experiment falls within a rejection region, the chance hypothesis supposedly responsible for the outcome is rejected. The reason for setting a rejection region prior to an experiment is to forestall what statisticians call "data snooping" or "cherry picking." Just about any data set will contain strange and improbable patterns if we look hard enough. By forcing experimenters to set their rejection regions prior to an experiment, the statistician protects the experiment from bogus patterns that could just as well result from chance.

***Figure 7.7*** *Archer shooting at a target*

A little reflection, however, makes clear that a pattern need not be given prior to an event to eliminate chance and implicate design. Consider the following cipher text:

nfuijolt ju jt mjlf b xfbtfm.

Initially this looks like a random sequence of letters and spaces; initially you lack any pattern for rejecting chance and inferring design.

But suppose next that someone comes along and tells you to treat this sequence like a Caesar cipher,[12] moving each letter one notch down the alphabet to decrypt the cipher. Now the sequence reads,

methinks it is like a weasel.

Even though the pattern is given after the fact, it is still the right sort of pattern for eliminating chance and inferring design. In contrast to statistics, which always identifies its patterns before an experiment is performed, cryptanalysis must discover its patterns after the fact. In both instances, however, the patterns are suitable for ruling out chance and inferring design.

Patterns thus divide into two types, those that in the presence of complexity warrant a design inference and those that despite the presence of complexity do not warrant a design inference. The first type of pattern we call a *specification*, the second a *fabrication*. Specifications are the non-*ad hoc* patterns that can be legitimately used to eliminate chance and warrant a design inference. In contrast, fabrications are the *ad hoc* patterns that cannot be legitimately used to warrant a design inference.

This distinction between specifications and fabrications hinges on the descriptive complexity of the underlying patterns, specifications having low descriptive complexity, fabrications having high or unknown descriptive complexity. In the case of the archer, if the target is fixed, it is easily described as "the target located at that particular place on the wall (e.g., 20 feet from the left, 10 feet from the top)." Such a description is short enough. Movable targets, however, are a different matter. One does not specify a target by saying "its bull's-eye is where an archer will hit the wall by shooting an arrow." Such a description is completely inadequate for reconstructing the target. To reconstruct the target in that case depends on the precise future action of the archer, which is absent from this description and from any background knowledge presupposed by it.

All the most widely cited examples of specified complexity employ, as they must, short descriptions of the underlying patterns. For instance, a bacterial flagellum is described as "a bidirectional motor driven propeller." Mount Rushmore is described as "the faces of four U.S. presidents: George Washington, Thomas Jefferson, Theodore Roosevelt, and Abraham Lincoln." The decisive signal in *Contact* is described as "the prime numbers from 2 to 101." In each of these cases the description is short, but the event described (e.g., the actual signal radio signal denoting prime numbers from 2 to 101) is highly improbable apart from the activity of a designing intelligence.

Such examples, however, raise a concern over the choice of language used in such descriptions. The worry is this: humans develop languages for efficiently describing the kinds of phenomena they want to specify. So, doesn't the definition of specified complexity, in looking to minimum length (and therefore efficient) descriptions, stack the deck? Doesn't our very choice of language bias us toward finding specified complexity (and therefore design) in certain phenomena as opposed to others? But in that case, how do we know that specified complexity is truly a marker of design and not merely an artifact of our choice of language?[13]

This objection begins with a valid observation but does not reason logically from it. It is certainly true that our ability to detect design via specified complexity depends on our background knowledge, and that includes whatever languages are at our disposal. Thus, if you hear a known language, you can readily discern whether you are listening to random gibberish or meaningful communication. On the other hand, if you hear a completely foreign language, you will be hard pressed to draw the same distinction. It takes intelligence to recognize intelligence, and in so doing we are limited to the languages at our disposal.

But this does not mean that the language we apply to a given range of phenomena was concocted on the basis of those phenomena and therefore biases us toward finding design in those phenomena when it actually isn't there. Consider the bacterial flagellum described as a "bidirectional motor driven propeller." That description came not from looking at molecular machines inside cells, but from human engineers who invented such machines before biologists ever suspected that they formed parts of cells. If anything, the ability of our language to efficiently describe the phenomena of molecular biology, far from suggesting that the specified complexity found there is an artifact of language, should lead us to think that the design in molecular biology is real. On this view, the language of engineering describes molecular machines in biology because these are in fact engineered systems.[14]

## 7.4 CLIMBING MOUNT IMPROBABLE WITHOUT DESIGN?

Proponents of intelligent design are not alone in seeing specified complexity as the key to determining whether design actually is present in biological systems. Take Oxford biologist Richard Dawkins, as vocal an opponent of intelligent design as one will find.[15] In his book *The Blind Watchmaker,* he attempts to explain why biological systems are not designed despite exhibiting specified complexity: "Complicated things have some quality, specifiable in advance, that is highly unlikely to have been acquired by random chance alone. In the case of living things, the quality that is specified in advance is . . . the ability to propagate genes in reproduction."[16] Dawkins is here attributing specified complexity to biological systems. Nevertheless, from their specified complexity Dawkins does not conclude that they are designed. Why? Because the probabilities are, in his estimation, not small enough. For instance, in regard to the origin of life, he writes:

> We can accept a certain amount of luck in our explanations, but not too much. . . . In our theory of how we came to exist, we are allowed to postulate a certain ration of luck. This ration has, as its upper limit, the

> number of eligible planets in the universe. . . . We [therefore] have at our disposal, if we want to use it, odds of 1 in 100 billion billion as an upper limit (or 1 in however many available planets we think there are) to spend in our theory of the origin of life. This is the maximum amount of luck we are allowed to postulate in our theory. Suppose we want to suggest, for instance, that life began when both DNA and its protein-based replication machinery spontaneously chanced to come into existence. We can allow ourselves the luxury of such an extravagant theory, provided that the odds against this coincidence occurring on a planet do not exceed 100 billion billion to one.[17]

Dawkins is here quite right that getting too lucky is not a good scientific explanation and that how much luck one can allow in a scientific explanation depends on how many opportunities there are to be lucky. These opportunities, or what Dawkins calls "ration of luck," are what in section 7.2 we termed "probabilistic resources." Probabilistic resources determine whether a seemingly improbable event really is improbable once all the opportunities for it to occur and to be specified are factored in. Accordingly, if the odds against a coincidence exceed one's probabilistic resources (for Dawkins, 100 billion billion planets in the case of life's origin), then a chance-based explanation is inadequate.

Indeed, Dawkins's entire defense of evolutionary theory consists in showing that the odds against forming complex specified systems are really not as bad as one at first suspects. That's why he wrote *Climbing Mount Improbable.*[18] In this book, Dawkins compares the emergence of biological complexity to climbing a mountain. He calls it Mount Improbable because if you had to get all the way to the top in one fell swoop (that is, achieve a massive increase in biological complexity all at once), it would be highly improbable. But Mount Improbable does not, according to him, have to be scaled in one leap.

Evolutionary theory purports to show how Mount Improbable can be scaled in small incremental steps. Thus, according to Dawkins, Mount Improbable always has a gradual serpentine path leading to the top that can be traversed in baby-steps. But such a claim requires verification. It might be a fact about nature (i.e., about inherent discontinuities in biological configuration space) that Mount Improbable is sheer on all sides and getting to the top from the bottom via baby-steps is effectively impossible. A gap like that would reside in nature herself and not in our knowledge of nature.

Dawkins rejects this possibility. Although biological systems may look vastly improbable, he argues that they are really not once one factors in the Darwinian mechanism of natural selection and random variation. This mechanism, according to Dawkins, takes seemingly improbable evolutionary events and renders them probable and therefore explainable apart from intelligence. Accordingly, the Darwinian mechanism of random variation and natural selection becomes biology's premier divide-and-conquer strategy for producing specified complexity.

As Dawkins characterizes the Darwinian mechanism, each generation in an organism's evolutionary history constitutes a round of random variation and natural selection, and therefore a baby step in the organism's evolutionary path. The Darwinian mechanism proceeds by trial and error, with natural selection providing the trial and random variation providing the error. As with all trial-and-error mechanisms, the Darwinian mechanism hinges on slow, gradual improvements. Insofar as it succeeds, it does so by numerous divisions and numerous small conquests. This is how evolution is supposed to scale Mount Improbable.

In essence, the Darwinian mechanism explains specified complexity by explaining it away—by breaking the improbability associated with an evolving system into a sequence of manageable probabilities and thus overcoming that improbability one evolutionary baby step at a time. In the remainder of this chapter we will show why at least some of the improbabilities associated with biological systems cannot be overcome in this way. Instead, we will show why specified complexity points not just to the appearance but to the actual design of biological systems.

## Monkeys Typing Shakespeare

To make materialistic accounts of evolution appear plausible, its proponents sometimes invoke "deep time." Take, for instance, the following remark by Nobel prize-winning biologist George Wald on the origin of life via chemical evolution:

> Time is in fact the hero of the plot. The time with which we have to deal is of the order of two billion years. What we regard as impossible on the basis of human experience is meaningless here. Given so much time, the "impossible" becomes possible, the possible probable, and the probable virtually certain. One has only to wait: time itself performs miracles.[19]

Given enough time, Wald argues, anything can happen. And that's right, provided the universe is old enough and big enough. But even with a universe

*continued on next page*

billions of years old and billions of light years in diameter, in fact very little of consequence can happen by chance.

To illustrate this point, in 1913 the French mathematician Émile Borel argued that a million monkeys typing ten hours a day would be extremely unlikely to reproduce the books in the world's libraries.[20] Borel's point was to illustrate the extreme unlikelihood of certain events and the inability of the known physical universe, given its duration and material resources, to produce such events.

How many monkeys and how much time would in fact be required to produce the world's books? Let's not focus on all the world's books or even on all the works of a given author, such as Shakespeare, or even on the entirety of Shakespeare's best known play, *Hamlet*. Rather, let's focus on the famous soliloquy of Hamlet that starts with "To be or not to be, that is the question." How many monkeys and how much time would be required for chance to produce just that soliloquy?

Suppose you had a million monkeys randomly typing on a million typewriters at a rate of a million keystrokes per second for a million years. To make things easy on the monkeys, let's assume that they only type capital Roman letters, spaces, commas, periods, and apostrophes. In other words, they only use the twenty-six letters of the alphabet plus four additional characters. That's thirty characters instead of the entire character set on a typical typewriter keyboard. In that case, the monkeys would have only a fifty-fifty chance of typing the first six words from that soliloquy:

TO BE, OR NOT TO BE

These monkey-at-a-typewriter examples illustrate an important principle: it takes time and matter for chance to produce information, and without enough time and matter, chance can only go so far.[21] In fact, work by MIT quantum computational theorist Seth Lloyd shows that in the known physical universe, chance can effectively produce only 400 bits of prespecified information (400 bits is a string of 400 zeros and ones).[22] In terms of our reduced set of 30 characters, that amounts to strings consisting of 82 characters. Accordingly, the longest initial segment of Hamlet's soliloquy that the entire universe could by chance produce in its multibillion year history is

TO BE, OR NOT TO BE, THAT IS THE QUESTION.
WHETHER 'TIS NOBLER IN THE MIND TO SUFFER[23]

Clearly, chance isn't going to get very far here. Indeed, all the chance in the known physical universe will never get Émile Borel's monkeys to randomly type even the first two lines of Hamlet's soliloquy, much less all of *Hamlet*, still less all the works of Shakespeare, much less still all the books in the world's libraries. Moreover, in these discussions of monkeys at typewriters, let's not forget to ask, Where did the typewriters come from? And the language? And those that communicate and understand the language? And the monkeys?[24]

## Natural Selection at the Monkey's Shoulder

The odds are overwhelmingly against monkeys randomly typing even two lines of *Hamlet*, much less the entire works of Shakespeare. But what if each monkey has someone stand at its shoulder and erase every error it makes (thereby keeping only what Shakespeare actually wrote). According to Darwinist and education lobbyist Eugenie Scott, this is exactly what natural selection does:

> [Suppose] you got a million monkeys sitting there typing on their machine. If you want to make this an analogy that makes sense from the standpoint of evolution, you've got a million technicians standing behind them with a very large vat of white out and every time the monkey types the wrong letter, you correct it. That's what natural selection basically does. It's not just the random production of variation.[25]

But with technicians like this backing up the monkeys, there's no need for an army of monkeys. Given even one such technician, a single randomly typing monkey will, in short order, type all of *Hamlet* and even all the works of Shakespeare. Within Darwinism, natural selection is supposed to fulfill the role of Scott's technicians. Thus natural selection is said to monitor the course of evolution, get rid of evolution's "mistakes," and thereby ensure that evolution moves along efficiently and doesn't get stuck in dead-ends.

Although Scott's error-correction approach to overcoming randomness sounds plausible, it is in fact deeply confused. In the case of a monkey at a typewriter,

*continued on next page*

what exactly are the qualifications of the technician standing at the monkey's shoulder doing the erasing? Indeed, how does the technician know what to erase? The whole point of having monkeys at a typewriter is to account for the emergence of Shakespeare's works without the need to invoke an intelligence (like Shakespeare) that already knows Shakespeare's works. In other words, the whole point was to get Shakespeare's works without Shakespeare. But that's not what is happening here. *Clearly, the only way to erase errors in the typing of Shakespeare's works is to know Shakespeare's works in the first place.* Indeed, the very concept of error presupposes that there is a right way that things ought to be. That's the problem: Eugenie Scott's technicians, to do their work, need already to know the works of Shakespeare.

When Eugenie Scott calls for a technician to stand over a monkey's shoulder and correct its mistakes, she commits the fallacy of begging the question or arguing in a circle. In other words, Scott presupposes the very thing she needs to establish as the conclusion of a sound scientific argument. Indeed, scientific rigor demands that we ask who in turn is standing over the technician's shoulder and instructing the technician what is and is not a mistake in the typing of Shakespeare. If the technician's assistance to the monkey is to mirror natural selection, then the technician needs to help the monkey without knowing or giving away the answer. And yet that's exactly what the technician is doing here.

Bottom line: Monkeys cannot type Shakespeare apart from Shakespeare!

## 7.5 THE APPEARANCE OF DESIGN

In Chapter 6, we saw that there are no detailed, testable, step-by-step Darwinian accounts for the evolution of any irreducibly complex biochemical machine such as the bacterial flagellum. We also saw that without the bias of speculative Darwinism coloring our conclusions, we are naturally inclined to see such irreducibly complex systems as the products of intelligent design. All our intuitions certainly point in that direction. Even opponents of intelligent design realize this. That's why Richard Dawkins writes, "Biology is the study of complicated things that give the appearance of having been designed for a purpose."[26] That's also why Francis Crick writes, "Biologists must constantly keep in mind that what they see was not designed, but rather evolved."[27] Yet for Dawkins, Crick, and fellow Darwinists, the appearance of design in biology cannot be trusted. Accordingly, any intuitions that lead us to see actual design in biological systems are supposedly leading us astray.

But intuitions need not lead us astray; they can also lead us aright. In fact, they often lead us to truths that might otherwise elude us. How, then, do scientists tell apart sound intuitions that lead us aright from faulty intuitions that lead us astray? The problem for science with intuitions is that they are informal and imprecise. Hence, to determine whether intuitions are leading us astray or aright, scientists attempt to flesh out intuitions with precise formal analyses. Darwinists claim to have done just that. Thus, they purport to have shown where our intuitions about design in biology break down and how the Darwinian selection mechanism can bring about the appearance of design in biology. But Darwinists have demonstrated no such thing. As we've seen, Darwin's theory fails to account for the emergence of irreducibly complex molecular machines.

It follows that we need once again to take seriously our intuitions that such systems (e.g., the bacterial flagellum) are in fact designed. The challenge, then, for the design theorist is to provide precise formal analyses showing that our intuitions about design in biology are indeed justified and, specifically, how various biological systems exhibit the key marker for detecting design outlined in the previous sections of this chapter, namely, specified complexity. To be sure, design theorists also want to explain how, in engineering terms, the design in living things came to be implemented. But given that the technology in living things so far exceeds human engineering capabilities, explaining how the design in biology got there is a research project more for the future than for the present.

What, then, does such a formal, design-theoretic analysis of irreducibly complex systems look like? How does it demonstrate that such systems are indeed complex and specified, therefore exhibit specified complexity, and thus are in fact designed? The details here are technical, but the general logic by which design theorists argue that irreducibly complex systems exhibit specified complexity is straightforward: for a given irreducibly complex system and any putative evolutionary precursor, show that the probability of that precursor evolving by the Darwinian mechanism into the irreducibly complex system is small.

In such analyses, specification is never a problem: in each instance, the irreducibly complex system, any evolutionary precursor, and any intermediate between the precursor and the final irreducibly complex system are always specified in terms of their biological function. Moreover, in such analyses, probabilities need not be calculated exactly (in general, they cannot be). It's enough to establish reliable upper bounds on the probabilities and show that they are small. A rule of thumb here is that if the probability of evolving a precursor into a plausible intermediate is small, then the probability of evolving that precursor through the intermediate into the irreducibly complex system will be even smaller.

Darwinists object to this approach to establishing the specified complexity of irreducibly complex biochemical systems. They contend that design theorists, in taking this approach, have merely devised a "tornado-in-a-junkyard" strawman. The image of a "tornado in a junkyard" comes from astronomer Fred Hoyle. Hoyle imagined a junkyard with all the pieces for a Boeing 747 strewn in disarray and then a tornado blowing through the junkyard and producing a fully assembled 747 ready to fly.[28] Darwinists object that this image has nothing to do with how Darwinian evolution produces biological complexity. Accordingly, in the formation of irreducibly complex systems such as the bacterial flagellum, all such arguments are said to show is that these systems could not have formed by purely random assembly. But, Darwinists contend, evolution is not about randomness. Rather, it is about natural selection sifting the effects of randomness.

To be sure, if design theorists were merely arguing that pure randomness cannot bring about irreducibly complex systems, there would be merit to the Darwinists' tornado-in-a-junkyard objection. But that's not what design theorists are arguing. The problem with Hoyle's tornado-in-a-junkyard image is that, from the vantage of probability theory, it made the formation of a fully assembled Boeing 747 from its constituent parts as difficult as possible. But what if the parts were not randomly strewn about in the junkyard? What if, instead, they were arranged in the order in which they needed to be assembled to form a fully functional 747. Furthermore, what if, instead of a tornado, a robot capable of assembling airplane parts were handed the parts in the order of assembly? How much knowledge would need to be programmed into the robot for it to have a reasonable probability of assembling a fully functioning 747? Would it require more knowledge than could reasonably be ascribed to a program simulating Darwinian evolution?

Design theorists, far from trying to make it difficult to evolve irreducibly complex systems such as the bacterial flagellum, strive to give the Darwinian selection mechanism every *legitimate* advantage for evolving such systems. The one advantage that cannot legitimately be given to the Darwinian selection mechanism, however, is prior knowledge of the system whose evolution is in question. That would be endowing the Darwinian mechanism with teleological powers (in this case, foresight and planning) that Darwin himself insisted it does not, and indeed cannot, possess if evolutionary theory is effectively to dispose of design. Yet, even with the most generous allowance of legitimate advantages, the probabilities computed for the Darwinian mechanism to evolve irreducibly complex biochemical systems such as the bacterial flagellum always end up being exceedingly small.[29]

What makes these probabilities so small is the difficulty of coordinating successive evolutionary changes apart from teleology or goal-directedness. In the Darwinian mechanism, neither selection nor variation operates with reference to future goals (such as the goal of evolving a bacterial flagellum from a bacterium lacking this structure). Selection is natural selection, which is solely in the business of conferring immediate benefits on an evolving organism. Likewise, variation is random variation, which is solely in the business of perturbing an evolving organism's heritable structure without regard for how such perturbations might benefit or harm future generations of the organism. In the absence of goal-directedness, evolution faces a number of daunting hurdles that render the formation of irreducibly complex molecular machines highly improbable and should lead us to question whether the appearance of design in biology is merely an appearance. We turn next to these hurdles.

## 7.6 HURDLES EVOLUTION MUST OVERCOME

In attempting to coordinate the successive evolutionary changes needed to bring about irreducibly complex molecular machines, the Darwinian mechanism encounters a number of hurdles. These include the following:[30]

1. *Availability.* Are the parts needed to evolve an irreducibly complex biochemical system such as the bacterial flagellum even available?

2. *Synchronization.* Are these parts available at the right time so that they can be incorporated when needed into the evolving structure?

3. *Localization.* Even with parts that are available at the right time for inclusion in an evolving system, can the parts break free of the systems in which they are currently integrated (without harming those systems) and be made available at the "construction site" of the evolving system?

4. *Interfering Cross-Reactions.* Given that the right parts can be brought together at the right time in the right place, how can the wrong parts that would otherwise gum up the works be excluded from the "construction site" of the evolving system?

5. *Interface Compatibility.* Are the parts that are being recruited for inclusion in an evolving system mutually compatible in the sense of meshing or interfacing tightly so that, once suitably positioned, the parts work together to form a functioning system?

6. *Order of Assembly.* Even with all and only the right parts reaching the right place at the right time, and even with full interface compatibility, will they be assembled in the right order to form a functioning system?

7. *Configuration.* Even with all the right parts slated to be assembled in the right order, will they be arranged in the right way to form a functioning system?

To see what's at stake in overcoming these hurdles, imagine you are a contractor who has been hired to build a house. If you are going to be successful at building the house, you will need to overcome each of these hurdles. First, you have to determine that all the items you need to build the house (e.g., bricks, wooden beams, electrical wires, glass panes, and pipes) exist and thus are *available* for your use. Second, you need to make sure that you can obtain all these items within a reasonable period of time. If, for instance, crucial items are back-ordered for years on end, then you won't be able to fulfill your contract by completing the house within the appointed time. Thus, the availability of these items needs to be properly *synchronized.* Third, you need to transport all the items to the construction site. In other words, all the items needed to build the house need to be brought to the location where the house will be built.

Fourth, you need to keep the construction site clear of items that would ruin the house or interfere with its construction. For instance, dumping radioactive waste or laying high-explosive mines on the construction site would effectively prevent a usable house from ever being built there. Less dramatically, if excessive amounts of junk found their way to the site (items that are irrelevant to the construction of the house, such as tin cans, broken toys, and discarded newspapers), it might become so difficult to sort through the clutter, and thus to find the items necessary to build the house, that the house itself might never get built. Items that find their way to the construction site and hinder the construction of a usable house may thus be described as producing *interfering cross-reactions.*

Fifth, procuring the right sorts of materials required for houses in general is not enough. As a contractor, you also need to ensure that they are properly adapted to each other. Yes, you'll need nuts and bolts, pipes and fittings, electrical cables and conduits. But unless nuts fit properly with bolts, unless fittings are adapted to pipes, and unless electrical cables fit inside conduits, you won't be able to construct a usable house. To be sure, each part taken by itself can make for a perfectly good building material capable of working successfully in some house or other. But your concern here is not with some house or other but with the house you are actually building. Only if the parts at the construction site are adapted to each other and interface correctly will you be able to build a usable house. In short, as a contractor you need to ensure that the parts you are bringing to the construction site not only are of the type needed to build houses in general but also share *interface compatibility* so that they can work together effectively.

Sixth, even with all and only the right materials at the construction site, you need to make sure that you put the items together in the correct order. Thus, in building the house, you need first to lay the foundation. If you first erect the walls and then try to lay the foundation under the walls, your efforts to build the house will fail. The right materials require the right *order of assembly* to produce a usable house. Seventh, and last, even if you are assembling the right building materials in the right order, the materials need also to be arranged appropriately. That's why, as a contractor, you hire masons, plumbers, and electricians. You hire these subcontractors not merely to assemble the right building materials in the right order but also to position them in the right way. For instance, it's all fine and well to take bricks and assemble them in the order required to build a wall. But if the bricks are oriented at strange angles or if the wall is built at a slant so that the slightest nudge will cause it to topple over, then no usable house will result even if the order of assembly is correct. In other words, it's not enough for the right items to be assembled in the right order; in addition, as they are being assembled, they need to be properly *configured*.

Now, as a building contractor, you find none of these hurdles insurmountable. That's because, as an intelligent agent, you see the big picture. You can look ahead to where you're going and what your final product will be. You can therefore coordinate all the tasks needed to overcome these hurdles. You have an architectural plan for the house. You know what materials are required to build the house. You know how to procure them. You know how to deliver them to the right location at the right time. You know how to secure the location from vandals, thieves, debris, weather, and anything else that would spoil your construction efforts. You know how to ensure that the building materials are properly adapted to each other so that they work together effectively once put together. You know the order of assembly for putting the building materials together. And, through the skilled laborers you hire (i.e., the subcontractors), you know how to arrange these materials in the right configuration. All this *know-how* results from intelligence and is the reason you can build a usable house.

But the Darwinian mechanism of random variation and natural selection has none of this know-how. All it knows is how to randomly modify biological structures and then to preserve those random modifications that happen to be useful at the moment (usefulness being measured in terms of survival and reproduction). The Darwinian mechanism is an instant gratification mechanism. If the Darwinian mechanism were a building contractor, it might put up a wall because of its immediate benefit in keeping out intruders from the construction site even though by building the wall now, no foundation could be laid later and, in consequence, no usable house could ever be built. That's how the Darwinian mechanism works, and that's why it is so limited. It is a trial-and-error tinkerer for which each act of tinkering needs to maintain or enhance present advantage or select for a newly acquired advantage. It cannot make present sacrifices to achieve future as-yet unrealized benefits.

Imagine, therefore, what it would mean for the Darwinian mechanism to clear these seven hurdles in evolving a bacterial flagellum (see section 6.3). We start with a bacterium that has no flagellum, no genes coding for proteins in the flagellum, and no genes homologous to genes coding for proteins in the flagellum. Such a bacterium is supposed to evolve, over time, into a bacterium with the full complement of genes needed to put together a fully functioning flagellum. Is the Darwinian mechanism adequate for coordinating all the biochemical events needed to clear these seven hurdles and thereby evolve the bacterial flagellum? To answer yes to this question is to attribute creative powers to the Darwinian mechanism that are implausible in the extreme.

To see this, let's run through these seven hurdles in turn, assessing the challenge each poses to the Darwinian evolution of the bacterial flagellum. Let's start with availability: can the Darwinian mechanism overcome the *availability hurdle*? To overcome this hurdle, the Darwinian mechanism needs to form novel proteins. (The bacterial flagellum, if it evolved at all, evolved from a bacterium without any of the genes, exact or homologous, for the proteins constituting the flagellum.) Now, the Darwinian mechanism may be capable, in some cases, of modifying existing proteins or recruiting them wholesale for new uses (see, for instance, the work of Meier *et al.* cited at the end of section 7.8). But, as a general purpose mechanism for producing novel proteins, the evidence goes against the Darwinian mechanism. As we shall see in section 7.9, research into the folding characteristics of protein domains indicates that certain classes of proteins are highly unlikely to evolve by Darwinian processes. Accordingly, for the Darwinian mechanism to generate the novel proteins (as well as the novel genes coding for them) required in the evolution of the bacterial flagellum seems unlikely.

Consider next the *synchronization hurdle*. Darwinian evolution has a long time to work, and may not be affected by short-term deadlines (though astrophysics imposes long-term deadlines, as with the Sun turning into a red giant in about 5 billion years, causing it to expand and burn up everything in its path, including the Earth[31]). So it may not be crucial when a specific protein or anatomical structure becomes available for evolution—unless it becomes available so prematurely that it decays before it can be put to use. But note that development is not so forgiving. As an organism develops from a fertilized egg to an adult, it needs specific building blocks at the correct times or it will die. Although evolution may be relatively insensitive to the synchronization hurdle, development is not.

The *localization hurdle,* on the other hand, seems considerably more difficult for the Darwinian mechanism to clear. The problem here is that items originally assigned to certain systems need to be reassigned and recruited for use in a newly emerging system. This newly emerging system starts as an existing system that then gets modified with items previously incorporated in other systems. But how likely is it that these items break free and get positioned at the construction site of another system, thereby transforming it into a newly emerging system with a novel or enhanced function? Our best evidence suggests that this repositioning of items previously assigned to different systems is improbable and becomes increasingly improbable as more items need to be repositioned simultaneously at the same location. There are two reasons for this. First, the construction site for a given biochemical system tends to maintain its integrity, incorporating only proteins pertinent to the system and keeping out stray proteins that could be disruptive. Second, proteins don't break free of systems to which they are assigned as a matter of course; rather, a complex set of genetic changes is required, such as gene duplications, regulatory changes, and point mutations.

The *interfering cross-reaction* hurdle intensifies the challenge to the Darwinian mechanism posed by the previous hurdle. If the bacterial flagellum is indeed the result of Darwinian evolution, then evolutionary precursors to the flagellum must have existed along the way. These precursors would have been functional systems in their own right, and in their evolution to the flagellum would have needed to be modified by incorporating items previously assigned to other uses. These items would then need to have been positioned at the construction site of the given precursor. Now, as we just saw with the localization hurdle, there is no reason to think that this is likely. Indeed, foreign proteins floating around in places where they are not expected to be tend to get broken down and recycled (e.g., by protein scavengers known as proteasomes). But suppose the construction site becomes more open to novel proteins (thus lowering the localization hurdle and thereby raising the probability of overcoming it). In that case, by welcoming items that could help in the evolution of the bacterial flagellum, the construction site would also welcome items that could hinder its evolution. It follows that to the degree that the localization hurdle is easy to clear, to that degree the interfering cross-reaction hurdle is difficult to clear, and vice versa.

The *interface-compatibility hurdle* raises yet another difficulty for the Darwinian mechanism. The problem is this. For the Darwinian mechanism to evolve a system, it must redeploy parts previously targeted for other systems. But that's not all. It also needs to ensure that those redeployed parts mesh or interface properly with the evolving system. If not, the evolving system will malfunction and thus no longer confer a selectable advantage. The products of Darwinian evolution are, after all, kludges. In other words, they are systems formed by sticking together items previously assigned to different uses. Now, if these items were built according to common standards or conventions, there might be reason to think that they could work together effectively. But natural selection, as an instant gratification mechanism, has no inherent capacity for standardizing the products of evolution. Yet without standardization, what evolution can manufacture and innovate becomes extremely limited.

Think of cars manufactured by different auto makers—say, a Chevrolet Impala from the United States and a Honda Accord from Japan. Although these cars will be quite similar and have subsystems and parts that perform identical functions in identical ways, the parts will be incompatible. You can't, for instance, swap a piston from one car for a piston in the other or, for that matter, swap bolts, nuts, and screws from the two vehicles. That's because these cars were designed independently according to different standards and conventions. Of course, at the Chevrolet plant that builds the Impala, there will be standardization ensuring that different parts of the Impala and other Chevrolet models have compatible interfaces. But across automobile manufacturers (e.g., Chevrolet and Honda), there will be no (or very little) standardization to which the construction of parts must adhere. In fact, common standards and conventions that facilitate the interface compatibility of distinct functional systems points not just to the design of the systems but also to a common design responsible for the standardization.

But the Darwinian mechanism is incapable of such common design. As an instant gratification mechanism, its only stake is in bringing about structures that constitute an immediate advantage to an evolving organism. It has no stake in ensuring that such structures also adhere to standards and conventions that will allow them to interface effectively with other structures down the line. For instance, evolutionists sometimes argue that the bacterial flagellum evolved from a microsyringe known as the type III secretory system (TTSS for short—see the general notes to section 6.4 for the background on this model). In this model, a pilus or hairlike structure gets redeployed and attached to the TTSS, eventually to become the whiplike tail that moves the bacterium through its watery environment. Yet before the pilus attached to the TTSS, these two systems had to evolve independently. Consequently, short of invoking sheer blind luck, there is no reason to think that these systems should work together—any more than there is to think that independently designed cars would have swappable parts. This weakness of Darwinian theory can be tested experimentally: take an arbitrary TTSS and pilus and determine the extent of the genetic modifications needed for the pilus to extrude through the TTSS's protein delivery system (which is how the pilus is supposed to interface with the TTSS). At present, there are no sound theoretical or experimental reasons to think that the Darwinian mechanism can overcome the interface compatibility hurdle.

For the Darwinian mechanism to clear the *order-of-assembly hurdle* is also a stretch. The Darwinian mechanism works by accretion and modification: it adds novel parts to already functioning systems as well as modifies existing parts in them. In this way, new systems with enhanced or novel functions are formed. Now, consider what happens when novel parts are first added to an already functioning system. In that case, the earlier system becomes a subsystem of a newly formed supersystem. Moreover, the order of assembly of the subsystem will, at least initially (before subsequent modifications), be the same as when the subsystem was a standalone system. In general, however, just because the parts of a subsystem can be put together in a given order doesn't mean those parts can be put together in the same order once it is embedded in a supersystem. In fact, in the evolution of systems such as the bacterial flagellum, we can expect the order of assembly of parts to undergo substantial permutations (certainly, this is the case with the model for the evolution of the bacterial flagellum from the TTSS). How, then, does the order of assembly undergo the right permutations? For most biological systems, the order of assembly is entrenched and does not permit substantial deviations. The burden of evidence is therefore on the Darwinist to show that for an evolving system, the Darwinian mechanism coordinates not only the emergence of the right parts but also their assembly in the right order. Darwinists have shown nothing of the sort.

Finally, we consider the *configuration hurdle*. In the design and construction of human artifacts, this hurdle is one of the more difficult to overcome. On the other hand, in the evolution of irreducibly complex biochemical systems such as the bacterial flagellum, this is one of the easier hurdles to overcome. Here's why. In the actual assembly of the flagellum and systems like it, the biochemical parts do not come together haphazardly. Rather, they self-assemble in the right configuration when chance collisions allow specific, cooperative, local electrostatic interactions to lock the flagellum together, one piece at a time. Thus, in the evolution of the bacterial flagellum, once the interface-compatibility and order-of-assembly hurdles are cleared, so is the configuration hurdle. There's a general principle here: for self-assembling structures, such as biological systems, configuration is a byproduct of other constraints (such as interface compatibility and order of assembly). But note, this is not to say that the configuration of these systems comes for free. Rather, it is to say that the cost of their configuration is included in other costs.

## 7.7 THE ORIGINATION INEQUALITY

The seven hurdles just described should not be construed as merely subjective or purely qualitative challenges to the Darwinian mechanism. It is possible to assess objectively and quantitatively the challenge these hurdles pose to the Darwinian mechanism. Associated with each hurdle is a probability:

$p_{avail}$ The probability that the types of parts needed to evolve a given irreducibly complex biochemical system become available (the *availability probability*).

$p_{synch}$ The probability that these parts become available at the right time so that they can be incorporated when needed into the evolving system (the *synchronization probability*).

$p_{local}$ The probability that these parts, given their availability at the right time, can break free of the systems in which they are currently integrated and be localized at the appropriate site for assembly (the *localization probability*).

$p_{i\text{-}c\text{-}r}$ The probability that other parts, which would produce interfering cross-reactions and thereby block the formation of the irreducibly complex system in question, get excluded from the site where the system will be assembled (the *interfering-cross-reaction probability*).

$p_{i\text{-}f\text{-}c}$ The probability that the parts recruited for inclusion in an evolving system interface compatibly so that they can work together to form a functioning system (the i*nterface-compatibility probability*).

$p_{o\text{-}o\text{-}a}$ The probability that even with the right parts reaching the right place at the right time, and even with full interface compatibility, they will be assembled in the right order to form a functioning system (the *order-of-assembly probability*).

$p_{config}$ The probability that even with all the right parts being assembled in the right order, they will be arranged in the right way to form a functioning system (the *configuration probability*).

Note that each of these probabilities is conditional on the preceding ones. Thus, the synchronization probability assesses the probability of synchronization *on condition that* the needed parts are available. Thus, the order-of-assembly probability assesses the probability that assembly can be performed in the right order *on condition that* all the parts are available at the right time and at the right place without interfering cross-reactions and with full interface compatibility. In consequence, the probability of an irreducibly complex system arising by Darwinian means cannot exceed the following product (note that because the probabilities are conditional on the preceding ones, in forming this product no unwarranted assumption about probabilistic independence is being slipped in here):

$$p_{\text{avail}} \times p_{\text{synch}} \times p_{\text{local}} \times p_{\text{i-c-r}} \times p_{\text{i-f-c}} \times p_{\text{o-o-a}} \times p_{\text{config}}.$$

If we now define $p_{\text{origin}}$ (the *origination probability*) as the probability of an irreducibly complex system originating by Darwinian means, then the following inequality holds:

$$p_{\text{origin}} \leq p_{\text{avail}} \times p_{\text{synch}} \times p_{\text{local}} \times p_{\text{i-c-r}} \times p_{\text{i-f-c}} \times p_{\text{o-o-a}} \times p_{\text{config}}.$$ [32]

This is the *origination inequality.*

Because probabilities are numbers between zero and one, this inequality tells us that if even one of the probabilities to the right of the inequality sign is small, then the origination probability must itself be small (indeed, no bigger than any of the probabilities on the right). It follows that we don't have to calculate all seven probabilities to the right of the inequality sign to ensure that $p_{\text{origin}}$ is small. It also follows that none of these probabilities needs to be calculated exactly. It is enough to have reliable upper bounds on these probabilities. If any of these upper bounds is small, then so is the associated probability and so is the origination probability. And if the origination probability is small, then the irreducibly complex system in question is both highly improbable and specified (all these irreducibly complex systems are specified in virtue of their biological function). It follows that if the origination probability is small, then the system in question exhibits specified complexity; and since specified complexity is a reliable empirical marker of actual design, it follows that the system itself is designed.

The origination inequality shows that intelligent design is a testable scientific theory. Take, for instance, the interface-compatibility probability. Scientists can bring together existing biochemical systems (anything from individual proteins to complex biochemical machines) and determine experimentally the likelihood that and the degree to which their interfaces are compatible. So too, they can take apart existing biochemical systems, perturb them, and then try to put them back together again. To the degree that these systems tolerate perturbation, they are candidates for Darwinian evolution. Conversely, to the degree that these systems are sensitive to perturbation, they are inaccessible to Darwinian evolution. Experiments like this can be conducted on actual biochemical systems or using computer simulations that model biochemical systems.

For instance, in simulating how gene duplications might facilitate protein evolution, Michael Behe and David Snoke have outlined a strategy for estimating interface-compatibility probabilities.[33] Specifically, they calculate the size of populations and number of generations for duplicated genes to evolve and produce novel proteins capable of meshing with an evolving irreducibly complex system. With very modest assumptions about the number of genetic changes that duplicated genes must undergo to produce novel interface-compatible proteins, they find that large populations (circa $10^{20}$) and large numbers of generations (circa $10^8$) are required before such a protein gets fixed in the population.[34] Although Behe and Snoke don't do this, such waiting times can be converted to probabilities: a longer waiting time for fixation corresponds to a smaller probability of fixation in a given generation.[35] And since the waiting time for fixation that Behe and Snoke calculate is large, the probability of evolving a novel interface-compatible protein will be small. In light of the origination inequality, the work of Behe and Snoke therefore argues for the design of the bacterial flagellum.

In any case, with the interface-compatibility probability and the other probabilities in the origination inequality, there is no inherent obstacle to deriving reliable, experimentally confirmed estimates for them. Both Darwinists and design theorists have a stake in estimating these probabilities, and research attempting to do so is only now beginning. The origination inequality does not stack the deck either for or against the intelligent design of irreducibly complex biochemical systems. So long as each of its probabilities is shown to be large or remains unestimated, such a system fails to exhibit specified complexity. In this case, Darwinists would be in their rights to reject that such systems display marks of intelligent design. On the other hand, should any of the probabilities prove sufficiently small, then the system exhibits specified complexity. In this case, since specified complexity is a reliable empirical marker of intelligence, design theorists would be in their rights to insist that such systems do in fact display marks of intelligent design. Thus, we see that the origination inequality makes for a level playing field in deciding between Darwinian and intelligent design theories.

### Keeping Darwin's Theory Honest

Darwinists tacitly embrace the origination inequality whenever they invoke high probability events to support their theory. For instance, in arguing that antibiotic resistance in bacteria results from the Darwinian mechanism and not from intelligent design, Darwinists attempt to show that the probability of the genetic changes needed for antibiotic resistance is large (e.g., that the same bacteria exposed to the same antibiotic regime reliably experience the same resistance-conferring mutations).[36] But having embraced the origination inequality when it confirms their theory, Darwinists tend to shun it when it disconfirms their theory. Indeed, one will be hard-pressed to find a Darwinist who concedes that some low-probability transformation constitutes an obstacle to Darwinian processes. Obviously, there's a double-standard at play here.

Darwinists, going right back to Darwin himself, have been guilty of this double-standard, stacking the deck in favor of their theory and against intelligent design. In the *Origin of Species* Darwin stated a test that, unlike the origination inequality, gives his theory no chance to fail and intelligent design no chance to succeed. Here is his test: "If it could be demonstrated that any complex organ existed, which could not possibly have been formed by numerous, successive, slight modifications, my theory would absolutely break down. But I can find out no such case."[37] This test seems to make a sweeping concession, but in fact imposes conditions that are impossible to meet. The test is therefore no test at all. As philosopher Robert Koons points out,

> How could it be proved that something could not possibly have been formed by a process specified no more fully than "numerous, successive, slight modifications"? And why should the critic [of Darwin's theory] have to prove any such thing? The burden is on Darwin and his defenders to demonstrate that at least some complex organs we find in nature really can possibly be formed in this way, that is, by some specific, fully articulated series of slight modifications.[38]

An honest test of Darwin's theory requires that its defenders first propose specific evolutionary pathways leading to the biological structures that the theory is supposed to explain (e.g., irreducibly complex biochemical systems such as the bacterial flagellum). Only with such proposals in hand can one begin to test the theory. Darwin's defenders, by contrast, take any absence of such proposals as simply showing that evolutionary biology has yet to determine the right evolutionary pathways by which Darwinian processes produced the systems in question.

Who is right? By now it's clear that neither party to this controversy is going to give way any time soon. From the vantage of the design theorist, the Darwinist has artificially insulated Darwinian theory and rendered it immune to disconfirmation because the universe of unknown Darwinian pathways can never be exhausted. From the vantage of the Darwinist, on the other hand,

*continued on next page*

nothing less than an in-principle exclusion and exhaustion of all conceivable Darwinian pathways suffices to shift the burden of evidence onto the Darwinist. To an outsider, with no stake in the outcome of this controversy, the asymmetry of these positions will be obvious. Intelligent design allows the evidence of biology both to confirm and to disconfirm it. Darwinism, by contrast, assumes no parallel burden of evidence—it declares itself the winner against intelligent design by default.[39]

This unwillingness of Darwinism to assume any burden of evidence is unworthy of science. Science, if it is to constitute an unbiased investigation into nature, must give the full range of logically possible explanations a fair chance to succeed. In particular, science may not by arbitrary decree rule out logical possibilities. Evolutionary biology, by unfairly privileging Darwinian explanations, has settled in advance which biological explanations must be true as well as which must be false apart from any consideration of empirical evidence. This is not science. This is arm-chair philosophy. Even if intelligent design is not the correct theory of biological origins, the only way science could discover that is by admitting design as a live possibility rather than by ruling it out in advance. Darwin unfairly stacked the deck in favor of his theory. Notwithstanding, elsewhere in the *Origin of Species*, he wrote: "A fair result can be obtained only by fully stating and balancing the facts and arguments on both sides of each question."[40] That balance is now shifting away from Darwinism and toward intelligent design.

## 7.8 NOT TOO COMPLEX, NOT TOO SIMPLE, JUST RIGHT

So far we have discussed in general terms the probabilistic hurdles that obstruct the evolution of irreducibly complex biochemical systems. Can we now put precise numbers to these probabilities? Although critics of intelligent design sometimes claim that this is impossible, their skepticism is unfounded. Certainly, it is more difficult to assign probabilities to the evolution of biological structures than to progressions of coin tosses. The probabilities involved with coin tossing are easy to calculate; those involved with real-world biological structures are not. But science is in the business of tackling difficult problems, and estimating the probabilities needed to evolve biological structures is well within its job description.

In estimating the probabilities that appear in the origination inequality, it is important to choose manageable systems (i.e., systems that are not so complex that we can't handle them with our scientific methods). In fact, one might say that Darwinists have traditionally hidden behind the complexities of biological systems to shelter their theory from critical scrutiny. Choose a biological system that is too complex, and one can't even begin to calculate the probabilities associated with its evolution. Consider the eye. A widely held myth in the biological community is that Darwin's theory has explained the evolution of the vertebrate eye. In fact, the theory hasn't done anything of the sort.

Rather, Darwinists have identified many different eyes exhibiting varying degrees of complexity, everything from the full vertebrate eye at the high end of the complexity scale to a mere light sensitive spot at the low end. But merely identifying eyes of varying complexity and then, as it were, drawing arrows from less complex to more complex eyes to signify evolutionary relationships does nothing to explain how increasingly complex eyes actually evolved.[41] The gaps between eyes of differing complexity become unbridgeable chasms once one begins to ask about the actual changes in genes, embryological development, and neural wiring that had to take place to evolve full-fledged vertebrate eyes from simpler precursors. Yet, instead of conceding that these gaps constitute disconfirming evidence against their theory, Darwinists strike a pose of invincibility: "Prove us wrong," they say in effect. "Show that it didn't happen that way." By so shifting the burden of proof, Darwin's theory trumps alternative theories every time.

Michael Behe's great coup was to identify a class of simpler biological systems for which it is easier to assess the probabilistic hurdles that must be overcome for them to evolve. With the human eye, one is dealing with billions of cells, especially when one factors in the neural machinery (e.g., visual cortex) that needs to complement it. But with irreducibly complex biochemical systems, one is looking at molecular structures inside the cell that reside at the very threshold of life—go below this threshold, and one is in the realm of prebiological chemical building blocks. A bacterial flagellum, for instance, is a molecular machine that consists of proteins. Proteins in turn consist of amino acids, which may arise in the absence of life (see section 8.4). Yet proteins themselves, as functional sequences of amino acids, are only known to arise in cellular contexts and thus in the presence of life.

But even with irreducibly complex biochemical systems (vastly simpler though they are than individual retinal cells, to say nothing of the eye itself), complexities quickly mount and become unwieldy. The bacterial flagellum consists of only forty or so distinct proteins, but those proteins get repeated in some cases hundreds or even thousands of times. The flagellar tail, for instance, consists of tens of thousands of protein subunits (i.e., copies of a given protein). The bacterial flagellum itself is therefore a vastly complex assemblage of many thousands of protein parts.

Estimating probabilities for the evolution of the bacterial flagellum thus becomes difficult, especially because Darwinists never identify detailed evolutionary pathways for such irreducibly complex biochemical systems. With sufficiently detailed pathways in hand, probabilities at each step along a path could, in principle, be estimated. Absent such pathways, however, we are limited to more general probabilistic considerations in assessing the improbability of evolving the flagellum. For instance, an evolving flagellum must overcome the problem of interface compatibility, in which proteins new to the structure must mesh suitably with it if they are to be successfully incorporated and evolution is to take place (see the discussion of this point in sections 7.6 and 7.7).

In estimating probabilities for the evolution of biochemical systems by Darwinian processes, we therefore need to analyze structures even simpler than the flagellum. The place to look is the improbability of evolving individual proteins. In terms of the origination inequality, this means looking at the availability probability (i.e., the probability of individual proteins becoming available for inclusion in an evolving irreducibly complex system) and showing that it is small. This research, however, has yet to be done for the proteins of the bacterial flagellum or for those of other irreducibly complex protein machines.[42] As a consequence, forming hard estimates for the improbability of evolving the flagellum and other irreducibly complex systems remains an open problem.

Nonetheless, hard estimates for the improbability of evolving certain individual proteins are available. Proteins reside at just the right level of complexity and simplicity to determine, at least in some cases, their improbability of evolving by Darwinian processes. To make this determination correctly, however, requires care. Suppose existing proteins are evolving into other proteins (their evolution being mediated, of course, through genetic changes since genes encode proteins—see Chapter 2). What determines their evolvability is not merely how sparsely proteins are distributed among all possible amino acid sequences, but also, and more importantly, to what degree proteins of one function are isolated from proteins of other functions.

To illustrate the evolvability of proteins, think of them as English sentences. Both employ a fixed alphabet: in the case of proteins, twenty L-amino acids; in the case of English sentences, twenty-six letters (it's convenient here also to add a space). To evolve one protein into another by a Darwinian process, modest amino-acid changes must gradually transform one protein into another so that each intermediate along the way not only has largely the same sequence of amino acids as its immediate predecessor and successor, but also folds properly and therefore can have a biological function. Likewise, to evolve one English sentence into another by a Darwinian process, modest letter changes must gradually transform one sentence into another so that each intermediate along the way not only has largely the same sequence of letters as its immediate predecessor and successor, but also is a meaningful English sentence.

Consider, for instance, the "evolution" of THE CAT SAT ON THE MAT. Because we are modeling Darwinian evolution, this sentence would have to evolve gradually. For definiteness, let's say that a single new letter or space can be added to, dropped from, or changed in a given sentence at each evolutionary step. This requirement ensures that the evolution of sentences is gradual. But gradualism of the evolutionary process is not enough. Because we are modeling Darwinian evolution, each intermediate must also preserve meaning (this, in biology, corresponds to preservation of function).[43] Thus, at no point in the evolution of THE CAT SAT ON THE MAT can the sentences derived from it become gibberish. What follows is one possible evolutionary path issuing from this starting sentence (note that altered or inserted characters are indicated with an underline, dropped characters are indicated with an underline on either side):

THE CAT SAT ON THE MAT
THE CATS SAT ON THE MAT
THE CATS EAT ON THE MAT
THE CATS BEAT ON THE MAT
THE BATS BEAT ON THE MAT
THE BATS BEAT ON THE BAT
THE BATS BEAT ON THE BOAT
THE BAT _EAT ON THE BOAT
THE BASS EAT ON THE BOAT
THE BASS EATS ON THE BOAT
THE BOSS EATS ON THE BOAT

What are we to make of this example? Is THE CAT SAT ON THE MAT evolvable? Evolvability is a matter of degree, so a simple yes or no won't adequately answer this question. Given the changes permitted at each step (i.e., addition, deletion, or alteration of a character), it's evident that one can evolve sentences that mean something quite different from the original. But is every meaningful sentence in the English language evolvable by a gradually changing sequence of meaningful intermediates from this original sentence? Or, as a Darwinist might say, is THE CAT SAT ON THE MAT a universal common ancestor for all other English sentences? That seems unlikely, though the analysis to verify this claim has yet to be done. To do so would require a computer taking the original sentence and running from it all the various evolutionary pathways consistent with the changes permissible at each evolutionary step. In addition, it would require a set of human eyes to ensure that meaning is preserved at each evolutionary step.[44]

Replace "meaning" with "function," and protein evolution parallels the evolution of English sentences. How evolvable, then, are proteins? It's well known that proteins tolerate numerous amino-acid substitutions.[45] In fact, single amino-acid changes that disrupt function are rare.[46] That's because what determines a protein's function is its overall shape (known as its tertiary structure), and many different amino-acid sequences are compatible with the same shape. A protein's shape and function therefore cannot be reduced to its precise sequence of amino acids (known as its primary structure) or even to its pattern of regular bonds (known as its secondary structure).

A protein's shape and function are a consequence of how it folds, and what stabilizes a fold is the pattern of interactions among its amino acids that are favorable to holding the structure together. Moreover, it's possible to arrange the same pattern of favorable interactions in a large number of ways. Thus, in principle, proteins could be highly evolvable as to their primary structure and yet be completely unevolvable as to their tertiary structure. (This would be like the sequence of letters in THE CAT SAT ON THE MAT changing radically over numerous small steps, and yet at each step retaining the same meaning.) Since tertiary structure determines biological function, such a form of evolution would be no evolution at all, producing no novel structures or functions and giving natural selection nothing to select.

Is all protein evolution like this? No. There is some evidence that protein domains (i.e., folding regions within a protein), if they are short, can evolve structurally dissimilar folds that in turn might induce new functions for the protein itself. For instance, research by Sebastian Meier and his collaborators on a very short protein domain (circa 25 residues) showed that changing a single residue could convert the protein to a new structure.[47] But this example seems exceptional. The amino-acid change in this case altered a disulfide bond, which is much stronger than the hydrogen bonds typical of many proteins. Moreover, a stretch of 25 amino-acid residues is so short that it cannot, as many proteins do, stabilize itself through the formation of a hydrophobic core (i.e., those amino acids in a protein that flee water and associate with one another). To talk about folding such a short protein domain is like talking about knotting a string that is too short to be knotted.

Folding becomes increasingly difficult and complicated for proteins consisting of longer strings of amino acids. Like folding a protruding satellite structure into a compact ball for placement into a rocket payload, there are certain key hinging actions that must occur in a particular sequence to get a large unfolded protein to fold into the right shape. Thus, it is not surprising that longer proteins often require assistance (from other proteins, known as chaperones) to fold. Short proteins, on the other hand, are much simpler to fold. Instead of requiring layer upon layer to be folded in a particular sequence, a short protein does something more like a simple constrained collapse for which the ordering of events isn't so important. It is therefore interesting that Meier and his collaborators found a way to modify the fold of a very short amino-acid sequence by changing a single residue, but generalizing this result to larger proteins is unwarranted.[48]

## 7.9 VARIATION AND SELECTION OUT OF SYNC

Does any experimental evidence confirm that larger proteins may be unevolvable? Such evidence exists. Research by molecular biologist Douglas Axe confirms that a domain of circa 150 residues of the protein TEM-1 β-lactamase is unevolvable by Darwinian processes. (Note that although this domain is much larger than the 25-residue domain studied by Meier *et al.*, 150 residues is pretty average as domains go. An average protein has several hundred amino acids.) Axe chose this protein carefully. As he noted, β-lactamases protect "bacteria from the effects of penicillin-like antibiotics," and thus provide "a simple means of selecting functional variants over a wide range of thresholds."[49]

This ability of β-lactamases to confer antibiotic resistance allowed Axe to apply a technique in his biological research that is widely employed in the field of evolutionary computing. Evolutionary computing, a branch of computer engineering, attempts to model Darwinian processes computationally by running evolutionary algorithms. Such algorithms solve problems by optimizing fitness measures. It can happen, however, that certain solutions to a problem are already known. In that case, one wants to keep the evolutionary algorithm away from those solutions and searching for new solutions. To do this, one rewards novelty, explicitly increasing the fitness of novel solutions (often this is done by assigning a penalty to old solutions).[50]

Axe therefore randomized amino acids in a 150-residue domain of TEM-1 β-lactamase, forcing variant domains to differ from the original in at least 30 amino-acid positions. He then rewarded bacteria with these increasingly distant variant domains provided the domains folded sufficiently to afford the bacteria minimal protection against penicillin-like antibiotics (note that dosages of antibiotics were under experimental control, so Axe was able to fine-grade just how much functionality for antibiotic resistance these variant domains retained). In thus forcing divergence and rewarding function, Axe used natural selection to run the evolutionary process away from the wild-type TEM-1 β-lactamase, for which the domain in question had a rock-stable fold.

Because of the importance of hydrophobic interactions to protein folding (and therefore to its tertiary structure and function), Axe focused especially on amino-acid sequences with the pattern of hydrophobic interactions characteristic of his β-lactamase domain. Axe argued that amino-acid sequences with this pattern of hydrophobic interactions, or "hydropathic signature" as he called it, are so physically specific that they either have the same fold (i.e., tertiary structure) as his β-lactamase domain or don't fold at all. Thus, in particular, amino-acid sequences with this hydropathic signature cannot assume a different functional fold. Axe used this constraint on protein folding as a way of producing variation consistent only with the original fold.

Next, Axe estimated that for amino-acid sequences with this hydropathic signature, only one in $10^{64}$ formed a "working domain."[51] Axe defined a working domain as one that, even if somewhat unstably folded and therefore handicapped, could still provide some minimal antibiotic resistance and thus could still be preserved by natural selection. Nonworking domains are thus of no biological significance, invisible to natural selection except as rubbish that might be selected against. How small is one in $10^{64}$. Here is that probability spelled out:

$$\frac{1}{\begin{array}{l}10{,}000{,}000{,}000{,}000{,}000{,}000{,}000{,}000{,}000{,}000{,}000{,}\\000{,}000{,}000{,}000{,}000{,}000{,}000{,}000{,}000{,}000\end{array}}$$

This number is exceedingly small as probabilities go. French mathematician Émile Borel, whom we met earlier in this chapter on the topic of monkeys typing Shakespeare, proposed one in $10^{50}$ as a universal probability bound against which chance-based mechanisms are powerless.[52] Axe's number is therefore well below Borel's universal probability bound.

In light of Axe's experiments, what does this one in $10^{64}$ improbability mean? This number identifies the improbability of evolving by Darwinian processes a working protein domain with the same pattern of hydrophobic interactions characteristic of his β-lactamase domain. Thus, for all practical purposes, there's no way this domain could have evolved by Darwinian processes. To see that this interpretation of Axe's results is warranted, it will help to revisit the example of linguistic evolution from section 7.8 in which THE CAT SAT ON THE MAT evolved into THE BOSS EATS ON THE BOAT. What allowed the first sentence to evolve into the second, even though they are substantially different both structurally and semantically, is that selection was able to track and exploit variation. Specifically, small changes in letters (of the sort permissible at each evolutionary step) could induce changes in meaning that in turn could be selected.[53]

The ability of selection to track and exploit variation in this example may seem so obvious as not to need stating, but it raises a less obvious question: What could go wrong with the sources of variation so that selection cannot track and exploit it? Think of it this way. In evolutionary scenarios, variations don't just happen willy-nilly. Rather, they happen according to well-defined mechanisms that operate on the evolving system at a certain level of the system's organization. The level of organization at which variation occurs let us call the *unit of variation.* Likewise, in evolutionary scenarios, selection doesn't just happen willy-nilly. Rather, it preferentially selects descendants of an evolving system according to how the environment judges certain features at a certain level of the system's organization more or less fit. The level of organization at which selection is effective let us call the *unit of selection.*[54] Could the units of variation and selection be so out of sync that evolution grinds to a halt?

In our earlier example of linguistic evolution, where THE CAT SAT ON THE MAT evolved into THE BOSS EATS ON THE BOAT, the unit of variation (in this case, alterations of individual letters) *coordinates* with the unit of selection (in this case, the meaning of sentences). Axe's research, by contrast, points up what happens when the unit of variation *fails to coordinate* with the unit of selection. To illustrate such a failure of coordination, modify THE CAT SAT ON THE MAT example as follows: imagine that the unit of variation is not individual letters but individual pixels on a screen. Assume, as would ordinarily be the case, that many pixels are required to form each letter. In that case, if only a few pixels can change at each evolutionary step (consistent with Darwinian gradualism), then THE CAT SAT ON THE MAT would never evolve into any sentence with a different meaning.

To see this, consider that even though many individual pixels may change via such an evolutionary process, the letters comprising those pixels would be remarkably stable and their meaning, taken in sequence, would not change. That's because pixel-driven variation does not coordinate with meaning-driven selection. Pixel-driven variation of THE CAT SAT ON THE MAT will lead either to no noticeable change in letters (thus preserving meaning), or to letters that are fuzzy, still recognizable, but unchanged (thus again preserving meaning), or to letters that become so fuzzy as to be unrecognizable (thus degrading meaning and being selected against). In addition, pixel-driven variation might introduce stray marks (thus again degrading meaning and being selected against). But what would not happen with pixel-driven variation is that the letters themselves should gradually transform into other letters so that each evolutionary intermediate consists only of recognizable letters and always preserves meaning. The lack of coordination between unit of variation and unit of selection voids this possibility.

Axe's work with β-lactamase points up the same problem. Indeed, the breakdown between unit of variation and unit of selection in both instances is virtually parallel. Pixel-driven (as opposed to alphabet-driven) variation changes a distribution of pixels but leaves unaffected higher level alphabetic structures whose arrangement (in linear sequence) determines meaning and is the unit of selection. Likewise, neo-Darwinian variation (which ultimately stems from random changes of genes that encode proteins) changes a linear sequence of amino acids (i.e., the protein's primary structure) but leaves unaffected the pattern of hydrophobic interactions whose associated tertiary structure determines biological function and is the unit of selection. In both instances, unit of variation and unit of selection are out of sync.

How can we see that amino-acid-driven variation and biological-function-driven selection are out of sync in the evolution of domains with the hydropathic signature of Axe's β-lactamase? Axe showed that an arbitrary domain with that hydropathic signature is extremely unlikely (one in $10^{64}$) to have any biologically significant catalytic activity in protecting against penicillin-like antibiotics. In other words, for a randomly chosen amino-acid sequence with that hydropathic signature, it is overwhelmingly likely that it won't be a working domain. Should such a nonworking domain appear, it will render β-lactamase ineffective at conferring antibiotic resistance. Thus selection pressure, in this case resulting from the application of penicillin-like antibiotics, will tend to eliminate bacteria with these nonworking domains.

On Darwinian principles, however, Axe's β-lactamase domain had to evolve from some distinct precursor domain and do so by a sequence of intermediates that, at each evolutionary step, were functional and therefore selectable. Indeed, prior to assuming that signature, the precursor domain had a different signature, fold, and function (if not, there would be no point talking about the evolution of this precursor into the β-lactamase domain because these domains would then be structurally and functionally equivalent). How, then, did this precursor domain acquire the hydropathic signature of Axe's β-lactamase domain? As Darwin noted in his *Origin of Species*, "unless profitable variations do occur, natural selection can do nothing."[55] Variation and selection therefore must synchronize as follows: variation must produce its variants; selection must then scrutinize these variants, preserving those that are profitable and eliminating those that are not. This process of variation proposing and selection disposing then repeats.

Thus, when the precursor varies into a domain with the hydropathic signature of Axe's β-lactamase domain, selection has yet to act. But the unit of variation is amino acids and not patterns of hydrophobic interactions. Thus, in randomly varying amino acids before selection can act, Darwinian processes could have no preference between working and nonworking domains with this hydropathic signature (such a preference can result only from selection, but selection cannot operate until variation has done its work). Yet without such a preference, working domains with this signature have only a one in $10^{64}$ probability. And since these are the only domains with this signature that selection can preserve, Darwinian processes are no better at evolving a precursor domain into a working version of Axe's β-lactamase domain than simply sampling at random from domains with this hydropathic signature.

Bottom line: With variation and selection out of sync, Darwinian processes can do no better than pure random sampling. And with an improbability of one in $10^{64}$, random sampling is hopeless. Axe's β-lactamase domain is therefore unevolvable by Darwinian processes. How firm is this conclusion? Axe's work on the unevolvability of particular proteins by Darwinian processes is the best to date. Whether his research (which is ongoing) ends up carrying the day and convincing the biological community that the design of certain proteins is real remains to be seen. At the very least, Axe's work demonstrates that there is a debate of real scientific merit about the design of proteins. More significantly, his work demonstrates that the best evidence is now on the side of his β-lactamase domain standing beyond the reach of Darwinian processes, exhibiting specified complexity, and therefore being designed.

The burden of evidence is therefore now shifted to the Darwinist who wants to dispute the design of Axe's β-lactamase domain. Nor will it do to speculate idly that Axe's research might have missed some unknown Darwinian pathway by which his domain might have evolved. To base an argument for the power of Darwinian processes on such unknown possibilities constitutes an argument from ignorance: "Axe's research does not rule out all conceivable Darwinian pathways, so they must exist." Science is called to a higher standard of evidence.

But what if Darwinists do end up finding a plausible evolutionary precursor and a fully articulated evolutionary pathway to the β-lactamase domain studied by Axe? To be sure, that would shift the burden of evidence back on the design theorist—but only for this system. To establish that proteins exhibit specified complexity and therefore are designed, it is not necessary to show that *all* proteins exhibit specified complexity or that Darwinian processes are completely incapable of effecting any evolution of proteins. Rather, it is enough to show that *even a single* protein exhibits specified complexity and is unevolvable by Darwinian processes. The Darwinist is committed to all aspects of life being devoid of actual design. Thus, to refute this view, the design theorist need merely show that some aspect of life clearly exhibits design.

Though Axe's protein domains are not as easily visualized as Behe's irreducibly complex biochemical systems, Axe's research closes a loophole to Darwinian processes that Behe's systems left open (which is not to say that Behe has been refuted—far from it; the loophole consists solely in the imagination of Darwinists, not in any empirical evidence from biology). Take the bacterial flagellum. Embedded in the flagellum (a motor-driven propeller) is the type-three secretory system (a microsyringe that pumps toxins). Indeed, for most of Behe's systems, one can identify subsystems that might perform a function on their own. Accordingly, Darwinists point to these subsystems as possible evolutionary precursors to the systems in which they are embedded (e.g., Darwinists point to the type-three secretory system as a precursor to the flagellum—see sections 6.3 and 6.4 to understand why this simple act of pointing does not constitute evidence for Darwinism or counterevidence against intelligent design). But the domains studied by Axe have an integrity that admits no functional subdivisions and thus leaves Darwinian evolution with no plausible precursors.

*The general notes on the CD included with this book expand on the material in this chapter and throw light on the following discussion questions.*

## 7.10 DISCUSSION QUESTIONS

1. List all the special sciences you can think of that depend on the detection of intelligent agents. How do these sciences detect intelligent agents? Is there a common methodology that they employ? What is the role of specified complexity in these sciences? Is there any legitimate reason for excluding biology from those sciences that allow for the possibility of design detection?

2. Consider four compact disks: one with your favorite album, one with random noise, one with your favorite computer-game software, and one that's blank. In what way(s) is design present or absent from these CDs? Can the laws of physics and chemistry alone distinguish these CDs?

3. What is specified complexity? What conditions need to be satisfied for an object, event, or structure to exhibit specified complexity? What conclusion should we draw about a biological system if it exhibits specified complexity?

4. What are probabilistic resources? Define the two types of probabilistic resources. How do they differ? Give examples of each. How are probabilistic resources relevant to the typing by monkeys of Shakespeare's works?

5. Do irreducibly complex biochemical systems exhibit specified complexity? If so, what does this tell us about the design of these systems? Is there a general logic by which design theorists argue that irreducibly complex systems exhibit specified complexity? If so, what is it? Is this logic sound?

6. What is the tornado-in-a-junkyard objection against attributing specified complexity to irreducibly complex biochemical systems? Does the logic by which design theorists argue that irreducibly complex systems exhibit specified complexity in fact fall prey to this objection? Explain your answer.

7. Describe the seven hurdles that the Darwinian mechanism must overcome to evolve irreducibly complex biochemical machines. How are these hurdles represented in the origination inequality? Explain why the interface-compatibility hurdle is especially difficult for Darwinian processes to overcome. How does the origination inequality render intelligent design testable?

8. Did the vertebrate eye evolve by Darwinian processes? What is the evidence that the eye evolved by such processes? How have Darwinists used the immense complexity of the eye to short-circuit any criticism of its evolution by Darwinian processes? More generally, how can the extreme complexity of biological systems shield Darwinism from critical scrutiny?

9. What is a unit of variation and unit of selection? If Darwinian processes are to lead to any interesting change in a given evolutionary scenario, must unit of variation and unit of selection be suitably coordinated? What happens when they fail to be coordinated? Give a fresh example—one not in the text—of such a breakdown in coordination.

10. Summarize the research of Douglas Axe on TEM-1 β-lactamase. How did the 150-residue protein domain that was the focus of his research achieve just the right balance between complexity and simplicity for assessing whether this system exhibits specified complexity?

CHAPTER EIGHT

# The Origin of Life

## 8.1 WHAT NEEDS TO BE EXPLAINED?

The origin of life is as difficult a problem as exists in contemporary science. But what exactly is the problem? And what would it mean to solve the problem? The mathematician George Polya once quipped, "If you can't solve a problem, then there is an easier problem you can solve: find it."[1] This can be good advice if the easier problem illuminates the original problem. But if the easier problem lulls us into thinking that we have solved, or are on the verge of solving, the original problem when in fact we have no clue about how to solve it, then Polya's advice is bad. And in the case of life's origin, Polya's advice is bad indeed.

Most of origin-of-life research consists in conveniently redefining the problem of life's origin to make it so easy that origin-of-life researchers think they have a reasonable shot at solving it when in fact they do not. In so redefining the problem, however, they sidestep the real problem of life's origin. Most of origin-of-life research is as relevant to the real problem of life's origin as rubber-band powered propeller model planes are to the military's most sophisticated stealth aircraft. Real life is so much more sophisticated than any of the supposed precursors to life posited in conventional origin-of-life research that there is no reason to think that this research provides any substantive insight into life's actual origin. What, then, is the real problem of life's origin? It is to explain the origin of cells as we currently find them on planet earth in their full jaw-dropping complexity (see the next sidebar and figure 8.1).

## The Cell as an Automated City

(Adapted from Michael Denton[2])

Magnified several hundred times with an ordinary microscope, as was available in Darwin's day, a living cell is a disappointing sight. It looks like an ever-changing and apparently disordered collection of blobs and particles that unseen turbulent forces continually toss in all directions.

To grasp the reality of life as revealed by contemporary molecular biology, we need to magnify the cell a billion times. At that level of magnification, a typical eukaryotic cell (i.e., cell with a nucleus) is more than ten miles in diameter and resembles a giant spaceship large enough to engulf a sizable city. Here we see an object of unparalleled complexity and adaptive design.

On the surface are millions of openings, like the portholes of a ship, opening and closing to allow a continual stream of materials to flow in and out. As we enter one of these openings, we discover a world of supreme technology and bewildering complexity. We see endless highly organized corridors and conduits branching in every direction from the perimeter of the cell, some leading to the central memory bank in the nucleus and others to assembly plants and processing units.

The nucleus itself is a vast chamber a mile in diameter resembling a geodesic dome. Inside we see, all neatly stacked together in ordered arrays, coiled chains of DNA, thousands and even millions of miles in length. This DNA serves as a memory bank to build the simplest functional components of the cell, the protein molecules. Yet proteins themselves are astonishingly complex pieces of molecular machinery. An average protein consists of several hundred precisely ordered amino acids arranged in a highly organized 3-dimensional structure.

Robot-like machines working in synchrony shuttle a huge range of products and raw materials along the many conduits to and from all the various assembly plants in the outer regions of the cell. Everything is precisely choreographed. Indeed, the level of control implicit in the coordinated movement of so many objects down so many seemingly endless conduits, all in unison, is mind-boggling.

As we watch the strangely purposeful activities of these uncanny molecular machines, we quickly realize that despite all our accumulated knowledge in the natural and engineering sciences, the task of designing even the most basic components of the cell's molecular machinery, the proteins, is completely beyond our present capacity. Yet the life of the cell depends on the integrated activities of numerous different protein molecules, most of which work in integrated complexes with other proteins.

In touring the cell, we see that nearly every feature of our own advanced technologies has its analogue inside the cell:

- Information processing, storage, and retrieval.
- Artificial languages and their decoding systems.
- Error detection, correction, and proof-reading devices for quality control.
- Elegant feedback systems that monitor and regulate cellular processes.
- Digital data embedding technology.
- Signal transduction circuitry.
- Transportation and distribution systems.
- Automated parcel addressing ("zip codes").
- Assembly processes employing prefabrication and modular construction.
- Self-reproducing robotic manufacturing plants.

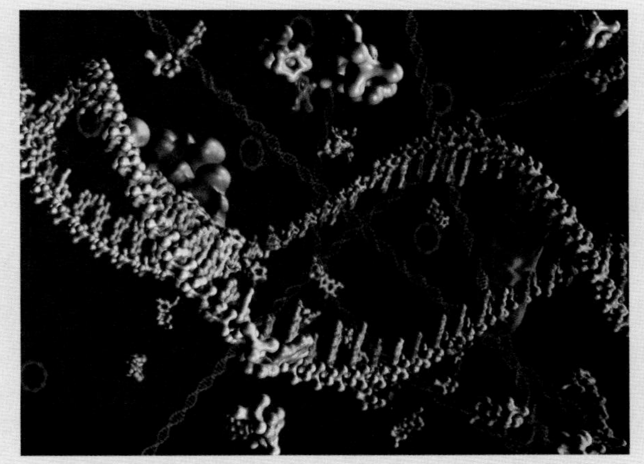

***Figure 8.1*** *This image, from a computer animation in* Unlocking the Mystery of Life (www.illustramedia.com), *illustrates information processing inside the cell (i.e., the transcription of DNA into mRNA). The animation shows why it is entirely appropriate to compare the cell with an automated city.*[4]

Nanotechnology of this elegance and sophistication beggars all feats of human engineering. Except for the hope of a materialistic solution to this problem, the inference to design would be immediate.[3] This chapter shows why that hope is unrealistic and why design is again a live option in explaining life's origin.

Some origin-of-life researchers object that in identifying life's origin with the origin of eukaryotic cells, we have set the bar too high. As cells with nuclei, eukaryotic cells are the most complicated cells we know. Accordingly, these researchers will claim that eukaryotic cells needed themselves to evolve and therefore are not the first living forms that origin-of-life studies need to explain. Implicit in this objection is the divide-and-conquer strategy of evolutionary biology that we have seen repeatedly in this volume. According to this strategy, to explain a complex system, one needs to explain how it could have evolved from a simpler system. This strategy, in effect, merely restates Polya's dictum: solve hard problems by finding easy problems and solving them. Evolutionary biology, which presupposes the presence of life and attempts to explain its diversification, is fully committed to this strategy. But so are materialistic approaches to the origin of life, which attempt to identify a likely sequence of chemical steps by which life may have formed under plausible prebiotic conditions. Hence the preoccupation in origin-of-life research with "prebiotic" or "chemical evolution."

Whatever its merits as a general principle for problem solving, the divide-and-conquer strategy has proven singularly ineffective in resolving the origin-of-life problem. It's true that eukaryotic cells are the most complicated cells we know. But the simplest life forms we know, the prokaryotic cells (which lack a nucleus, such as bacteria), are themselves immensely complex. Moreover, they are every bit as high-tech as the eukaryotic cells. If eukaryotes are like state-of-the-art laptop computers, then prokaryotes are like state-of-the-art cell phones. Or, to revert to our previous analogy of an automated city, if an average eukaryotic cell is ten miles in diameter at a billion magnification, then an average prokaryotic cell is, on average, a mile in diameter. Thus, if eukaryotes are cities, prokaryotes are towns—yet they are towns that rival the technological sophistication of the larger cities. Not only do eukaryotic and prokaryotic cells perform all the basic functions traditionally associated with life (e.g., reproduction, growth, metabolism, homeostasis, well-defined internal organization, maintenance of boundaries, stimulus-response repertoire, and goal-directed interaction with the environment), but they do so by utilizing many of the same basic structures.

For instance, the genetic code and the synthesis of proteins (by reading messenger RNA off DNA and then feeding it to ribosomes) is essentially the same in all cells, prokaryotic and eukaryotic. Ribosomes, the engines of protein synthesis, are themselves immensely complicated biochemical machines consisting of at least fifty separate proteins and RNA subunits. In fact, the very simplest prokaryotic cell requires hundreds of genes to handle its basic tasks of living.[5] Thus, even if it were possible to explain the origin of eukaryotic cells via the evolution of prokaryotic cells,[6] the origin of prokaryotic cells, which have essentially the same high-tech information processing capabilities of eukaryotic cells, would still need to be explained.

But this level of complexity in even the simplest cells raises a far more vexing problem for materialistic origin-of-life research, committed as it is to explaining the origin of life through a sequence of chemical steps that neither individually nor as a whole required intelligent input. Standard dating of the earth places its origin at about 4.5 billion years ago. For the first half billion or so years of its existence, the earth was too hot and tempestuous for any life form to exist on it at all. Then, at the moment when the earth cooled sufficiently to permit life, *boom!*, within a hundred or so million years, prokaryotic life as we know it appeared suddenly and abundantly (the best current estimates for its first appearance are 3.8 to 3.9 billion years ago). Moreover, there is no evidence whatsoever of earlier, more primitive life forms from which these early prokaryotes evolved. Origin-of-life researchers sometimes argue that prokaryotic forms were so successful that they consumed any evidence of their evolutionary precursors. But the fact remains that we have no evidence of early life forms other than these.

At a minimum, the problem of life's origin is therefore to explain the origin of prokaryotic life and, in particular, its DNA-based protein synthesis apparatus. This is a huge unresolved problem. Karl Popper explains why:

> What makes the origin of life and of the genetic code a disturbing riddle is this: the genetic code is without any biological function unless it is translated; that is, unless it leads to the synthesis of the proteins whose structure is laid down by the code. But, as [Jacques] Monod points out, the machinery by which the cell (at least the nonprimitive cell which is the only one we know) translates the code "consists of a least fifty macromolecular components *which are themselves coded in DNA*." Thus the code cannot be translated except by using certain products of its translation. This constitutes a really baffling circle: a vicious circle, it seems for any attempt to form a model, or a theory, of the genesis of the genetic code.[7]

Popper wrote these words in the 1970s, but their impact is as powerful today. For the remainder of this chapter we will examine attempts to come to terms with this problem.

## The Origins of Life, Second Genesis, and Astrobiology

Origin-of-life researchers refer increasingly not to the origin of life (singular) but to the origins of life (plural). The presumption here is that life in the universe had many origins, and that in unduly focusing on the origin of

*continued on next page*

earth-based cellular life, we are discriminating against all the other life forms that are supposed to exist elsewhere in the universe or that are supposed to predate and ultimately to have evolved into cellular life on earth. In this respect, what is known as "second genesis" has become the holy grail of origin(s)-of-life research, namely, the discovery, or invention in the lab, of an entirely new life form whose architecture and machinery is completely different from that of DNA-based cellular life.

Astrobiology has become the new umbrella discipline for making sense of all these alternative life forms. NASA, which is the principal funder of astrobiology, defines this discipline as follows on its website:

> Astrobiology is the study of life in the universe. It investigates the origin, evolution, distribution, and future of life on earth, and the search for life beyond earth. Astrobiology addresses three fundamental questions: How does life begin and evolve? Is there life beyond earth and how can we detect it? What is the future of life on earth and in the universe?[8]

There is a sense in which astrobiology is a perfectly valid discipline and the origins of life and second genesis are perfectly valid concepts. The problem, however, is that for now they apply to exactly one class of objects, namely, cellular life as we know it on earth. Indeed, there is not a shred of evidence for any other form of life, whether here on earth, on Mars, or elsewhere in the universe.

Astrobiology is essentially what used to be called exobiology. Whereas astrobiology covers all life in the universe, exobiology covers all life outside of earth. With astrobiology, we have a population size of one: at least one place in the universe is known where life exists—earth. With exobiology, we have a population of size zero: earth is excluded and no other habitat for life in the universe is known. Exobiology is therefore a field of science for which no evidence exists, and astrobiology is a field of science for which all the evidence derives from conventional biology.

In practice, the origins of life (plural), second genesis, and astrobiology denote not so much scientific enterprises as exercises in hype. Until and unless there is solid evidence of full-fledged life forms (i.e., forms that reproduce, grow, metabolize, process information, interact with and adapt to changing environmental conditions, etc.) distinct from the DNA-based life here on earth, the use of these terms is highly misleading, suggesting that we have scientific confirmation for these alternative life forms when in fact we have nothing of the sort. Keep this in mind the next time you see astrobiology taught as science and see intelligent design's status as science questioned.

## 8.2 OPARIN'S HYPOTHESIS

In 1924, Russian biochemist Alexander Oparin proposed a purely materialistic approach to life's origin that has set the tenor for origin-of-life research ever since.[9] According to him, the first cell originated from nonliving matter, but not all at once. Instead, it arose gradually in stages.[10] Oparin argued that simple chemicals combined to form organic compounds, such as amino acids. These in turn combined to form large, complex molecules, such as proteins. And these aggregated to form an interconnecting network inside a cell wall.

***Figure 8.2*** *A. I. Oparin, who hypothesized about a gradual route from nonlife to life.*

According to Oparin, the atmosphere of the early earth was very different from the present one. Energy sources such as heat from volcanoes or lightning were said to act on simple carbon compounds in the atmosphere, transforming them into more complicated compounds. In the earth's early seas these newly formed compounds were said to come together to form microscopic clumps, the forerunners of the first living cells on earth.

In 1928 the English biochemist J. B. S. Haldane put forward essentially the same idea.[11] He thought that ultraviolet light from the sun caused simple gases in the earth's primitive atmosphere (e.g., carbon dioxide, methane, water vapor, and ammonia) to transform into organic compounds, turning the primitive ocean into a hot, dilute "soup" (see figure 8.3). Out of this soup came virus-like particles that eventually evolved into the first cells.

Oparin and Haldane thus laid the groundwork for a theory of *prebiotic or chemical evolution*, according to which life began in a sea of chemicals, sometimes called the *prebiotic soup*. To this day, their view is known as the Oparin hypothesis or the Oparin-Haldane hypothesis. Despite many modifications to the hypothesis in later years, it exemplifies the standard contemporary evolutionary approach to the origin of life.

According to the Oparin hypothesis, chance by itself could not drive the interactions of chemicals and compounds needed to form complex biomolecules and from there life. Instead, the hypothesis posits that some internal tendency of matter toward self-organization gives rise to the ordered structures we see in life. Let us now examine this hypothesis in more detail, noting carefully the assumptions on which it is built.

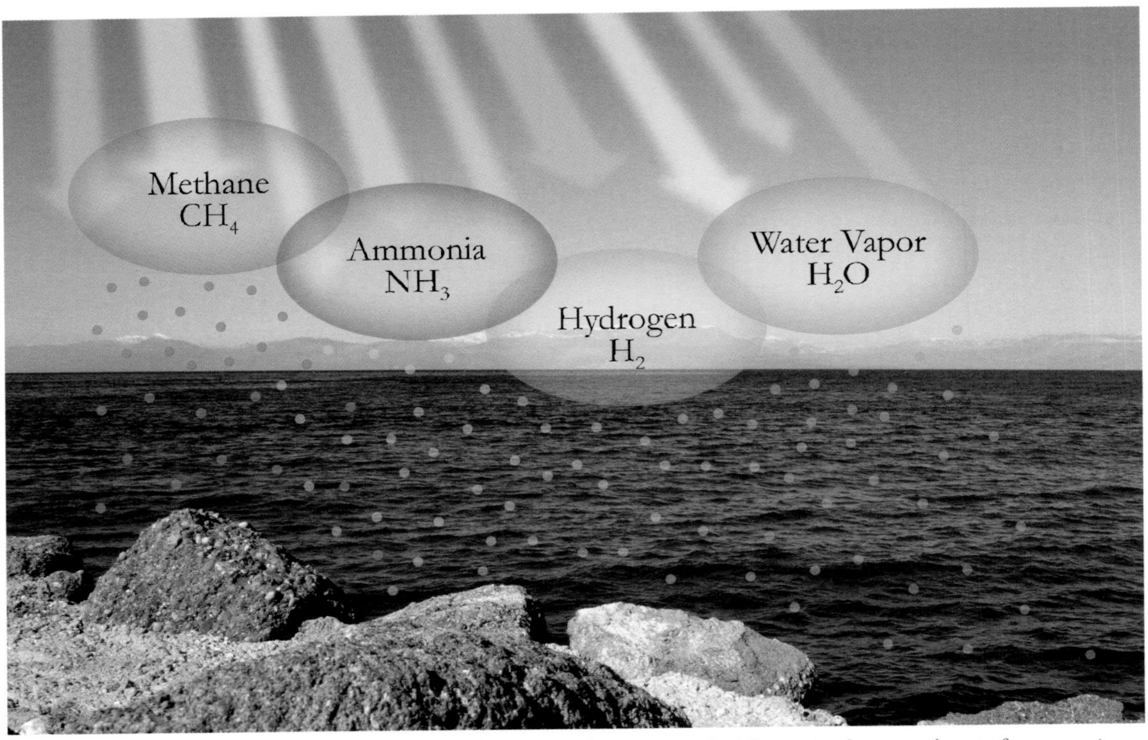


***Figure 8.3*** *As envisioned by Oparin's hypothesis, ultraviolet radiation reacted with gases in the atmosphere to form organic compounds on the early earth.*

**Assumption #1: Reducing Atmosphere.** ***The earth's early atmosphere contained abundant free hydrogen and little or no oxygen.***
According to Oparin, the first cells arose gradually over millions of years. He believed that conditions at the surface of the early earth allowed a massive accumulation of organic compounds before life began. Such an accumulation would improve the probability that all the right compounds could have come together and combined into a cell.

This accumulation of organic compounds could not have occurred, however, if the earth's atmosphere contained significant amounts of free oxygen ($O_2$), because oxygen destroys organic compounds by reacting with them in a process called *oxidation.* Oparin believed that the early atmosphere must have been composed of hydrogen-rich gases such as methane ($CH_4$), ammonia ($NH_3$), hydrogen ($H_2$), and water vapor ($H_2O$), but not oxygen (see figure 8.4). Such an atmosphere is called a *reducing atmosphere.*

Furthermore, Oparin believed that the first cells were *anaerobic* (able to survive without oxygen) and *heterotrophic* (unable to make their own food), obtaining many of their essential nutrients instead from the surrounding water. These anaerobic heterotrophs were thus said to obtain their energy by fermentation, a method of releasing energy from organic molecules in the absence of free oxygen.

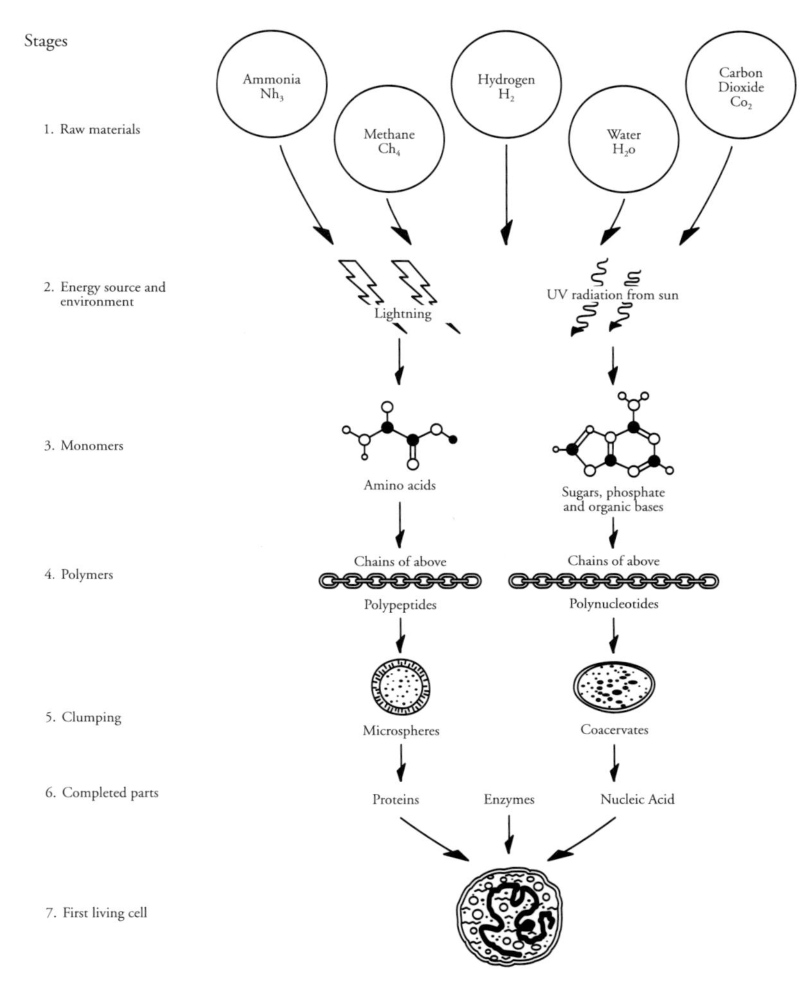


**Figure 8.4** *Oparin's vision of chemical evolution*

How did Oparin justify his belief in a reducing atmosphere? Oparin argued that because hydrogen (H) is the most common element in the universe, early in the history of the universe it would readily have combined with other light elements to form hydrogen-based compounds such as methane and water. All the free oxygen would therefore have been used up.

**Assumption #2: Preservation.** ***Simple organic compounds that formed in the primeval soup were somehow preserved so that the energy responsible for their formation did not also destroy them.***

Atmospheric gases could have reacted to form more complex compounds only if energy was available to cause them to react. Ultraviolet light from the sun, cosmic rays, electrical energy from lightning bolts, heat, and radioactivity might have provided the necessary energy. According to Oparin's hypothesis, the available energy converted the atmospheric gases into more complicated compounds such as sugars, amino acids, and fatty acids. But the same energy sources could also have broken down the molecular structure of these compounds. Oparin therefore assumed that these compounds were somehow protected from such destructive effects and collected in the earth's primitive oceans to form a soup out of which life could form.

**Assumption #3: Concentration.** ***Biological compounds accumulated in sufficiently high concentration so that they could combine with each other to form the large, complex molecules needed for life.***
Oparin proposed that as simple organic compounds accumulated in the primitive oceans, they were not so diluted as to exclude biologically significant interactions. Rather, material forces were able to set aside and concentrate these compounds, thereby enabling them to combine with each other to form more complex substances such as proteins, nucleic acids (RNA and DNA, whose double-helix structure was unknown to Oparin when he proposed his hypothesis), polysaccharides (long chains of sugar molecules), and lipids (fats).

At first, these large biological molecules were far simpler than their biochemical counterparts today, but gradually they became more and more complex. Eventually, catalytic proteins with the ability to accelerate reaction rates emerged. They were the forerunners of the first enzymes. In light of this assumption, scientists came to believe that amino acids reacted with each other to form larger amino-acid chains (polypeptides), that nucleotide bases combined with each other to form long DNA and RNA sequences, and that simple sugars combined with other simple sugars to form complex sugars (polysaccharides).

**Assumption #4: Uniform Orientation.** ***Only "left-handed" or L-amino acids combined to produce the proteins of life, and only the "right-handed" or D-sugars reacted to produce polysaccharides and nucleotides.***
Sometime early in their history, cells developed a preference for a certain type of amino acid and a certain type of sugar. Two amino acids can be chemically identical, having the same chemical make up. Nevertheless, they can differ in their three-dimensional shape in the same way that the left hand differs from the right. In other words, they can be mirror images of each other. The proteins that make up living things are composed of only left-handed amino acids (except glycine which is neither right- nor left-handed). Yet, in the absence of life, left- and right-handed forms are equally probable in nature. Sugars also occur in mirror images. The genetic material of living organisms (DNA and RNA) is composed of nucleotide bases along a sugar-phosphate backbone. These sugars are all right-handed. Here again, in the absence of life, left- and right-handed forms are equally probable in nature.

**Assumption #5: Simultaneous Emergence of Interdependent Biomacromolecules (such as DNA and proteins).** ***The genetic machinery that tells the cell how to produce proteins and the proteins required to build that genetic machinery both originated gradually and were present and functioning in the first reproducing protocells.***
In living cells today, both DNA and protein depend on each other for their existence. Scientists differ about how one could have originated without the other already being present. Some maintain the genetic coding came first. Others maintain the functioning protein came first. Still others maintain that both DNA and proteins appeared simultaneously. In recent years the "RNA first" view has gained prominence. Also known as "ribozyme engineering," this view holds that RNA with catalytic properties was the precursor to both DNA and proteins (see section 8.6). All such proposals for the materialistic origin of these biomacromolecules are highly speculative.

**Assumption #6: Functional Integration.** ***The highly organized arrangement of thousands of parts in the cell's chemical machinery, which is needed to achieve the cell's specialized functions, originated gradually in coacervates or other protocells.***
In present-day life, biological macromolecules (such as DNA and proteins) are parts of a much more complicated functionally integrated system—the cell. According to Oparin, before the origin of cells, biological macromolecules combined to form complex microscopic aggregates called coacervates. Coacervates are organized droplets of proteins, carbohydrates, and other materials formed in a solution. They may have "competed" with one another for dwindling supplies of "food" molecules in the primitive oceans and thus been the forerunners of the first living cells. Oparin thought of this competition as a form of Darwinian natural selection resulting in the survival and domination of ever more complex and life-like coacervates until a true cell finally appeared. These first cells would have had cell membranes, complex metabolism, genetic coding, and the ability to reproduce; moreover, they would have dominated the primitive seas.

**Assumption #7: Photosynthesis.** ***The chemical process known as photosynthesis, which captures, stores, and utilizes the energy of sunlight to make nutrients, gradually developed within coacervates.***
Oparin speculated that the further development of primitive heterotrophic organisms (organisms unable to make their own food from inorganic starting materials) resulted in the formation of cells capable of photosynthesis. These were the first autotrophic organisms (organisms able to make their own food). According to Oparin, the driving force for this evolutionary development was the gradual decrease of organic nutrients from the primitive oceans (such molecules having been largely consumed by the heterotrophic organisms).

Oparin therefore proposed that spontaneous chemical events within the coacervates led to the formation of photosynthesis. Photosynthesis captures and processes energy from sunlight. According to Oparin, photosynthesis supplied the energy needs of primitive cells. Since photosynthesis releases free oxygen ($O_2$) into the air, Oparin argued that free oxygen was not available on the earth until after the emergence of photosynthesis. Only then could autotrophic organisms develop that used oxygen in respiration and thus did not have to depend on fermentation to supply organic nutrients.

In summary, Oparin visualized a gradual origin of life from nonliving organic chemicals by a long process spread over hundreds of millions of years and without the aid of intelligent agency. Instead of cells appearing suddenly through spontaneous generation (as was believed still in Darwin's day—see the general notes to this section), the Oparin hypothesis breaks the origin of life down into seven stages. Moreover, because transitions from one stage to the next could, at least to some extent, be tested experimentally, the hypothesis itself could be tested.

Thus, Oparin's hypothesis was more than just a revival of spontaneous generation, the idea that full-blown cellular life can emerge abruptly from nonlife. Rather, it employed both a divide-and-conquer and a self-organizational strategy to explain the cell's complexity. It used divide-and-conquer to break the problem of life's emergence into seemingly manageable steps. Moreover, at each step it appealed to self-organizational properties of matter to bring about the next needed advance in the progression from nonlife to life.

## 8.3 THE MILLER-UREY EXPERIMENT

One of the advantages of Oparin's hypothesis was that it could to some extent be tested. Not directly tested, of course, because we cannot observe a past event such as the origin of life. But scientists can construct hypothetical scenarios for events that might have happened and then set up laboratory experiments to see if similar events could occur today. These are called *simulation experiments.* They are designed to simulate what might have happened on the early earth when life began. Results from such simulation experiments could then assess the reasonableness of Oparin's hypothesis. Although the hypothesis itself was sketchy, Oparin and his followers were confident that it was only a matter of time before laboratory experiments would fill in the details of how life first developed.

Oparin proposed that life arose from chemical reactions among simple gases in the atmosphere: methane, ammonia, hydrogen, and water vapor. These reactions would be activated by various forms of energy present on the earth before life began—by lightning, heat from volcanoes, kinetic energy from earthquakes, and light from the sun. When atmospheric gases encountered this energy, they would react to form organic compounds such as amino acids, fatty acids, and sugars.

How has this hypothesis been tested in the laboratory? Scientists have taken the simple gases suggested by Oparin, enclosed and mixed them together in a glass apparatus, and then subjected them to various energy sources, such as ultraviolet light (to simulate sunlight) and electrical discharges (to simulate lightning). Such experiments are called *primitive atmosphere simulation experiments*, and many have been performed since the early 1950s. These experiments attempt to reproduce likely conditions on the early earth. As such, they cannot provide direct observation of life's origin. The purpose of these experiments is to determine what compounds might reasonably have been formed on the primitive earth prior to the emergence of life and whether any of these compounds are biologically relevant.

In 1953, Stanley Miller and Harold Urey reported the first such experiment. At the time, Miller was a graduate student at the University of Chicago working with Urey (who had won the Nobel Prize in chemistry in 1934). When Miller began his graduate work, nobody had yet carried out experiments to see if the primitive atmosphere assumed by Oparin would indeed produce organic compounds necessary for life. Miller and Urey were interested in doing just that. To duplicate the conditions that Oparin assumed to exist on the primitive earth, Miller and Urey designed the apparatus shown in figures 8.5 and 8.6. They boiled water in a round-bottomed flask, thereby saturating the atmosphere of the flask with water vapor. They then eliminated any trace of $O_2$. Next they piped in methane and hydrogen gases, and then they generated ammonia gas by heating dissolved ammonium hydroxide ($NH_4OH$) in the water at the bottom of the flask. During the experiment, the water in the flask boiled, driving the gases in a clockwise direction through the apparatus. At the top of the apparatus was a 5-liter glass sphere containing two electrodes connected

***Figure 8.5*** *Stanley Miller, shown here with the apparatus used in his original origin of life experiments as a young graduate student at the University of Chicago.*

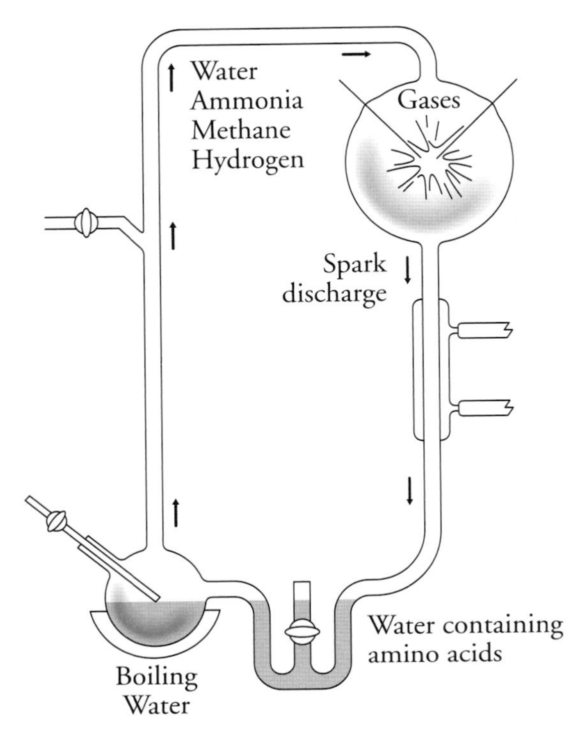


***Figure 8.6*** *Miller's apparatus. The laboratory equipment used by Stanley Miller to perform his early prebiotic simulation experiments.*

to a source of electricity. As the gases passed between these electrodes, they were subjected to 50,000-volt sparks.

Leaving the spark chamber, the gases passed through a cooling device that condensed the water vapor and any nonvolatile (nongaseous) organic compounds that formed in the glass sphere. This solution then collected in a trap at the bottom of the apparatus. Miller analyzed the gooey, tar-like substance formed in the flask and identified several of the amino acids found in proteins today, including glycine, alanine, aspartic acid, and glutamic acid. He also found several nonbiological amino acids, as well as urea and some simple organic compounds such as formic acid, acetic acid, and lactic acid.

Since Miller's early work in the 1950s, other biological compounds have been detected in similar atmospheric simulation experiments. The list now includes many essential organic compounds found in living things. Imagine the excitement in the scientific community as the results of these experiments were first published. The possibility of humans creating life in the laboratory seemed just around the corner. These experiments seemed to prove that many of the chemical building blocks of life could have formed naturally under conditions assumed to have existed on the primitive earth (one of the main assumptions being a reducing atmosphere).

Experimental evidence thus seemed to support the first stage in Oparin's hypothesis. In consequence, Oparin's views on chemical evolution gained credibility and new adherents. But when scientists sought to go beyond the simplest building blocks of life, they were quickly frustrated. The step from simple compounds to the complex molecules of life, such as proteins and DNA, has proved exceedingly difficult. Thus far, this step has resisted all efforts by scientists working on the problem.

The reason for this failure is that the needed chemical reactions do not occur. Some chemical reactions occur quite readily whereas others do not. As long as no oxygen is present, the reactions that produce the simple building blocks of life occur readily in the laboratory, but the chemical reactions required to form proteins and DNA do not. In fact, these biomacromolecules haven't been produced in any simulation experiment to date. In addition, the assumptions underlying primitive atmosphere simulation experiments have proven problematic. Such experiments should, after all, simulate realistic prebiotic conditions (i.e., conditions that might reasonably be expected to have obtained on the early earth). Yet many don't.

## Primitive Undersea Simulation Experiments

Our best evidence for the earth's early atmosphere suggests that it was inhospitable to the origin of life because of the presence of free oxygen. Some researchers, therefore, speculate that life originated at undersea hydrothermal vents where oxidation may have played less of a role in hindering the formation of life's building blocks. Oceanographer Jack Corliss discovered these vents in 1977. Despite the intense heat released by these vents from the earth's interior, life abounds around them. Corliss hypothesized that this is where life might have first arisen. Corliss's proposal runs counter to earlier speculation, such as that of Oparin, which took the sun as the crucial energy source driving the origin of life. For Corliss, heat from the earth's interior and high pressure from the ocean depths provided the crucial energy driving life's origin.[12]

Corliss's proposal has attracted much attention. Organic chemist Günter Wächtershäuser has tried to fill in the details, speculating that metal sulfide minerals such as iron pyrite, which are abundant in seafloor rocks, might play a catalytic role in the chemical reactions needed to form biologically relevant compounds (a task that living cells perform with enzymes).[13] Researchers at the Carnegie Institution have been testing Corliss's proposal through high-pressure, high-temperature simulation experiments that attempt to recreate ocean conditions at the hydrothermal vents.[14]

As with atmospheric simulation experiments, researchers have found that biologically relevant compounds (including amino acids) can be formed in such undersea simulation experiments. To prevent such compounds from degrading, however, these experiments control for the effects of free oxygen. Yet in doing so, they may not be simulating authentic prebiotic conditions. If oxygen was present in the earth's primitive atmosphere, then seawater synthesis experiments do not redress the problem of oxidation of organic

*continued on next page*

compounds *by atmospheric oxygen.* Current oceans are in equilibrium with the atmosphere, and seawater contains enough oxygen ($O_2$) to oxidize the current levels of dissolved organic matter many times over along with the vast majority of organic particles that fall through it. The only reason we find dissolved organic matter in seawater today is that biological processes constantly add to it. Stop the supply, and existing organic matter will disintegrate. There is no good reason to think that the situation would have been different on the early earth had oxygen been present: the oceans will not save organic matter synthesis simply because the compounds are in water.

All the problems associated with atmospheric simulation experiments therefore remain. In particular, high-pressure, high-temperature undersea simulation experiments do nothing to address the fundamental problem that plagues all origin-of-life scenarios, namely, how to arrange the basic building blocks of life (amino acids, nucleic acids, lipids, etc.) into the highly organized, information-rich structures required for life. Additionally, undersea simulation experiments introduce a new problem. Water at high pressure and high temperature, especially if shooting out of vents, is turbulent. Even with the building blocks for life in place, the turbulence of such watery environments will hinder the further organization of these building blocks into biomacro-molecules. Yes, the origin of life requires energy. But this energy needs to be directed, and the haphazard motion of water at high temperature and high pressure impedes such direction.[15] Primitive undersea simulation experiments therefore create more difficulties than they resolve.

## 8.4 PROBLEMS WITH OPARIN'S ASSUMPTIONS

**Contra Assumption #1: The Presence of Free Oxygen.** All experiments simulating the atmosphere of the early earth have excluded free oxygen (i.e., oxygen in a dissolved or gas state not bound up with other substances). That's because oxygen acts like a wrench in a gear-box, actively hindering the chemical reactions that produce organic compounds. Moreover, if any such compounds did happen to form, free oxygen would quickly destroy them in a process called oxidation. That's why many food preservatives are antioxidants—they protect food from the effects of oxidation.

As a consequence, the standard story of chemical evolution assumes that there was no oxygen present in the earth's atmosphere at the origin of life. Yet scientists now have strong geological evidence that significant amounts of oxygen were present in the earth's atmosphere from the earliest ages (and thus at the time the first life was supposed to be forming).[16] For instance, many minerals react with oxygen (as in the rusting of iron), and the resulting oxides are found in rocks dated earlier than the origin of life.

If oxygen had been present in the earth's early atmosphere, it would have been impossible for organic compounds to have formed and accumulated the way they did in Miller's experiment. Such experiments also leave an unresolved paradox. Because of oxidation, the presence of oxygen would have prevented the accumulation of organic compounds on the early earth. Yet without oxygen, as assumed by Oparin and in the Miller-Urey experiment, organic compounds may not have accumulated either. Significant levels of oxygen would have been necessary to produce ozone. Ozone shields the earth from levels of ultraviolet radiation lethal to biological life. Since life did in fact flourish on the early earth, a realistic simulation of the early earth's atmosphere may need to include oxygen.

Another problem is that the Oparin hypothesis and the Miller-Urey experiment assumed that the earth's early atmosphere was hydrogen-rich. Yet geochemists concluded in the 1960s that the earth's primitive atmosphere was derived from volcanic outgassing, and consisted primarily of water vapor, carbon dioxide, nitrogen, and only trace amounts of hydrogen. Because the earth's gravity is too weak to retain light hydrogen gas in the atmosphere, most of the volcanic hydrogen would have been lost to space. With no hydrogen to react with the carbon dioxide and nitrogen, methane and ammonia could not have been major constituents of the early atmosphere.

Geophysicist Philip Abelson concluded in 1966: "What is the evidence for a primitive methane-ammonia atmosphere on earth? The answer is that there is no evidence for it, but much against it."[17] In 1975, Belgian biochemist Marcel Florkin announced that "the concept of a reducing primitive atmosphere has been abandoned," and the Miller-Urey experiment is "not now considered geologically adequate."[18] Sidney Fox and Klaus Dose conceded in 1977 that a reducing atmosphere did "not seem to be geologically realistic because evidence indicates that . . . most of the free hydrogen probably had disappeared into outer space and what was left of methane and ammonia was oxidized."[19] Since 1977 this view has become a near-consensus among geochemists. As Jon Cohen wrote in Science in 1995, many origin-of-life researchers now dismiss the 1953 experiment because "the early atmosphere looked nothing like the Miller-Urey simulation."[20]

What if the Miller-Urey experiment is repeated with a more realistic mixture of water vapor, carbon dioxide, and nitrogen? Fox and Dose reported in 1977, and Heinrich Holland reiterated in 1984, that no amino acids are produced by sparking such a mixture.[21] In 1983, Miller reported that he and a colleague were able to produce a small amount of the simplest amino acid, glycine, by sparking an atmosphere containing carbon monoxide and carbon dioxide instead of methane as long as substantial free hydrogen was present.[22] But he conceded that glycine was about the best they could do in the absence of methane. John Horgan summarized the state of this research for *Scientific American* in 1991: an atmosphere of carbon dioxide, nitrogen and water vapor "would not have been conducive to the synthesis of amino acids."[23]

So Oparin's assumption that the early earth had a reducing atmosphere conducive to the pre-biotic synthesis of amino acids (Assumption #1) appears to be incorrect.

**Contra Assumption #2: Reversible Reactions.** A second obstacle faced by any theory of chemical evolution can be stated as a paradox. Some chemicals react quite readily with one another. They connect easily, like the north and south poles of two magnets. Others resist reacting. Getting them to react is like forcing the magnets' north ends together. To drive such a chemical reaction forward requires energy (heat, for example, or electricity). But—and here is the paradox—energy also breaks chemical compounds apart.

Energy is therefore a two-edged sword, building up complex molecules from simpler parts but also breaking down developing molecules. In the formation of life, it is therefore crucial that the destructive effects of energy not outpace the constructive effects. For life to form, there has to be a proper balance, or equilibrium, between these two opposing tendencies, the tendency for energy to build things up and its tendency to break things down. Yet, when we take into account the destructive effects of energy on the early earth, the equilibrium state of the "primeval soup" would favor not complex molecules but simple ones that have no inherent capacity for spontaneously organizing themselves into the machinery of a living cell.

Even the preservation of simple organic molecules on the early earth is not as straightforward as one might think. Origin-of-life researchers are quick to note that the Murchison meteorite carried numerous organic compounds to earth. Thus, it is thought, if chemical pathways to organic compounds fail on the early earth, organic matter can always hitch a ride on meteorites to get here. But, can enough organic matter, and of the right sort, be supplied to earth in this way to jump-start life? As we shall see in the next point, organic matter produced under plausible prebiotic conditions tends to be useless for building up biological complexity.

In any case, the evidence for the early earth as a preserver of simple organic molecules is hardly compelling. Atmospheric simulation experiments, such as the one by Miller and Urey, neglect the destructive effects of energy. In such experiments, amino acids and other organically relevant products that form are siphoned off through a trap to protect them from breaking down. In the trap, they are preserved from the destructive effects of the electrical discharges. But suppose the amino acids and other products had been continually exposed to such discharges, as on the early earth. In that case, they would have disintegrated as soon as they formed, and Miller could not have detected them with his apparatus (see figure 8.7). Such traps do not realistically correspond with any protective mechanism reasonably presumed to have existed on the early earth.

***Figure 8.7*** *The same energy that would cause amino acids to form short polypeptides in a primordial soup would, with additional exposure, also cause those developing polypeptides to break down.*

Oparin's assumption that simple organic compounds were somehow preserved on the early earth (Assumption #2), though plausible, therefore remains to be proven.

**Contra Assumption #3: Interfering Cross-Reactions.** Many reactions needed to form biologically important compounds have been observed under (and often only under) artificial laboratory conditions. At the same time, many reactions that occur in nature work against the formation of biologically important compounds. Amino acids, for instance, do not readily react with each other. They do, however, readily react with other substances, such as sugars. But this creates a problem. If amino acids formed on the early earth, they would not float around in lakes and ponds simply waiting for the correct amino-acid partners to show up to form proteins. Instead, they would combine with other compounds in all sorts of cross-reactions, tied up and unavailable for any biologically useful function (see figure 8.8). This explains why, in prebiotic simulation experiments, the predominant outcome is large yields of nonbiological sludge.[24]

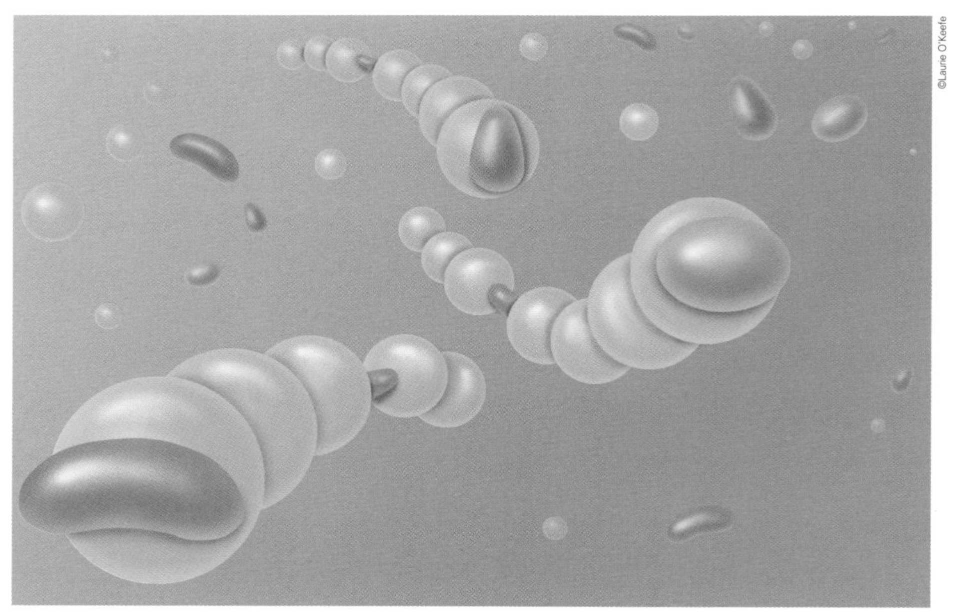


***Figure 8.8:*** *Cross-reactions. In a primordial soup, nothing would prevent amino acids from reacting with innumerable other ingredients of the primordial soup, thus tying them up and ending their potential to become useful protein molecules.*

For Oparin's account of life's origin to hold up, simple biological compounds would need to concentrate in sufficient quantities before they could combine to form the complex biomacromolecules necessary for life. This is Oparin's assumption of "concentration." But there is no evidence that nature can concentrate simple biological compounds so that they can later arrange themselves into the large complex molecules needed for life. Nature provides no lay-away plan for setting aside biologically important compounds for future use.

In an organic soup full of diverse chemicals, amino acids would not tend to combine with amino acids or sugars with sugars. Rather, interfering cross-reactions would tie up such biologically relevant chemical compounds and render them biologically useless.[25] Primitive atmosphere simulation experiments confirm this point. No biopolymers (chains of amino acids or nucleotides) useful to life have been found in such experiments except for some very small peptides. Most of what is produced in such experiments, however, is nonbiological sludge—hydrogen-poor, insoluble materials known as tars.[26] This shows that interfering cross-reactions occur under even the most favorable experimental conditions, casting real doubt on Oparin's hypothesis of concentration.

Further doubt arises because there is no geological evidence of any significant prebiotic accumulation of organic matter. It is often claimed that all such evidence would have disappeared as soon as life arose and organisms began to feed on that matter. But clay deposits at the time of life's origin have been found in abundance and would have retained large amounts of hydrocarbons and nitrogen-rich compounds from the prebiotic soup if they had been present. The surface of the clay has tiny cavities that would have imprisoned these molecules, where they would still be evident today. Thus, if the prebiotic soup had really existed, we would expect to find such surviving traces of it in the oldest rocks. In fact, we don't find any such traces.[27]

Oparin's assumption that simple organic compounds concentrated in sufficient quantities to form the biomacromolecules needed for cellular life (Assumption #3) therefore appears to be incorrect.

**Contra Assumption #4: Racemic Mixtures.** Amino acids, sugars, proteins, and DNA are not simply bundles of chemicals. They exhibit very specific three-dimensional structures. When synthesized in the laboratory, they may have the right chemical constituents but still exhibit the wrong three-dimensional form. For example, amino acids appear in two forms or *chiralities*. These are mirror images of each other just as a right glove is the mirror image of a left glove. The two forms are referred to as right- and left-handed amino acids. Living things use only left-handed amino acids (L-amino acids) in their proteins (and thus are said to be *homochiral*). Right-handed ones (R-amino acids) don't "fit" the metabolism of the cell any more than a right-handed glove would fit onto a left hand. If just one right-handed amino acid finds its way into a protein, the protein's ability to function is diminished and often completely destroyed.

Although the amino acids that make up the proteins of living cells are all "left-handed" (the L-form), in simulation experiments such as the one by Miller and Urey, the amino-acid products found in their apparatus are always a racemic mixture, that is, 50 percent left-handed and 50 percent right-handed (the L- and D-forms of the amino acid alanine are shown in figure 8.9). No one knows why amino acids in living things only occur in one of their two possible forms. Yet if life originated by purely material processes, then they must be capable of concentrating the L-form of amino acids in a specific location.

Is there any evidence for such a capacity in nature? Perhaps the closest thing is work by geochemist Robert Hazen and fellow researchers. They placed four calcite crystals in a racemic mixture of the amino acid aspartic acid. Two of the crystals were smooth, and two had microscopic terraces. The researchers found that one of the terraced crystals had a slight excess of right-handed amino acids and that the other had a similar excess of left-handed amino acids. The most extreme disproportion that they found was 55-45 instead of the usual 50-50. By contrast, the smooth crystals did not distinguish between the L- and D-forms of aspartic acid.[28]

Even so, this is hardly compelling evidence that purely material processes can segregate L-amino acids from D-amino acids. Hazen's research applies to only one amino acid—aspartic acid. Moreover, his calcite crystals merely induced a slight disproportion in the ratio of L- to D-amino acids. Life as we know it requires pure concentrations of the L-form of amino acids—a substantial admixture of the D-form, as in Hazen's experiments, will not do. As a consequence, Noam Lahav of the Hebrew University in Jerusalem does not regard Hazen's research as providing a general mechanism by which L-amino acids can be segregated from R-amino acids.[29]

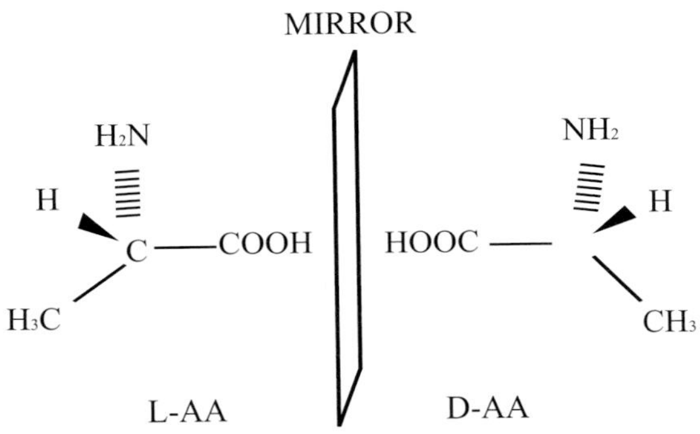


***Figure 8.9*** *Amino acids come in two forms "right-handed" (D-) and "left-handed" (L-). Both forms of the amino acid alanine are shown here. While the proteins of living things are composed of only the L-forms, both forms are found in prebiotic simulation experiments."*

Like amino acids, sugars come in two forms (or chiralities) but only appear in one form in living things. When sugars are synthesized in the laboratory, the result is an equal (racemic) mixture of both L- and D-forms, like a pile of left- and right-handed gloves. Yet living things include only "right-handed" (D-) sugars. How did living things come to prefer only one form? Scientists have conducted experiments to try to determine how these preferences might have arisen by material mechanisms. So far, they have found no mechanism that will produce only the correct three-dimensional structure. When material mechanisms (such as Hazen's calcite crystals) are applied to such mixtures to segregate L- from D-forms, the result is at best a slight disproportion of one form over the other. In this and other ways, life shows characteristics that are alien to anything known to be produced under ordinary material conditions.

Oparin's assumption that biologically relevant molecules such as amino acids and sugars could by purely material means assume the right uniform orientation (Assumption #4) therefore also appears to be incorrect.

**Contra Assumption #5: The Synthesis of Polymers.** Once sugars and amino acids of the right chirality (or handedness) are concentrated in the right place, the real work of constructing life begins: L-amino acids need to be combined in the right sequence to form proteins and nucleotide bases attached to a (D-) sugar-phosphate backbone need to be combined in the right sequence to form DNA and RNA. Proteins, DNA, and RNA are long polymer chains whose biological function depends on the precise sequence of monomers that make up the chains (an average protein is several hundred amino acids in length; DNA can be millions of nucleotide bases in length).

Synthesizing polymers with the right sequences involves two things: (1) getting adjacent monomers to join together properly and (2) getting properly joined monomers to assume a functionally meaningful order. On the first point, because of the ways that amino-acid and nucleotide monomers react chemically, they can join in numerous ways. Amino acids, for instance, can react with each other to form a variety of chemical bonds. Yet only one, called the *peptide bond*, occurs in functional proteins. This is not to say that amino acids joined by peptide bonds will form a functional protein—that depends on the ordering of amino acids. But a necessary condition for a chain of amino acids to form a protein is that peptide bonds join all the amino acid residues and that the chain be linear. Branching chains, in which amino acid residues are joined by peptide bonds yet form side chains, are also possible but will not work. The arrangement of amino acids in proteins is not branched but linear.

To see what's at stake, consider an analogy based on the following sequence of letters:

**ANERUT**

Ordinarily, when we arrange letters in a sequence, we orient them vertically from the top down. This orientation corresponds, in our analogy, to amino acids joined exclusively by peptide bonds. As we arrange letters in sequence, however, we could also rotate them. Thus, we might not only rearrange the order of the previous letters but also rotate them:

Such rotations correspond to nonpeptide bonds in a linear sequence of amino acids.[30]

Rotating the letters in odd ways like this disrupts any meaning that the sequence might have if the letters were given their proper vertical orientation (for instance, is the first letter a "Z" or an "N"?). Likewise, an amino acid sequence that would form a functional protein if all the bonds were peptide bonds would have its function disrupted by nonpeptide bonds. Thus, for the letters A, N, E, R, U, and T to form a meaningful word, the letters must be both properly oriented and correctly ordered:

Likewise, to form a functional protein, amino acids must be both properly linked (by peptide bonds) and correctly ordered (to ensure that the resulting amino-acid sequence folds).

Nucleotide sequences face exactly the same constraints: in the polymerization of polynucleotides, which is the primary step in the formation of DNA and RNA, 3-5 phosphodiester linkages are the only ones that cellular life allows even though 2-5 linkages dominate in the absence of cellular life (and thus in prebiotic environments). It follows that without some catalyst that promotes peptide bonds (for amino-acid sequences) or 3-5 phosphodiester linkages (for nucleotide sequences), there can be no materialistic route to proteins, DNA, and RNA. But the only catalysts we know capable of handling this task are enzymes and other protein-based products (e.g., the ribosome), and these in turn presuppose the entire DNA-RNA-protein machinery. This machinery, however, is precisely what origin-of-life research may not presuppose but rather must explain.

To describe the requirements for obtaining functional amino-acid or nucleotide sequences does not explain how such sequences originated. According to the Oparin hypothesis, such sequences assembled themselves gradually through an undefined process of self-organization. But this answer is unconvincing. The probability of obtaining one specific protein in an undirected search of amino-acid sequences is practically zero. Consider a small protein consisting of 100 amino-acid subunits. How many different sequences of the basic 20 amino acids are possible in a chain of 100 subunits? The answer is $20^{100}$, or approximately $10^{130}$ (1 followed by 130 zeros). The improbability of finding such a small protein by a blind or random search of protein sequence space is therefore 1 in $10^{130}$. The number $10^{130}$ is so enormous that even with billions and billions of years there have not been enough opportunities for the known physical universe to sort through every combination of amino acids to find the specific combination of amino acids for even one such protein (see the sidebar "Monkeys Typing Shakespeare" in section 7.4).

These numbers can be whittled down, but not drastically. A chain of amino acids tolerates substantial variation at sites along the chain without disrupting the protein's function. MIT biochemist Robert Sauer has applied a technique known as *cassette mutagenesis* to several proteins to determine how much variation among amino acids can be tolerated at any given site. His results show that even with this variation taken into account, the probability of forming a 100-subunit functional protein is only 1 in $10^{65}$—this is still a vanishingly small probability (there are estimated to be $10^{65}$ atoms in our galaxy).[31]

Building on Sauer's work, Douglas Axe, while at Cambridge University, developed a refined mutagenesis technique to measure the sequence specificity of enzymes such as barnase. Axe's research indicates that previous mutagenesis experiments actually underestimated the functional sensitivity of proteins to amino-acid sequence change because they presupposed (incorrectly) the context independence of function-preserving changes at individual amino-acid sites.[32] The probability of achieving a functional 100-subunit sequence of amino acids by undirected search is therefore considerably smaller than the 1 in $10^{65}$ improbability calculated by Sauer.[33]

These estimates of improbability are extremely conservative, giving prebiotic chemistry every unfair advantage to succeed. Modern organisms use just 20 basic amino acids. But what evidence we have for a "prebiotic soup" (e.g., from simulation experiments and from studying amino-acid-rich meteorites) indicates that it would have contained many more than these 20. The Murchison meteorite, for instance, has been found to contain over 70 amino acids.[34] Any realistic origin-of-life scenario attempting to account for the polymerization of amino acids is therefore likely to contain at least 50 amino acids, each with right- and left-handed forms. In a sequence of amino acids, that's 100 amino-acid possibilities at each location. Thus, for an arrangement of 100 amino acids, there are $100^{100} = 10^{200}$ possibilities. And that assumes only linear arrangements and only peptide bonds. Even this estimate, however, is conservative. A more realistic estimate would include not only these additional amino acids but also reactions with countless competing compounds (sugars, aldehydes, etc.). The number of possibilities, and the corresponding improbabilities, quickly become staggering.

How do origin-of-life researchers handle such improbabilities? Do they attribute specified complexity to these systems and therefore infer design? No. Typically they respond, as in section 7.5, by charging that design theorists have merely devised a tornado-in-a-junkyard strawman, calculating the probability of forming functional protein types by purely random assembly. Instead, so they claim, evolutionary mechanisms exist that substantially raise the probability of the formation of these systems. But what is the evidence for the existence and efficacy of such mechanisms? It is nil.

To be sure, origin-of-life researchers have proposed speculative scenarios in which mechanisms of chemical evolution supposedly bring about biologically significant polymers (see sections 8.5 through 8.8). But these scenarios don't even begin to address the true complexities of life. Moreover, origin-of-life researchers have failed to test and confirm these scenarios experimentally. Oparin boldly assumed the chemical pathways he needed to account for the origin of life, but then left the crucial demonstration of those pathways to others. Not only are we still waiting for such a demonstration, but, as we have just seen, the chemical pathways we do know conspire against the formation of functional polymers such as proteins and DNA.

Oparin's assumption that protein and nucleotide polymers could gradually form by purely material processes acting on monomers (Assumption #5) therefore also appears to be incorrect.

**Contra Assumption #6: The Humpty-Dumpty Problem.** Take a clean, sterile Eppendorf tube (a small test tube with a tapered bottom that is often used by biochemists). Place into it a small amount of sterile salt solution, known as a buffer, at just the right temperature and pH. Place into the buffer a living cell, but puncture it with a sterile needle so that its contents leak out into the solution. (The amount of buffer can be kept as small as possible to avoid diluting the cell's contents too much.) You now have all the molecules needed to make a living cell—not just the building blocks, but all the fully assembled macromolecules such as DNA and proteins. Moreover, you have them in just the right proportions, under just the right conditions, and without any interfering substances. Yet, despite all such efforts to facilitate the transition from nonlife to life, the molecules will not form into a living cell—no matter how long you wait or what you do to them given present knowledge and technology. What's more, if our technology ever gets to the place where it can reassemble punctured cells, that will be evidence not for the power of chemical self-assembly but for design.

We've just sketched what may be called the Humpty-Dumpty experiment. Origin-of-life researchers, playing the role of "all the king's horses and all the king's men," have been spectacularly unsuccessful in putting Humpty-Dumpty back together again. But if reassembly has proven difficult, think of how much more difficult it is to assemble

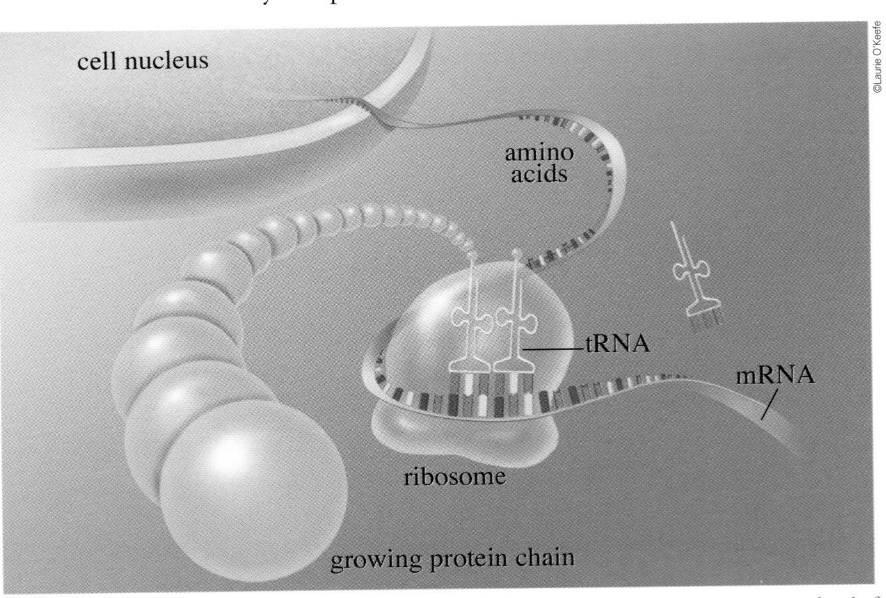


***Figure 8.10*** *Protein synthesis. A view from inside the cell of the assembly of a protein molecule from the DNA pattern brought by messenger RNA to the ribosome's assembly site.*

Humpty-Dumpty from scratch, as the Oparin hypothesis assumes and origin-of-life researchers hope to demonstrate. To put the problem in perspective, focus for the moment simply on the enzymes required for protein synthesis (the remarkable process by which proteins are synthesized is diagrammed in figure 8.10). Enzymes are proteins that specialize in carrying out chemical reactions quickly and efficiently. The enzymes in a cell work together like the parts of a finely crafted machine to perform the cell's metabolic functions. A typical enzyme consists of several hundred amino acids.

To build even a single protein in a cell, about sixty specific proteins acting as enzymes are needed. These include helicases (which unwind double-stranded DNA into single strands), topoisomerases (which control the supercoiling of DNA), RNA polymerases (which polymerize RNA with respect to a DNA template), aminoacyl-tRNA synthetases (which link amino acids to transfer RNA), and ribonucleases (which control the amount of unprotected RNA in the cell). These sixty enzymes need to act in concert with the cell's genetic machinery—its DNA. If even one of these enzymatic proteins necessary for protein synthesis is missing, the cell will not be able to form proteins. The enzymes involved in protein synthesis are therefore indispensable for life and, considered jointly, form an irreducibly complex system (see Chapter 6).

The precise division of labor in a cell's protein-making machinery underscores the enormity of the problem of evolving proteins under prebiotic conditions. Protein synthesis, which is absolutely essential for life as we know it, presupposes these highly specific enzymes. Moreover, not only must those sixty enzymes all exist at the same time; they must also occur together in the same tiny region of the cell. In other words, they have to be coordinated and targeted to the right location in the cell. But such coordination itself presupposes a vastly complex communication, transportation, and control system (and that system is itself enzymatically, and thus protein, driven).

Consequently, getting sixty specific enzymes all to congregate and act in precisely choreographed steps in the correct sequence and in the exact location still doesn't grasp the full enormity of the problem. The cellular environment for these proteins must itself be conducive to their coordinated function. And this presupposes that all basic cellular functions be in place from the start. These include: the storage, retrieval, and processing of the information in DNA; the manufacturing of proteins from RNA templates by ribosomes (ribosomes are themselves immensely complex, consisting of at least fifty protein and RNA subunits); the metabolic functions that extract energy from nutrients; and the very capacity to reproduce such vastly complex systems.

Oparin's assumption that the biomacromolecules could gradually be coordinated and integrated into a primitive life form by purely material processes (Assumption #6) therefore also appears to be incorrect.

**Contra Assumption #7: Harnessing the Sun.** Assumption #7 states that some cells eventually developed the ability to capture sunlight and transform it into usable energy by means of photosynthesis. No presently known material mechanisms account for how so complex a process as photosynthesis could have developed. In particular, no scientist has proposed any detailed step-by-step pathways by which the evolution of photosynthesis might be tested. To be sure, biologists have speculated how photosynthesis might have evolved.[35] But such speculation remains for now untestable. Indeed, the exquisite fine-tuning of photosynthetic complexes to capture solar light efficiently by exploiting quantum-theoretical effects and then by transferring that energy to reaction centers in the cell where it can be stored and later utilized bespeaks a level of design that far exceeds anything produced thus far from the field of photochemical engineering.[36]

Oparin's assumption that primitive cells could have developed photosynthesis by purely material processes (Assumption #7) therefore appears to be incorrect as well.

In summary, most simulation experiments do not realistically model the conditions that our best evidence suggests existed on the early earth. Given the probable composition of the primitive atmosphere (especially the likely presence of free oxygen), the fact of reversible reactions, the prevalence of interfering cross-reactions, and the absence of homochirality apart from life, there is no reason to think that conditions on the early earth favored the emergence of life's basic building blocks by purely material forces. The most probable scenario of early-earth history is not one that builds up biological complexity but one that breaks it down.[37] Moreover, there is neither a solid theory nor a promising experimental basis for the view that biomacromolecules could have emerged by purely material forces, much less that an interdependent and coordinated set of them could have gradually organized themselves into functionally integrated units with the complexity of actual cells. The problems raised in this section therefore challenge Oparin's entire framework for the origin of life.

## 8.5 THE PROTEINOID WORLD

In critiquing the Oparin hypothesis, we have thus far focused on the obstacles to making and arranging the building-blocks required to form complex biomolecular structures. Though these obstacles are daunting and, in our view, sufficient to invalidate the whole materialistic origin-of-life enterprise, origin-of-life researchers have not given up. Instead, they have come up with proposal after proposal for fleshing out Oparin's hypothesis in the hopes of overcoming these obstacles. Let us now turn to some of these proposals in detail. One of the earliest was the late Sidney Fox's proteinoid microspheres. Fox wrestled with how purely material forces could construct complex, information-rich biological structures from elementary constituents.

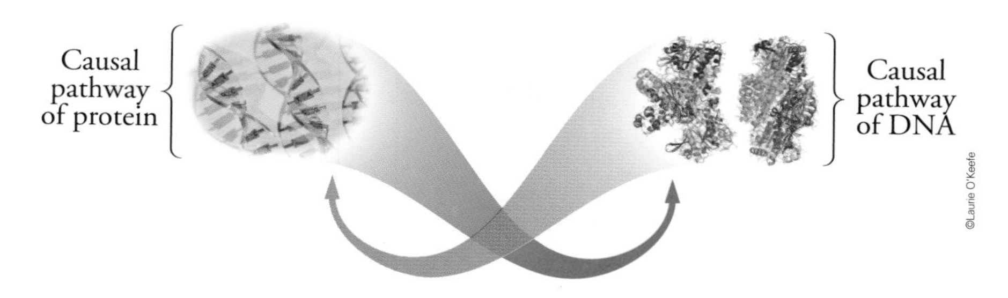


***Figure 8.11*** *Both proteins and DNA are absolutely dependent upon the prior existence of the other for their own existence.*

How, for instance, did amino acids come together to form the first proteins? Amino acids left to themselves tend to clump together, but not with the right bonds and not in the right order to form proteins. For that, they need enzymes that act as specific catalysts and DNA to serve as a sequence template. But these enzymes themselves are very special proteins that require DNA to code for their catalytic ability. Moreover, DNA itself requires enzymes for its production. How, then, did the first amino acids form enzymes when there were neither existing enzymes to serve as catalysts in linking the protein structure together nor DNA to code for the structure? (See figure 8.11.)

This is a classic case of "Which came first, the chicken or the egg?" To overcome this problem, scientists have attempted to link amino acids together into protein-like molecules without using either existing enzymes or DNA. Fox, while at the University of Miami, performed an experiment to that end, heating dry mixtures of amino acids at 160–180° C for several hours in a nitrogen atmosphere. He found that these amino acids joined together. Nevertheless, they did not form linear protein polymers. His polymers included nonpeptide bonds and were branched. To call them polypeptides would therefore have been incorrect. Fox therefore dubbed them "proteinoids." It would have been more accurate to call them "branched amino-acid polymers."

Fox's proteinoids faintly exhibited some of the properties of true proteins. For instance, the proteinoids catalyzed certain chemical reactions, making them go ever so slightly faster than they would otherwise. Fox contended that these proteinoids represented the beginnings of enzymatic activity on the primitive earth. Moreover, if proteinoids are dissolved in boiling water and the solution is cooled, the proteinoid molecules will clump together to form uniform microscopic spheres about the size of bacterial cells (see figure 8.12). Fox contended that these little spheres represent the first step toward cellular life on earth.

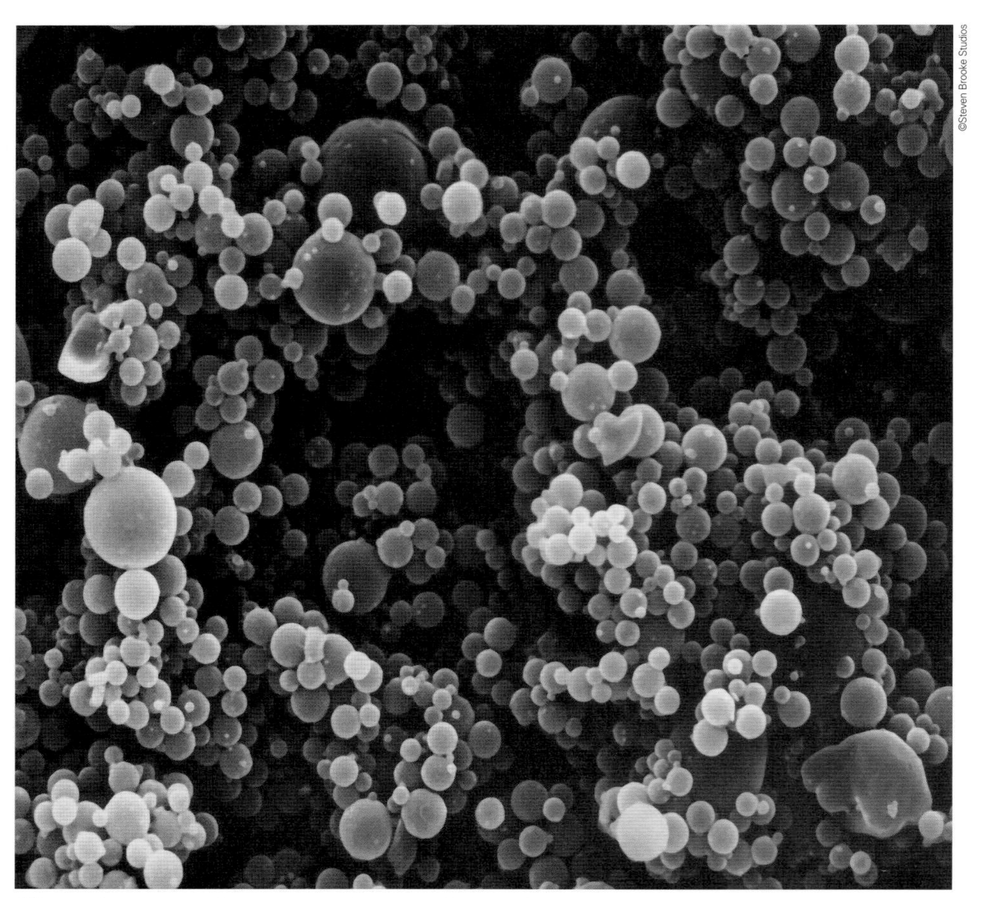
©Steven Brooke Studios

***Figure 8.12*** *Optical micrograph of proteinoid microspheres.*

The process of forming proteinoid microspheres from amino acids is thought to have taken place near volcanoes (this assumes, of course, that amino acids could have accumulated in sufficient numbers on the primitive earth—this assumption is itself problematic, as we saw in section 8.4). As Fox conceived it, amino acids formed when gases of the primitive atmosphere came in contact with molten rock from a volcano (at a temperature of about 1200° C). The amino acids then accumulated some distance from the center of the active volcano where surface temperatures ranged between 160 and 180° C. At such temperatures, the dry amino acids condensed into proteinoids. These were then protected from the destructive effects of heat by the cooling action of rainfall.

Once suspended in pools of water, the proteinoids would form microspheres. These in turn might acquire the ability to compete with each other for environmental resources (how they might do this without first acquiring the ability to metabolize nutrients and reproduce is unclear—the proteinoid microspheres that emerged from Fox's experiment were incapable of either). The most successful competitors might then survive, become dominant, and develop life-like qualities. The process could take millions of years, though Fox suggested it could take but hours to get to the protocell stage. Eventually, the first self-reproducing cells capable of fermentation would form, and life itself would gain its first foothold.

Fox's experiment seems to reach the threshold of life by transforming amino-acid building blocks, which can be produced in Miller-Urey type simulation experiments. Taken together, the two experiments seem to confirm Oparin's hypothesis: Miller-Urey type experiments supply the building blocks, and Fox's proteinoid microspheres assemble them into life-like wholes. Closer examination, however, reveals that Fox's experiment falls far short of providing a credible scenario for the origin of life.

There are several problems with Fox's experiment. First, Fox used mixtures containing only protein-forming L-amino acids. Where on the primitive earth, or elsewhere, could such a mixture have occurred? As we have seen, naturally occurring amino acids always include roughly the same number of L- and D-amino acids. Also, even with a mixture including only L-amino acids, interfering cross-reactions would have tied up the amino acids in nonbiological compounds, thus blocking the formation of proteinoids. Sugars, for instance, react with amino acids to form the nonbiological compound known as melanoidin. Because of such cross-reactions, which Oparin's model neglected, it is very unlikely that proteinoids could have formed under natural conditions on the primitive earth. As a consequence, many scientists regard Fox's use of selected and purified amino-acid mixtures as highly unrealistic.[38]

Second, the proposed sequence of events that supposedly occurred near volcanoes is dubious. The required combination of high and low temperatures, with rainstorms occurring at just the right time and place, seems unrealistically "choreographed." Moreover, the heat needed to form proteinoids would also have threatened to destroy them. And if the heat did not destroy them, the proteinoids would likely have disintegrated on their own before they could have played a role in the formation of life.

Third, unlike Fox's proteinoid microspheres, actual cells are surrounded by a complex cell membrane. Cell membranes are made of specialized fatty acids and phosphate groups that form a phospholipid bilayer, together with many specialized protein molecules. Unless the protein molecules are highly specific, they will not function properly in forming and maintaining the cell membrane. By contrast, proteinoid microspheres have an outside boundary that is much thicker than the membrane of actual cells. Moreover, the microspheres are composed entirely of proteinoids, very simple molecules compared to true proteins.

Last, and most significantly, even if the preceding problems could be resolved, so that proteinoid microspheres could readily be produced under realistic prebiotic conditions, massive differences exist between them and the very simplest living cells.[39] Unlike proteinoids, the simplest cell requires the highly organized arrangement of hundreds of precisely defined information-rich macromolecules (e.g., proteins and DNA) for its chemical machinery to accomplish all the specialized functions a cell needs to live and prosper. Fox's proteinoid microspheres provide no mechanism for the formation of these biomacromolecules and therefore leave their origin completely unexplained.

To recap, we have considered four problems with Fox's experiment: (1) his use of selected and purified amino-acid mixtures (only protein-forming L-amino acids); (2) his implausible scenario for the formation of proteinoid microspheres near volcanoes; (3) his identification of proteinoid microspheres with cell membranes even though the two are vastly different; and finally (4) his unfounded claim that the minimal catalytic activity of proteinoids might somehow lead to the biomacromolecules relevant for life and that these in turn might gradually organize themselves into actual cells. Each of these problems alone shows that Fox's experiment cannot adequately account for the origin of life. Together they show that Fox's proteinoid world is not a plausible route to life's origin.

## 8.6 THE RNA WORLD

The origin-of-life scenario that has attracted the most attention since the 1990s is known under the rubric "RNA world" or "ribozyme engineering." Primitive atmosphere simulation experiments such as those by Miller and Urey and proteinoid experiments such as those by Sidney Fox, even if they could explain the formation of life's elementary constituents (e.g., amino acids), nonetheless fail to explain how proteins could have formed on the early earth. Moreover, to elevate proteins as the master molecule from which all other biomacromolecules arise leads to a vicious circle: proteins presuppose DNA and DNA presupposes proteins. As a consequence, origin-of-life researchers have considered the possibility that proteins were not the first molecular building-blocks of life. DNA is not a good candidate, however, because it needs a whole array of complex proteins to make copies of itself.[40] Consequently, DNA could not have originated before proteins and could not have been the first step in the origin of life.

Another candidate is RNA. RNA is a close chemical relative of DNA that is used by all living cells to make proteins. In the 1980s molecular biologists Thomas Cech and Sidney Altman showed that RNA can sometimes behave like an enzyme—that is, like a protein that catalyzes chemical reactions. Keying off this discovery, molecular biologist Walter Gilbert suggested that RNA might be able to synthesize itself in the absence of proteins, and thus might have originated on the early earth before either proteins or DNA. This "RNA world" might then have been the molecular cradle from which living cells emerged. RNAs with catalytic properties became known as "ribozymes," and the search for such RNAs became known as "ribozyme engineering."

But this scenario is deeply problematic as well. No one has been able to demonstrate convincingly how RNA could have formed before living cells were around to make it. According to Scripps Research Institute biochemist Gerald Joyce, RNA is not a plausible candidate for the first building block of life "because it is unlikely to have been produced in significant quantities on the primitive earth."[41] RNA consists of a nucleotide base (either adenine, cytosine, guanine, or uracil), the five-carbon sugar ribose, and a phosphate group. The material processes that favor the formation of these nucleotide bases work against the formation of the corresponding sugar and phosphate group, and vice versa. Indeed, the experimental synthesis of RNA has to date occurred only under the most unrealistic prebiotic conditions. NYU biochemist Robert Shapiro describes one such experiment:

> [O]ne example of prebiotic synthesis [was] published in 1995 by *Nature* and featured in the *New York Times*. The RNA base cytosine was prepared in high yield by heating two purified chemicals in a sealed glass tube at 100 degrees Celsius for about a day. One of the reagents, cyanoacetaldehyde, is a reactive substance capable of combining with a number of common chemicals that may have been present on the early earth. These competitors were excluded. An extremely high concentration was needed to coax the other participant, urea, to react at a sufficient rate for the reaction to succeed. The product, cytosine, can self-destruct by simple reaction with water. . . . Our own cells deal with it by maintaining a suite of enzymes that specialize in DNA repair. The exceptionally high urea concentration was rationalized in the *Nature* paper by invoking a vision of drying lagoons on the early earth. In a published rebuttal, I calculated that a large lagoon would have to be evaporated to the size of a puddle, without loss of its contents, to achieve that concentration. No such feature exists on earth today.[42]

We hasten to add that there's no evidence that any such feature existed in the distant past either. But even if RNA could be produced under realistic prebiotic conditions, it would not survive long under the conditions thought to have existed on the early earth—it is simply too unstable. Joyce therefore concludes: "The most reasonable interpretation is that life did not start with RNA."[43] Although he still thinks that an

RNA world preceded the protein-DNA world, he believes that some kind of non-RNA life must have preceded RNA-life (that then evolved into DNA-life). According to Joyce, "You have to build straw man upon straw man to get to the point where RNA is a viable first biomolecule."[44]

The RNA world therefore offers no explanation for how such a "viable first biomolecule" might have arisen. But suppose some as-yet-unknown materialistic scenario could have produced such a molecule.[45] What would this molecule look like? And what would it be capable of evolving into? Would it be a single self-replicating RNA? If so, what happens next? Would more and more RNAs accumulate to form a coordinated system of self-replicating RNAs? And how would such a system evolve into the DNA-protein machinery that's standard issue for life as we know it? The RNA world leaves all such questions unanswered. Indeed, origin-of-life researchers have no clue about how to answer them. The RNA world—like the protein-first scenario in the Miller-Urey and Fox experiments—therefore presupposes a still more fundamental origin-of-life scenario.

## 8.7 SELF-ORGANIZING WORLDS

Given the daunting challenges facing a materialistic explanation of life's origin, one might think that origin-of-life researchers would be pessimistic about their prospects for resolving this problem. But one would be mistaken. Origin-of-life researchers tend to be optimistic about their prospects for ultimate success, at times even giving the impression that, except for a few minor details, the problem of life's origin has essentially been solved already. As an example of such optimism, consider the following remark by mathematician Ian Stewart:

> The origin of life no longer appears to be a particularly difficult problem. We know that—at least on this planet—the key ingredient is DNA. Life's basis is molecular. What we need is an understanding of complex molecules: how they might have arisen in the first place, and how they contribute to the rich tapestry of living forms and behavior. It turns out that the main scientific issue is *not* the absence of any plausible explanation for the origin of life—which used to be the case—but an embarrassment of riches. There are many plausible explanations; the difficulty is to choose among them. That surfeit causes problems for the question "How *did* life begin on earth?" but not for the more basic issue, which is "*Can* life emerge from nonliving processes?"[46]

Stewart's assessment of the origin-of-life problem is too cavalier. Indeed, the embarrassment of riches that Stewart cites should give one pause. It is usually hard enough to come up with even one good theory to account for a phenomenon. For instance, prior to Newton's theory of universal gravitation, there was no unified theory to account for the motion of stars, moons, and planets. An embarrassment of riches points not to the solution of a problem but to vain gestures at a solution.[47] Indeed, the very claim that "there are many plausible explanations" suggests that none is plausible. If any one of them were really plausible, we could expect to see a consensus among scientists that it really is plausible (to say nothing of it being the "true" or "correct" explanation). Instead, we find a vast proliferation of "plausible explanations," none of which is universally acclaimed and each of which has fatal flaws.

Whenever origin-of-life researchers accept plausibility rather than evidence as their standard for scientific truth, they in effect give up the search for what really happened or for what with reasonable probability could have happened. Plausibility, as Stewart and many origin-of-life researchers understand the term, implies no effort to estimate probability. Instead, they settle for what they can *imagine* was possible or could have happened. In this way, they substitute opinion and prejudice for experiments and data. Where is the science in all this? How is the plausibility that Stewart ascribes to the various materialistic accounts of life's origin anything other than an article of faith?

Even so, Stewart raises two questions that deserve closer scrutiny: (1) How did life begin on earth? and (2) Can life emerge from nonliving processes? Stewart regards the answer to the first question as a work in progress and the answer to the second as an unequivocal yes. But what is this second question really asking? If it is merely asking whether nonliving matter can be organized into living matter, the answer is clearly yes because, when broken down far enough, all matter consists of particles that can exist on their own apart from life. Indeed, cosmologists have determined that all matter on earth once belonged to stars, and stars in their heyday are too hot to permit life.

In saying that life can emerge from nonliving processes, Stewart is therefore saying more than that nonliving matter can be arranged to form living matter. Rather, he is saying that nonliving matter, without any outside assistance, has the ability to organize itself into living matter. Indeed, his reference to life's "emergence" denotes, within the origin-of-life community, the power of nonliving matter to organize itself into living matter not as some wildly improbable event but as an inherent feature of nonliving matter. Thus, origin-of-life researcher Harold Morowitz, in his book *The Emergence of Everything*, writes,

> The view of the emergence of biochemistry that we have been discussing represents a paradigm shift from what the reader may have encountered

in biology courses where it was assumed that random products of free-radical reactions lead to monomers, then to polymers, then to cells [cf. the RNA world discussed in section 8.6]. In the view elaborated here, selection rules lead to a core metabolism that then produces an ordered hierarchy of emergent structures and functions. These become increasingly complex, leading to the sophisticated chemistry of the universal ancestor. This is a very different view than may have been taught in standard introductory courses, but I believe that it is a *much more probable scenario.*[48]

Morowitz here contrasts his own "metabolism-first" approach to life's origin with the more conventional "genetics-first" approach that regards the key to life's origin as the synthesis of biologically functional polymers (as in the RNA world). Although the genetics-first approach has to date received the most attention from origin-of-life researchers, it suggests that at key points in life's origin, as monomers were being configured into polymers, vastly improbable events had to occur.[49] This is because functional polymers (unless they are extremely short) tend to be rare among the totality of possible polymers; moreover, the physical properties of monomers relevant to their linear arrangement exhibit no preference for one sequence over another.

For instance, with RNA, the sugar-phosphate backbone onto which the nucleotide bases get attached is completely indifferent to how those bases are ordered (much as a refrigerator does not care about the order in which children place magnetic letters on it). The ability of polymers to carry information results from the vast number of degrees of freedom that the laws of physics and chemistry permit *to the order in which the monomers may combine.* These laws prefer no sequence over any other. Yet precisely because of this freedom, there is no way for monomers to organize themselves spontaneously into biologically functional polymers except as vastly improbable events (unless they are extremely short, in which case they won't be improbable; compare sections 7.9 and 8.4).

The problem with vast improbabilities is this: If the origin of life is a vastly improbable event, then it is a freak—an exceedingly lucky event that happened but need not have happened and is far more likely not to have happened. But to say that life's origin is lucky does not constitute a scientific theory. A scientific theory of life's origin is possible only if life originated through a sequence of steps where each step is probable and the sequence as a whole is also probable. In that case, nonliving matter has an inherent propensity to organize itself into life. In other words, it constitutes a self-organizing world. Morowitz's metabolism-first approach, in which a core metabolism facilitates the emergence of increasingly complex structures and functions, is therefore an instance of a self-organizing world.

## Emergence: The New Alchemy?

The alchemists of old never explained exactly how gold could be produced from lead. Likewise, origin-of-life researchers who tout "emergence" never tell us exactly how "an ordered hierarchy [of] structures and functions"[50] could emerge from the chemical processes that might reasonably have obtained on the prebiotic earth. "X emerges" is an incomplete sentence. It needs to be amended by reading "X emerges from Y." And even in this form it remains incomplete until one precisely specifies Y and provides a detailed account of how Y could in fact produce X. Moreover, such an account needs to be backed up by evidence. Otherwise, to invoke emergence as an explanation constitutes a leap of faith.

A complete set of the building materials for a house does not account for a house—additionally what is needed is an architectural plan (drawn up by an architect) as well as assembly instructions (executed by a contractor) to implement the plan. Likewise, with the origin of life, it does no good simply to describe a plausible set of chemical precursors to life. A detailed account of how purely material forces could, under plausible prebiotic conditions, organize those precursors into a living organism needs to be specified as well. Origin-of-life researchers who characterize life as an emergent property but then don't fill in these details are like the alchemists of old, claiming the ability of matter to transform itself in remarkable ways, yet without identifying the precise causal pathways by which it could.

Besides Morowitz's self-organizing metabolism-first world, origin-of-life researchers have proposed many other self-organizing worlds. Here is a representative sample:[51]

- Stuart Kauffman, seeing self-catalyzing collections of chemicals as the key to life's origin, has proposed that "life emerged as a phase transition to collective autocatalysis once a chemical minestrone, held in a localized region able to sustain adequately high concentrations, became thick enough with molecular diversity."[52]

- Christian de Duve, in assuming that chemistry is prior to information, regards the origin of life as a cosmic imperative (i.e., as an outcome destined by the very nature of nature) and has proposed thioesters (certain sulfur compounds) as the basis for metabolism and thus as the key to life's origin.[53]

- Günter Wächtershäuser, looking to bacteria that consume iron and sulfur, has proposed the mineral pyrite (which is rich in iron and sulfur and is commonly known as "fool's gold") as a primary catalyst for the origin of life.[54]

- Michael Russell has proposed that life originated as convection currents produced by undersea volcanic vents dissolve iron and sulfur, setting up a temperature differential between competing fluid flows at which colloidal membranes made of iron sulfide can form.[55]

- David Deamer has proposed that the self-organization of lipids into bilayered vesicles, which resemble cell membranes, was a crucial factor in the origin of life.[56]

- Simon Nicholas Platts has proposed that life originated through the activity of polycyclic aromatic hydrocarbons (abbreviated PAHs) which, according to his model, supply "proto-informational templating materials" and thus serve as progenitors for the RNA world.[57]

- Graham Cairns-Smith has proposed a clay-template theory for the origin of life in which self-replicating clays form templates for subsequent carbon-based life.[58]

All such proposals are highly speculative, address only one or a few of the most elementary aspects of the origin of life, are thin on detail even in those aspects of the origin of life that they do address, have little if any experimental support, and require massive intervention from investigators to achieve any interesting results (thereby failing to reflect realistic prebiotic conditions).

Harold Morowitz, for instance, has for years been trying to get prebiotic chemicals to organize themselves into the citric acid cycle.[59] This cycle, as with other metabolic cycles, occurs in all cells and exhibits a form of self-replication (in metabolic cycles, one chemical catalyzes another, which catalyzes another, and so on, until the process circles back to the first chemical, producing more of it than was originally there and thus replicating the cycle). For Morowitz, the emergence of the citric acid cycle is a key step in the origin of life. Yet, to this day, he cannot obtain the citric acid cycle except with the help of enzymes,[60] which presuppose life as we know it and therefore properly speaking do not belong in the origin-of-life researcher's tool-kit.

Or take David Deamer's "lipid world." Deamer has demonstrated experimentally that carbon-based molecules taken from the Murchison meteorite (which presumably represents a pristine prebiotic environment) spontaneously organized themselves into small spheres the size of microbes. Moreover, the spheres consisted of lipid bilayers as found in cell membranes. So here we see another key step in the origin of life attributable to purely material factors. Or do we? "Biologists have been quick to point out," notes Robert Hazen, that "the vesicles produced in Deamer's work are a *far cry* from actual cell membranes, which feature a mind-boggling array of protein receptors that regulate the flow of molecules and chemical energy into and out of the cell."[61] Many other pieces of life's puzzle would need to fall into place before we could say with any assurance that Deamer's lipid world plays a pivotal role in the origin of life.

Even if we bend over backwards to be as charitable as possible to these self-organizing worlds, they remain deeply problematic. Suppose they were to succeed on their own terms. What exactly would that show? What exactly would have organized itself? And how would such products of self-organization be significant to the larger story of life's origin? The origin-of-life community has already conceded that catalytic RNAs cannot organize themselves from their elementary constituents. Ditto for DNAs and proteins. At best, self-organizing worlds therefore yield primitive replicators that look nothing like DNA, RNA, or protein. But in that case, why should we think that such replicators have any relevance to the origin of cellular life? Metabolism-first worlds, for instance, never describe the route by which primitive metabolisms achieve the ability to arrange nucleotides into functional sequences (as required by the RNA world).

A spectacular success for Wächtershäuser's iron-sulfur world model would be to demonstrate how a metabolic cycle could organize itself by employing only authentic prebiotic materials under authentic prebiotic conditions (and hence without excessive investigator interference). Although Wächtershäuser and his colleagues have accomplished nothing like this,[62] imagine that they had. What form would such "primitive metabolic life" take? It would be a coating on a mineral—perhaps a glaze or patina. Yet why should we think that such a coating might play a pivotal role in life's origin? We have no compelling reason to draw such a conclusion short of a detailed evolutionary path from such a coating to cellular life. But no such paths are known (not even paths terminating in the much simpler RNA world are known).

In thinking that Wächtershäuser's iron-sulfur world played a pivotal role in bringing about the actual world of cellular life, we might just as well think that waterwheel technology played a pivotal role in bringing about supercomputer technology. The supercomputer owes nothing to the waterwheel. This lack of connection is also evident with Wächtershäuser's model as well as with the others considered in this section. To think that cellular life emerged from any of these "worlds"—in the absence of hard evidence—is wishful thinking. And such evidence is notoriously lacking. Notwithstanding, origin-of-life researchers continue to regard self-organizing worlds as deeply relevant to the origin of life. To understand why, we need to turn to our next topic: molecular Darwinism.

## 8.8 MOLECULAR DARWINISM[63]

Origin-of-life researchers have one remaining strategy for explaining how life originated by purely material forces, namely, to apply the Darwinian mechanism not to organisms but to molecules. "Molecular Darwinism," as we may call it, is thereby supposed to explain the origin of life. What if researchers could identify a simple self-replicating molecule or set of molecules? Suppose such a molecular system is so simple that it could arise from one of the self-organizing worlds described in the last section. Might it not be possible for the Darwinian mechanism to do the rest? Why shouldn't natural selection and random variation act at the molecular level, bootstrapping that initial replicator all the way up to a full-blown cell?

There is good reason to be skeptical about the effectiveness of natural selection before the advent of cellular life. Origin-of-life researchers do not need to explain the origin of simple molecular replicators, but rather the origin of a self-replicating system of macromolecules that can perform the very specific functions universally associated with life as we know it (material transport, metabolism, energy conversion, signal transduction, information processing, sequestration from the environment, etc.). Origin-of-life research is constantly lowering the bar for what may count as first life. Not replication per se, but self-replicating multimolecular systems of the kind found in actual cells are what need to be explained.[64]

For instance, we have long known that a crystal seed placed in a supersaturated solution will cause crystal growth. In this way, the crystal seed may be said to self-replicate. Even so, self-replication here bears little resemblance to the self-replication of functionally integrated multimolecular systems found in cellular life. It is such systems that molecular Darwinism needs to explain. As a consequence, even if one were to concede that a relatively simple self-replicating molecule or set of molecules could arise by purely material processes (that is, without extensive investigator interference), the Darwinian mechanism of natural selection and random variation operating at the molecular level would still need to explain

- how these molecules could become enclosed in a channeled and gated membrane;
- how they could generate an ever-increasing assortment of biomacromolecules;
- how these biomacromolecules in turn could arrange themselves into a hierarchy of functionally integrated systems; and

- how this spiraling of molecular complexity could ultimately produce the DNA-RNA-protein machinery required to perform the specific functions that we associate with actual living cells.

Of course, one can imagine a simple molecular replicator, by natural selection acting over many generations, gradually producing the functionally integrated multimolecular systems that we now associate with cellular life. But imagination is cheap. Where is the evidence to support such a scenario? All experiments that attempt to simulate the production of biologically-relevant building blocks under realistic prebiotic conditions are subject to interfering cross-reactions between desired chemical products and undesired chemical byproducts (see section 8.4). These reactions produce biologically irrelevant tars and melanoids that do not evolve in biologically productive directions. Accordingly, what nature selects on its own, in plausible prebiotic chemical environments (and thus without extensive investigator interference), tends to be chemically inert, biologically irrelevant, and therefore a dead-end to further chemical evolution.

To be sure, origin-of-life researchers are able to circumvent the problem of interfering cross-reactions by carefully designing experiments and extensively constraining or manipulating their outcomes. For instance, researchers can simply remove undesirable chemical reaction products or use purified starting materials. The problem of interfering cross-reactions is therefore averted, but at the cost of a new problem: such actions involve input of information by design and correspond to no known material process.

The work of Julius Rebek is emblematic of the problems facing molecular Darwinism. Rebek, while still at MIT before joining Scripps, synthesized amino adenosine triacid ester (AATE), which consists of two components, pentafluorophenyl ester and amino adenosine.[65] Moreover, he placed these AATE molecules in an organic solvent that preserves rather than degrades them. Both his choices of solvent and molecules represent an artificial constraint on the environment that has no natural analogue.

Even granting this artificial set-up, the evolution of Rebek's molecules hardly describes a plausible route for life's origin. To be sure, Rebek's molecules make copies of themselves and therefore replicate. But, as Gerald Joyce pointed out, they reproduce too accurately.[66] Without a sufficient amount of random variation, the Darwinian mechanism has nothing to select. In the extreme case, where things just keep producing identical copies of themselves, no evolution at all is possible. Still more problematic for Rebek's model is this: not only do Rebek's molecules carry far less information than is required for life, but they suggest no plausible pathway to the vast increase in information required for life.[67] Leslie Orgel sums up the significance of Rebek's work for the origin of life as follows: "What Rebek has done is very clever, but I don't see its relevance to the origin of life."[68]

For molecular Darwinism to succeed as an explanation of life's origin, it must explain why evolution proceeds in a *complexity-increasing* direction. Starting with a simple replicator that could have emerged from a self-organizing world, molecular Darwinism needs to account for its evolution toward ever increasing complexity. Yet ironically, molecular Darwinism suggests that evolution should proceed in a *complexity-decreasing* direction. Consider, for instance, Sol Spiegelman's work on the evolution of polynucleotides in a replicase environment.[69] Leaving aside that the replicase protein is supplied by the investigator (from a viral genome), as are the activated mononucleotides needed to feed polynucleotide synthesis, the problem here, and in experiments like it, is the steady diminution of biologically relevant information over the course of the experiment. As Brian Goodwin notes:

> In a classic experiment, Spiegelman in 1967 showed what happens to a molecular replicating system in a test tube, without any cellular organization around it. The replicating molecules (the nucleic acid templates) require an energy source, building blocks (i.e., nucleotide bases), and an enzyme to help the polymerization process that is involved in self-copying of the templates. Then away it goes, making more copies of the specific nucleotide sequences that define the initial templates. But the interesting result was that these initial templates did not stay the same; they were not accurately copied. They got shorter and shorter until they reached the minimal size compatible with the sequence retaining self-copying properties. And as they got shorter, the copying process went faster. So what happened with natural selection in a test tube: the shorter templates that copied themselves faster became more numerous, while the larger ones were gradually eliminated. This looks like Darwinian evolution in a test tube. But the interesting result was that this evolution went one way: toward greater simplicity. Actual evolution tends to go toward greater complexity, species becoming more elaborate in their structure and behavior, though the process can also go in reverse, toward simplicity. But DNA on its own can go nowhere but toward greater simplicity. In order for the evolution of complexity to occur, DNA has to be within a cellular context; the whole system evolves as a reproducing unit.[70]

There are costs to becoming complex. For something to replicate, it has to reproduce itself in all essential respects. The bigger it is and the more it does, the greater the burden of replication. Simpler replicators therefore have an advantage over more complicated ones—they have less to keep track of in replication. Moreover, as in the Spiegelman case, natural selection is able to exploit this advantage. Darwinism places no inherent premium on complexity. In fact, other things being equal, the Darwinian mechanism prefers simplicity over complexity.[71] To be sure, other things are not always equal. In the history of life, where cells already possess vast amounts of complexity and vast behavioral repertoires, complexity has clearly increased. But there is no evidence that the Darwinian mechanism, when applied to individual molecules or simple molecular systems, accounts for evolution in a complexity-increasing direction. Molecular Darwinism therefore fails to resolve the problem of life's origin.

Theodosius Dobzhansky, a key architect of the neo-Darwinian synthesis, remarked that "prebiological natural selection is a contradiction in terms."[72] Dobzhansky may have overstated things, but not by much. It is possible to apply the Darwinian mechanism of random variation and natural selection at the molecular level to simple self-replicating systems. These systems will evolve provided they don't replicate too exactly and therefore allow some variability of offspring. Yet, in every instance, these simple molecular replicators have proven highly contrived and vastly simpler than actual biological organisms (cellular life). Moreover, they show no sign of evolving into anything that remotely resembles actual biological organisms—not in complexity and not in variety of functions. The quasi-mystical view that the Darwinian mechanism is a ratchet that can take just about any replicator and transform it into something vastly more splendid and complex has neither evidence nor theoretical foundation.

## 8.9 THE MEDIUM AND THE MESSAGE

Origin-of-life researchers readily admit that they don't know how life actually got started. At the same time, many are quite confident that they know the broad contours for how life *could have* gotten started and that eventually they will figure out the details. Accordingly, they see themselves as reverse engineers. For reverse engineers, it's not crucial to follow the exact path by which an item was first produced. Rather, it is enough to find at least one plausible path by which it could have been produced. For origin-of-life researchers, it is therefore enough merely to establish a "proof of principle" or "proof of concept."[73] How life actually originated, though no doubt interesting, is largely irrelevant.

To appreciate that this is how origin-of-life researchers actually do view the problem of life's origin, consider the following two remarks by two well known origin-of-life researchers. The first is by Stuart Kauffman, the second by Leslie Orgel:

> Anyone who tells you that he or she knows how life started on the earth some 3.45 billion years ago is a fool or a knave. Nobody knows. Indeed, we may never recover the actual historical sequence of molecular events that led to the first self-reproducing, evolving molecular systems to flower

forth more than 3 million millennia ago. But if the historical pathway should forever remain hidden, we can still develop bodies of theory and experiment to show how life might realistically have crystallized, rooted, then covered the globe. Yet the caveat: nobody knows.[74]

"Anybody who thinks they know the solution to this problem [of the origin of life][73] is deluded," says Orgel. "But," he adds, "anybody who thinks this is an insoluble problem is also deluded." One possible approach to the problem of life's origins is to ask the question scientifically rather than historically—how *can* life emerge rather than how *did* life emerge. In order to address this, scientists try to determine experimentally what is chemically feasible and what could have occurred on the prebiotic earth.[76]

These quotations oversell the prospects for resolving the problem of life's origin in purely materialist terms. Given the vast array of obstacles facing a purely material origin of life—obstacles outlined in detail in the previous sections, and especially in section 8.4—why should anyone think that the problem of life's origin is soluble in terms of "what is chemically feasible and what could have occurred on the prebiotic earth"? Is there any evidence for this view? Or has it merely become an article of faith among origin-of-life researchers? In fact, the obstacles cited (e.g., racemic mixtures, interfering cross-reactions, the synthesis of polymers) suggest anything but a purely material origin of life. There is no evidence that material processes are able to overcome these obstacles. Moreover, to say that they must have this ability—because, after all, here we are—simply begs the question. The belief that material processes can overcome these obstacles is therefore without basis and indeed an article of faith.

To ascribe a purely material origin to life is to affirm that life is, without remainder, chemistry. Harvard chemist George Whitesides spoke on this very point to and for the 160,000-member American Chemical Society when in 2007 he received the society's highest honor, the annually awarded Priestley Medal. As part of this honor, Whitesides delivered an award address in which he discussed the origin of life: "This problem [of life's origin] is one of the big ones in science. It begins to place life, and us, in the universe. Most chemists believe, as do I, that life emerged spontaneously from mixtures of molecules in the prebiotic earth. How? I have no idea."[77] Even so, Whitesides added that if anyone is going to resolve this problem, it will be the chemists: "I *believe* that understanding the cell is ultimately a question of chemistry and that chemists are, in principle, best qualified to solve it. The cell is a bag—a bag containing smaller bags and helpfully organizing spaghetti—filled with a Jell-O of reacting chemicals and somehow able to replicate itself."[78] In this way Whitesides reduced the problem of life's origin to a problem in chemistry.

Such a reduction of life, however, is fundamentally misconceived. Chemistry provides the medium for life, but the information that this medium carries, if life is to exist at all, is a message that cannot be reduced to it. "Obviously, life is a chemical phenomenon," writes Paul Davies, "but its distinctiveness lies not in the chemistry as such. The secret of life comes instead from its informational properties; a living organism is a complex information-processing system."[79] The medium of chemistry carries the message of life, but the medium cannot generate the message—the information. To suggest otherwise is like saying that ink and paper have the power to organize themselves into pages of meaningful text.[80] They have no such power.

Meaningful texts owe their meaning not to the physics and chemistry of ink on paper but to the infusion of semantic information. Likewise, life owes its origin not to the physics and chemistry of life's basic building blocks but to the infusion of biologically significant functional information. Whitesides is an outstanding chemist. He goes too far, however, in seeing everything as a problem of chemistry. He therefore misses that the origin of information is not a problem of chemistry. Chemistry can be a carrier of information, but it cannot be its source. Whitesides's failure to see this is perhaps not surprising. Chemists typically do not concern themselves with the problem of the origin of information because their work presupposes a smart chemist ready to provide it! But smart chemists infusing information into their chemical experiments are intelligent designers. Such experiments, if they support the origin of life at all, support not a materialistic but an intelligent origin of life.

Biologists now recognize the crucial importance of information to understanding life and, especially, its origin. Caltech president and Nobel Prize-winning biologist David Baltimore, in describing the significance of the Human Genome Project, stated, "Modern biology is a science of information."[81] Origin-of-life researcher and fellow Nobel Prize winner Manfred Eigen has identified the problem of life's origin with the task of uncovering "the origin of information."[82] Biologists John Maynard Smith and Eörs Szathmáry have explicitly placed information at the center of developmental and evolutionary biology: "A central idea in contemporary biology is that of information. Developmental biology can be seen as the study of how information in the genome is translated into adult structure, and evolutionary biology of how the information came to be there in the first place."[83]

Given the importance of information to biology, the obvious question is how does biological information arise? What is its source? Whitesides sees the ultimate source of biological information as residing in chemistry. Nobel laureate and origin-of-life researcher Christian de Duve develops this point more fully. In *Vital Dust*, he lays out seven "ages" in the history of life. Only the first four concern us in this chapter. Note the order: The Age of Chemistry, The Age of Information, The Age of the Protocell, and The Age of the Single Cell.[84] Here is how de Duve describes the transition from the first to the second age:

> History is a continuous process that we divide, in retrospect, into ages—the Stone Age, the Bronze Age, the Iron Age—each characterized by a major innovation added to previous accomplishments. This is true also of the history of life. . . . First, there is the Age of Chemistry. It covers the formation of a number of major constituents of life, up to the first nucleic acids, and is ruled entirely by the universal principles that govern the behavior of atoms and molecules. Then comes the Age of Information, thanks to the development of special information-bearing molecules that inaugurated the new processes of Darwinian evolution and natural selection particular to the living world.[85]

De Duve is here claiming that the ordering of nucleotides marks the advent of The Age of Information and that universal chemical principles govern everything in The Age of Chemistry leading up to it. But how does he know that such chemical principles provide a suitable and sufficient backdrop for the origin of biological information? Chemical principles describe both local law-like interactions among particles and quantum mechanical transitions among particle states. Nothing about such principles, however, guarantees the formation of nucleotide sequences or any other information-bearing molecules. Indeed, the principles of chemistry, though obviously compatible with the existence of biological information, offer no positive theoretical grounds for its origin.

Empirical evidence likewise fails to confirm de Duve's "chemistry first" view of life's origin. In particular, it fails to confirm that chemical processes operating under realistic prebiotic conditions can bring about nucleotide sequences or other information-bearing molecules. De Duve's "thioester world" is, in this respect, as ill-supported as its competitors (see section 8.7). For it to be well supported, Duve would need to provide a reasonably probable, fully articulated chemical pathway from his thioester world to the world of nucleotide sequences. He has accomplished nothing of the sort. If he had, prebiotic chemistry and chemical evolution would be going concerns and this chapter would have had to be rewritten to reflect the vitality of materialistic origin-of-life research. In fact, to dispassionate eyes, materialistic origin-of-life research is moribund, unable to recapture the initial excitement of the Miller-Urey experiments over half a century ago because, even with the building blocks of life in hand, they never spontaneously arrange themselves into information-rich biological structures. In short, the medium of chemistry gives no evidence of generating the message of life.

## 8.10 THE GOD OF THE GAPS

If chemistry alone is unable to account for the transition from The Age of Chemistry to The Age of Information, what, then, could account for it? With no empirical evidence or theoretical basis for the ability of chemistry by itself to generate biological information, one might think that the origin-of-life research community should be open to a design hypothesis that posits an information source capable of accounting for information-rich biological structures. But one would be mistaken. In fact, origin-of-life researchers, and biologists generally, tend to reject this option, charging those who defend an intelligent origin of life with committing a god-of-the-gaps fallacy (i.e., the fallacy of introducing a nonmaterial cause where a material cause, though for the present unknown, will do).

For instance, in his book *The Language of God*, Francis Collins, head of the Human Genome Project, admits that "no serious scientist would currently claim that a naturalistic explanation for the origin of life is at hand." Yet he immediately adds, "that is true today, and it may not be true tomorrow." Collins worries that an information source not reducible to de Duve's universal chemical principles is simply a place-holder for ignorance (i.e., a god of the gaps). As such, it could be maintained only so long as, in Collins's words, "scientific understanding is currently lacking." As further justification for his concerns, Collins cites the history of science: "From solar eclipses in olden times to the movement of the planets in the Middle Ages, to the origins of life today, this 'God of the gaps' approach . . . may be headed for crisis if advances in science subsequently fill those gaps."[86]

When Collins discusses the origin of life elsewhere, however, he seems less concerned about the god of the gaps and more open to a design hypothesis: "Four billion years ago, the conditions on this planet were completely inhospitable to life as we know it; 3.85 billion years ago, life was teeming. That is a very short period—150 million years—for the assembly of macromolecules into a self-replicating form. I think even the most bold and optimistic proposals for the origin of life fall well short of achieving any real probability for that kind of event having occurred." Collins then asks whether the origin of life might be where a nonmaterial intelligence entered natural history. In answer, he says, "I am happy to accept that model." This is a remarkable concession. To be sure, he immediately qualifies it by warning against the god of the gaps. But he then qualifies this qualification by further highlighting the sheer magnitude of the origin-of-life problem: "this particular area of evolution, the earliest step, is still very much in disarray."[87]

Origin-of-life researcher Robert Hazen likewise worries that to posit an information source not reducible to chemistry commits a god-of-the-gaps fallacy. According to him, proponents of intelligent design argue that "life is so incredibly complex and intricate that it must have been engineered by a higher being." This in turn means that "no random natural process could possibly lead from non-life to even the simplest cell, much less humans." The problem for Hazen with this argument is that hypothesizing an information source not reducible to chemistry "ignores the power of emergence to transform natural systems without conscious intervention." But appealing to emergence, as we saw in section 8.7, is itself a stop-gap for ignorance. Hazen tacitly admits as much when, right after invoking emergence as the general solution to life's origin, he concedes, "True, we don't yet know all the details of life's genesis story, but why resort to an unknowable alien intelligence when natural laws appear to be sufficient?"[88]

Hazen's question contains two false assumptions: the sufficiency of natural laws and the unknowability of the designer. The claim that natural laws are sufficient to account for the origin of life is far-fetched. Natural laws work *against* the origin of life. Natural laws describe material processes that *consume* the raw materials of life, turning them into tars, melanoids, and other nonbiological substances that thereafter are completely useless to life. For the raw materials of life to avoid being consumed in this way, material processes must, as this chapter has documented in painstaking detail, overcome a daunting set of obstacles (see section 8.4). Yet material processes give no evidence of being able to overcome these obstacles. Rather, they give evidence of being predisposed to crash into them and never to get past them. Life has clear needs, and natural laws, far from supporting or even being neutral about those needs, directly undercut them.

Hazen is also mistaken that any intelligence able to account for the information in biological systems must be "unknowable." We know intelligences through what they do, that is, by attending to the objects and events they produce. Knowledge of intelligences gained by examining their products may be limited, but it can be accurate as far as it goes. Thus, if intelligence is responsible for the origin of life, the study of biological systems yields knowledge about it. At the very least, we can know that such an intelligence is highly skilled in nano-engineering. This is not to say that the intelligence behind biology is merely a nano-engineer. But it is not less than a nano-engineer.

The worry that some presently unknown materialistic explanation will one day dawn on scientists and thereby invalidate a design explanation of life's origin seems therefore overblown. Science is a bold enterprise. It takes risks and can afford to take risks because it is always in contact with empirical evidence and therefore can correct itself in light of new facts. The design hypothesis for the origin of life may ultimately be shown wrong, but so what? All scientific hypotheses place themselves in empirical harm's way and may be shown wrong. Thus, if a compelling materialistic account for the origin of life is one day found, it will overthrow the design hypothesis by rendering it superfluous (note that this very possibility shows that intelligent design is testable). On the other hand, the design hypothesis may well be right. Yet if it is, the only way to determine its rightness is by admitting the hypothesis as a live scientific option and then subjecting it to the severe scrutiny characteristic of science. As things stand, by rejecting the design hypothesis out of hand, origin-of-life researchers artificially prop up materialistic origin-of-life scenarios, withholding from them the scrutiny they deserve.

If appealing to unknown materialistic explanations is not a good reason for rejecting design, then neither is Francis Collins's appeal to the history of science. Collins compares invoking design to explain life's origin with invoking design to explain eclipses and the motion of planets. But when in times past people invoked the action of an intelligence to explain eclipses or the motion of planets, it was in ignorance of the relevant astronomical facts underlying these phenomena. We find ourselves in a radically different situation with regard to life's origin: by knowing the relevant facts of biochemistry and molecular biology, we are in a position to assess how difficult it is for the chemical building blocks of life to arise and then arrange themselves into the information-rich structures required for cellular life. So long as design hypotheses are based on knowledge rather than ignorance, they are scientifically legitimate.

Ironically, the history of science suggests that design explanations of life's origin may be *less* problematic than materialistic explanations. In Darwin's day, the cell was regarded as extremely simple—it was thought to be essentially a blob of Jell-O enclosed by a membrane. Given this apparent simplicity, the origin of cellular life did not seem to require the long gradual ascent up a hierarchy of complexity as with Oparin's hypothesis. Instead, cells were thought to form instantaneously from pre-organic matter (see the discussion of "spontaneous generation" in the general notes to section 8.2). Instantaneous spontaneous generation strikes us now as ludicrous given what biochemistry and molecular biology reveal about the complexities of the cell. But Darwin and his nineteenth-century disciples, in ignorance of the cell's true complexities, mistook it for simple and, as a consequence, attributed it to purely material factors, thereby eliminating design from the scientific discussion of life's origin. The history of science therefore reveals that scientific ignorance can lead to design explanations being rejected and materialistic explanations being accepted *for the wrong reasons.*

## 8.11 A REASONABLE HYPOTHESIS

What, then, are the right reasons, if any, for accepting a design explanation of life's origin? Reasons for accepting something come in two forms, positive and negative, and it's always good to have both. For instance, in answering a multiple-choice test, it is best to have not just negative reasons for rejecting the incorrect answers but also positive reasons for accepting the correct answers. The same holds in science. In the case at hand, we have compelling negative reasons for rejecting a materialistic explanation of life's origin. These include

- the complete absence of plausible chemical pathways from realistic prebiotic scenarios to the information-rich structures found in cells;
- a catalogue of obstacles that any such pathways must overcome, that they display no ability to overcome, and that derail prebiotic chemistry before it can even get close to a materialistic origin of life (see section 8.4); and
- the failure of self-organizational, Darwinian, and other materialistic principles to provide a coherent theoretical framework for the origin of life.

In short, neither evidence nor theory supports a materialistic origin of life.

Besides negative reasons for rejecting a materialistic origin of life, there are also positive reasons for accepting an intelligent origin of life. These include

- the engineering features of cellular systems;
- the irreducible and specified complexity of cellular systems; and
- a conservation principle from the mathematical theory of information showing that evolutionary processes must always take in at least as much information as they give out.

Let's examine these points briefly.

**Engineering.** Many of the systems inside the cell represent nanotechnology at a scale and sophistication that dwarfs human engineering. Moreover, our ability to understand the structure and function of these systems depends directly on our facility with engineering principles (both in developing the instrumentation to study these systems and in analyzing what they do). Engineers have developed these principles by designing systems of their own, albeit much cruder than what we find inside the cell. Many of these cellular systems are literally machines: electro-mechanical machines, information-processing machines, signal-transduction machines, communication and transportation machines, etc. They are not just analogous to humanly built machines but, as mathematicians would say, *isomorphic* to them, that is, they capture all the essential features of machines.[89]

The genetic-coding and protein-synthesizing machinery in cells is of particular interest for the origin of life. The chemical properties of amino acids, nucleotide bases, sugars, phosphates, etc. that make up this machinery are, in and of themselves, not adequate to build this machinery. In fact, the natural chemical tendencies of these compounds under realistic prebiotic conditions would have inhibited the formation of coded information, which is at the heart of this machinery. For instance, amino acids react with sugars, preventing the formation of DNA and RNA. To be sure, prebiotic chemistry is compatible with certain forms of order (repeating patterns, symmetry, etc.). Yet nowhere in nonbiological nature do we find coded information.

The coded information inside cells is mathematically identical (isomorphic) with the coded information in written language.[90] Since both written language and DNA have that telltale property of encapsulating information in specific sequences of characters, and since intelligence alone is known to produce written language, and since there is no known materialistic route to the coded information inside cells, it is an entirely reasonable inference to identify the cause of the coded information in DNA with an intelligent information source. More generally, we know only one cause capable of producing such high-tech machinery as exists in all cells, namely, intelligence. The engineering features of cells therefore count as a positive reason to accept intelligent design.

## Fallacies of Composition and Division

In emphasizing that cells employ molecular machines and that engineering principles are indispensable to understanding cells, design theorists are often accused of reducing life to mechanism, conceiving of living forms as machines rather than as organisms.[91] But this accusation commits a fallacy of composition, arguing incorrectly that what is true of the parts must be true of the whole.[92] Just because cells have machine-like aspects does not mean that they are machines. Indeed, the intelligent design community regards living forms as much more than machines. But because life's machine-like aspects clearly signal the action of intelligence, they receive special attention in the theory of life's intelligent design.

Conversely, in attempting, under realistic prebiotic conditions, to reproduce various components of cells (e.g., RNA sequences, as in the RNA world, or metabolic cycles, as in metabolism-first worlds), origin-of-life researchers may be tempted to, and sometimes in fact do, commit a fallacy of division, arguing

*continued on next page*

incorrectly that what is true of the whole must be true of the part.[93] Thus, they ascribe life to these components when in fact this ascription is undeserved and properly belongs to life taken as a whole. For instance, complexity theorist Norman Packard founded a start-up company called ProtoLife with the goal of synthesizing life from non-living materials.[94] The complex organic structures that this company actually endeavors to synthesize, however, fall far short of cells and do not deserve to be called "life" or, for that matter, the equally misleading "protolife."

Interestingly, ProtoLife describes its research as belonging to "the field of complex chemical system *design*."[95] Its actual research program is therefore more consistent with a design-theoretic as opposed to materialistic approach to life's origin.

**Irreducible and Specified Complexity.** The protein synthesizing machinery inside the cell is not just irreducibly complex but forms a nested hierarchy of irreducibly complex systems (i.e., it consists of irreducibly complex systems within irreducibly complex systems). For instance, ribosomes, which stitch amino acids into proteins by reading information off an RNA template, are immensely complex, requiring at minimum fifty or so proteins and RNAs. The ribosome contains an irreducible core and is therefore irreducibly complex. Yet it is part of a larger system of enzymes and polynucleotides that read genetic information off a DNA template to produce proteins. This larger system, which includes the ribosome, the sixty enzymes discussed in section 8.4, and more—each indispensable to all known cellular life—is itself irreducibly complex. Indeed, the cell is filled with systems and subsystems that are indispensable to cellular life and that are, in their own right, each irreducibly complex.

Because these systems are indispensable to cellular life as such, facile appeals to supposed evolutionary precursors will not do. Such a precursor would be an earlier system that evolves into a later system.[96] Accordingly, in the early days of the precursor, the later system would not have existed. But if the later system is the protein synthesizing machinery, what earlier system could have evolved into it? The situation here is quite different from the bacterial flagellum, which might have evolved from the simpler type III secretory system that is embedded in it (recall Chapter 6). The type III secretory system can exist as part of a viable cell in the absence of the bacterial flagellum. But where is the precursor to the cell's protein synthesizing machinery that could function on its own in the absence of that machinery? Bacteria can do quite well without flagella. But no instance of cellular life is known that can do "quite well" without protein synthesizing machinery.

The only life we know is protein-based and presupposes an elaborate nested hierarchy of irreducibly complex machinery that synthesizes proteins. If this machinery evolved from a precursor that did something else (i.e., that did something other than build proteins), there's no reason to think that one has a viable cell. Irreducible complexity of systems that are indispensable for life is therefore irreducible complexity on steroids. It is a souped-up form of irreducible complexity that rules out all appeals to precursors based simply on general evolutionary principles.

Short of a concrete proposal for how these systems might have evolved from simpler precursors that do not merely exhibit some function or other but also maintain viability of the cell, neither direct nor indirect Darwinian pathways offer a clue, much less evidence, for their evolution. The fact is that no plausible chemical pathways to these systems from realistic prebiotic scenarios have been discovered—nothing even close. We therefore have no evidence that these systems are within the scope of materialistic processes. On the other hand, we do know, as in the previous point about engineering, that intelligence is able to produce irreducibly complex systems. Accordingly, we have here a good positive reason for thinking that these systems are actually designed.

As for the specified complexity of such irreducibly complex systems, which are indispensable for life, in the absence of any gradual step-by-step chemical pathways, their improbability staggers the imagination. In consequence, one might be tempted to think that such pathways must exist—after all, life did get here. But that would be begging the question. Given what we actually know about these systems (as opposed to completely unsubstantiated speculations about how they might have evolved), our best estimates of probability suggest that these systems exhibit specified complexity and therefore are designed (see Chapter 7).

**Conservation of Information.** Many searches in science are needle-in-the-haystack problems, looking for small targets in large spaces. In such cases, blind search stands no hope of success. Success, instead, requires an assisted search. This assistance takes the form of information: searches require information to be successful. Think of an Easter egg hunt in which saying "warmer" or "colder" indicates that one is getting respectively closer to or farther from the eggs. Clearly, this additional information assists those who are searching for the Easter eggs, especially when the eggs are well hidden and blind search would be unlikely to find them. But how does one secure the information required for a search to be successful? To pose the question this way suggests that successful searches do not magically materialize but need themselves to be discovered by a process of search.

The question then naturally arises whether such a higher-level "search for a search" is ever easier than the original search. Work in the field of *evolutionary informatics* indicates that this is not the case. Evolutionary informatics, a branch of information theory that studies the informational requirements of evolutionary processes, shows that the information required to make a search successful obeys a conservation principle known as the *conservation of information.*[97] To say that information is conserved means that the information required to find a successful search is never less than the information required to make the original search successful. In consequence, the higher-level search for a search is never easier than the original lower-level search.

Conservation of information implies that information, like money or energy, is a commodity that obeys strict accounting principles. Just as corporations require money to power their enterprises and machines require energy to power their motions, so searches require information to power their success. Moreover, just as corporations need to balance their books and machines cannot output more energy than they input, so searches, in successfully locating a target, cannot give out more information than they take in. Conservation of information has far reaching implications for evolutionary theory (for both chemical and biological evolution), pointing up that the success of evolutionary processes in exploring biological configuration space always depends on preexisting information. In particular, evolutionary processes cannot create the information required for their success from scratch.

To get around this conclusion, evolutionary theorists sometimes deny that biological evolution constitutes a targeted search. For instance, Oxford biologist Richard Dawkins illustrates biological evolution with a computer simulation that employs a targeted search (the simulation is explicitly programmed to search for the target phrase METHINKS IT IS LIKE A WEASEL). But after giving this illustration, he immediately adds: "Life isn't like that. Evolution has no long-term goal. There is no long-distant target, no final perfection to serve as a criterion for selection."[98] Dawkins here fails to distinguish two equally valid and relevant ways of understanding targets: (1) targets as humanly constructed patterns that we arbitrarily impose on things to suit our interests and (2) targets as patterns that exist independently of us and therefore regardless of our interests. In other words, targets can be extrinsic (i.e., imposed on things from outside) or intrinsic (i.e., inherent in things as such).

In the field of evolutionary computing (to which Dawkins's METHINKS IT IS LIKE A WEASEL example belongs), targets are given extrinsically by programmers who attempt to solve problems of their choice and preference. But in biology, not only has life come about without our choice or preference, but there are only so many ways that matter can be configured to be alive and, once alive, only so many ways it can be configured to serve different biological functions. Most of the ways open to evolution (chemical or biological) are dead ends. Evolution may therefore be characterized as the search for the alternative "live ends." In other words, demographics—whatever facilitates survival and reproduction—sets the targets. Evolution, despite Dawkins's denials, is therefore a targeted search after all.

Because conservation of information shows that evolutionary processes must always take in at least as much information as they give out, Dawkins's attempt to see evolution as fundamentally a matter of building up complexity from simplicity cannot stand. For Dawkins, proper scientific explanation is "hierarchically reductionistic," by which he means that "a complex entity at any particular level in the hierarchy of organization" must be explained "in terms of entities only one level down the hierarchy."[99] Thus, according to Dawkins, "the one thing that makes evolution such a neat theory is that it explains how organized complexity can arise out of primeval simplicity."[100] This is also why Dawkins regards intelligent design as unacceptable: "To explain the origin of the DNA/protein machine by invoking a supernatural [*sic*] Designer is to explain precisely nothing, for it leaves unexplained the origin of the Designer. You have to say something like 'God was always there', and if you allow yourself that kind of lazy way out, you might as well just say 'DNA was always there', or 'Life was always there', and be done with it."[101]

Conservation of information shows that Dawkins's primeval simplicity is not as nearly simple as he thinks. Indeed, what Dawkins regards as intelligent design's predicament of failing to explain complexity in terms of simplicity now confronts materialistic theories of evolution as well. In *Climbing Mount Improbable*, Dawkins's argues that biological structures that at first blush seem vastly improbable with respect to pure randomness (i.e., blind search) become quite probable once the appropriate evolutionary mechanism (i.e., assisted search) is factored in to revise the probabilities.[102] But this revision of probabilities just means that blind search has given way to an assisted search. And the information that enables this assisted search to be successful now needs itself to be explained. Moreover, by conservation of information, that information is no less than the information that the evolutionary mechanism adds in outperforming pure randomness.[103]

The conservation of information therefore points to an information source behind evolution that imparts at least as much information to the evolutionary process as this process in turn is capable of expressing. In consequence, such an information source

- cannot be reduced to materialistic causes,
- suggests that we live in an informationally open universe, and
- may reasonably be regarded as intelligent.

The conservation of information therefore counts as a positive reason to accept intelligent design.

Is the intelligent origin of life therefore a reasonable hypothesis? This chapter has laid out strong negative reasons for rejecting a materialistic origin of life as well as strong positive reasons for accepting an intelligent origin of life. Nonetheless, for the design hypothesis to seem reasonable to scientists, it needs a tie-in to scientific praxis. Charles Sanders Peirce, one of the founders of American pragmatist philosophy, held that for a difference to be a difference it has to make a difference.[104] Scientists reading this chapter might therefore wonder what practical difference accepting a design hypothesis might make to their understanding and investigation of life's origin. To be sure, chemistry by itself seems completely hopeless for generating information-bearing molecules, much less their hierarchical arrangement inside a cell. But what are we to make of an intelligence that is the source of biological information? If this information source is a supernatural agent that miraculously intervenes to produce biological information, then what hope is there of studying the origin of life scientifically?

To answer this question, note first that an intelligence that brought life into existence need not be supernatural—it could be a teleological organizing principle that is built into nature and thus be perfectly natural.[105] Next, note that however an intelligence may have acted to originate life (whether by channeling ordinary material processes or by miraculously overriding them), design-theoretic research into life's origin can simply investigate the transformations of matter by which life *could* originate. Such an investigation is entirely parallel to what origin-of-life research already does, namely, attempt to show how life could originate and not necessarily how it did (recall section 8.9). The only difference is that whereas materialistic research into life's origin is always supposed to restrict itself to "realistic prebiotic conditions" (and thus to chemistry undirected by intelligence), design-theoretic research into life's origin lifts that restriction.

Truth be told, origin-of-life research, as currently practiced, already lifts that restriction. Consider how philosopher of biology Michael Ruse summarizes research into the RNA world: "At this point, no one has actually confirmed that RNA can start self-replicating, nor has such an event been reproduced experimentally. . . . At the moment, the hand of *human design and intention* hangs heavily over everything, but work is going forward rapidly to create conditions in which molecules can make the right and needed steps without constant outside help."[106] If the hand of human design and intention indeed hangs heavily over *everything* in origin of life research, how can the work to show that "constant outside help" is unnecessary be "going forward rapidly"? The need for outside help in origin-of-life research suggests not a materialistic but a designer-assisted origin of life. So long as origin-of-life research is constantly requiring outside help of a sort that would not have been available under "realistic prebiotic conditions," its progress in showing how life might have originated without outside help is nil.

Origin-of-life research promises with one side of the mouth to stay true to materialistic principles but with the other asks indulgence for constantly needing to resort to the hand of human design and intention. Such equivocations are rampant throughout the field. Michael Ruse's comments about the RNA world are a case in point. Here is another: One of us (WmAD) debated physicist James Trefil at Boston University in 2005 on the topic of intelligent design.[107] Trefil, who is associated with Harold Morowitz's origin-of-life research group at George Mason University (recall from section 8.7 that Morowitz works on metabolism-first worlds), claimed at the end of his presentation that he and his colleagues were but a few years from creating a life form in the lab.

To be sure, Trefil immediately qualified this claim by saying that such a life form would be quite different from life as we know it (i.e., it would not be a full-fledged cell). Yet the thing to note is that regardless of what Trefil and his associates create in the lab, it is *they* who will be doing the creating. That is, scientists will be in the lab using their intelligence to create something that they regard as alive. Whether what they come up with deserves to be called alive is here beside the point. The point is that by using their intelligence to do things that matter left to its own devices would never do, they are in fact doing design-theoretic research. For this reason, design theorist Paul Nelson recommends that origin-of-life researchers line their labs with mirrors to remind themselves of their role as intelligent designers in feeding their experiments with information and guiding them along paths that nature, operating under "realistic prebiotic conditions," would never have taken.

From a materialistic perspective, the origin of life is a difficult problem because it is difficult to find the right and needed steps by which a blind materialistic process could have created life. Find the right "chemical minestrone," as Stuart Kauffman says, and the origin-of-life problem goes away.[108] Materialistic origin-of-life researchers therefore expect that by trying out enough recipes for such a chemical minestrone, they will eventually hit on the right one and unlock the mystery of life's origin. From a design-theoretic perspective, by contrast, the origin of life is a difficult problem because biological designs vastly exceed human designs in technological sophistication, and we are only beginning to grasp the technology. Design theorists therefore expect that by improving their understanding of technology, especially the nano-technology inside the cell, they will gain increasing insight into life's intelligent origin. At the same time, design theorists admit that certain features of the cell may so outstrip human understanding that a full grasp of life's intelligent origin may forever elude humanity. Regardless of who is right, given the evidence cited in this chapter, the burden of proof has now shifted to the materialistic origin-of-life researcher, who can no longer merely assert that cellular life arose by purely materialistic processes but must now demonstrate in detail how this could have happened.

Will the origin-of-life community assume this burden of proof? Not until it is ready to question a prior commitment to materialism. Richard Dawkins is as ardent a materialist as one will find.[109] Yet he writes, "The illusion of purpose is so powerful that biologists themselves use the assumption of good design as a working tool."[110] Scientists are pragmatists and therefore loath to give up effective tools that help them in their research. Design, ironically, is one such tool. Yet, conditioned by a materialistic outlook that denies design, many scientists find it difficult to acknowledge design even when they are using it, imagining that it is merely an illusion. We would therefore turn Dawkins's statement around: *Biologists use the assumption of design with such success as a working tool precisely because design in biology is not an illusion but real.*

*The general notes on the CD included with this book expand on the material in this chapter and throw light on the following discussion questions.*

## 8.12 DISCUSSION QUESTIONS

1. What is Polya's dictum? How is it relevant to the origin of life? What role do cells play in life's origin? How much of origin-of-life research adequately addresses the full complexity of cells? Why is the cell appropriately described as an "automated city"? Can you think of a better metaphor to describe the cell? Why is so much of origin-of-life research concerned with "toy problems," that is, with simplistic problems that avoid the true complexities of life? Why is the origin of life such a difficult problem?

2. Describe the work of Francesco Redi and Louis Pasteur on spontaneous generation (see the general notes). Is there a problem with ascribing the origin of life to spontaneous generation? Has spontaneous generation as an explanation for the origin of life disappeared from scientific theorizing? What remnants, if any, remain?

3. Who was Alexander Oparin? What is the Oparin hypothesis? Why is it also called the Oparin-Haldane hypothesis? Oparin divided the origin of life into stages. Describe these stages as well as the underlying assumptions for moving from one stage to the next. How plausible are the transitions between stages? What obstacles must be overcome to move from one stage to the next?

4. What is a primitive atmosphere simulation experiment? The Miller-Urey experiment is the best known of these. Describe it. What support does this experiment lend to the Oparin hypothesis? What was the reaction in the scientific world when the Miller-Urey experiment was first conducted? How is it viewed now by the scientific community? Are primitive undersea simulation experiments superior to primitive atmosphere simulation experiments? What problem do undersea simulation experiments claim to resolve? Do they in fact resolve it? What new problem do they introduce?

5. Describe the "RNA world" or "ribozyme engineering" proposal for life's origin. Why do many researchers believe that catalytic RNAs preceded the DNA/protein machinery of life? What difficulties confront the RNA world, making it unlikely that it can successfully account for life's origin? How easy is it to form ribonucleotides under realistic prebiotic conditions? Once they are available, how easy is it to arrange them in functional sequences? Once that is done, is there a clear chemical path from the RNA world to the DNA/protein machinery? What are the prospects for such a path?

6. Describe the self-organizational approach to life's origin. List three of the key researchers in this area and briefly summarize their work. Are self-organizing worlds necessarily metabolism-first worlds, or are there self-organizing worlds that do not employ metabolism? Illustrate and explain. What advantage do metabolism-first worlds have over genetics-first worlds (e.g., the RNA world, for which the origin of genetic information is primary)? Have metabolism-first worlds resolved the problem of life's origin? If not, why not? Is there a clear chemical evolutionary path from self-organizing worlds to information-bearing molecules such as RNA? If not, describe what the prospects are for finding such a path.

7. What is molecular Darwinism? Even if there is reason to think that the Darwinian mechanism of natural selection and random variation operates successfully at the level of full organisms, why should one think that this same mechanism can operate successfully at the molecular level to bring about the first life? Evaluate the strengths and weaknesses of molecular Darwinism in accounting for the origin of life.

8. Describe Sol Spiegelman's classic experiment with replicating molecules. In this experiment, did chemical evolution proceed in a complexity-increasing direction as it is presumed to do in real life? Should we expect the Darwinian mechanism of natural selection and random variation to propel evolution (whether chemical or biological) in a complexity-increasing or complexity-decreasing direction? If in a complexity-decreasing direction, what does this say about the ability of Darwinian theory to explain biological complexity?

9. Have all materialistic attempts to explain the origin of life thus far failed? What appear to be the most promising avenues for explaining the origin of life materialistically? Is it inevitable that materialistic approaches to the origin of life must in the end succeed? Or does intelligent design deserve a place in the scientific discussion of life's origin? Explain. Are there positive reasons for accepting the intelligent origin of life? If so, what are they? Which of these reasons do you find most compelling? Explain.

10. Is saying that the origin of life had an intelligent cause the same as saying that it had a supernatural cause? Is it possible for the origin of life to have had an intelligent cause without that cause being supernatural? What is a miracle? Does the origin of life require a miracle? Could the origin of life have been intelligently caused without being miraculous?

# CHAPTER NINE Epilogue: The "Inherit the Wind" Stereotype

## 9.1 HOLLYWOOD'S VERSION OF THE SCOPES "MONKEY TRIAL"

Critics of intelligent design frequently portray anyone who is willing to consider alternatives to Darwinian evolutionary theory as a religiously motivated opponent of science. Using a stereotype epitomized in the Hollywood film *Inherit the Wind*, a fictional portrayal of the 1925 Scopes "Monkey Trial," many in the academy and media treat any challenge to Darwinism as a challenge to truth and rationality. Yet, it is the failure to examine evolution critically that poses the real challenge to truth and rationality.

Jerome Lawrence and Robert Lee wrote the play *Inherit the Wind* in the 1950s. It was produced on Broadway in 1955, but is best known as a 1960 black-and-white movie starring Spencer Tracy and Frederic March. A more recent version of the movie, made in 1999, starred Jack Lemmon and George C. Scott, but the 1960 version has been far more influential (you should be able to find it in the "classics" section of your local video rental store).

Like the Scopes trial, the play is set in 1925. In it, Bert Cates (the John Scopes character) is hounded by religious fundamentalists in the town of Hillsboro (which corresponds to Dayton, Tennessee) for teaching Darwin's theory of evolution. Henry Drummond (who corresponds to famous defense attorney Clarence Darrow) bravely offers to defend Cates. To counteract Drummond and suppress the spread of evolutionary ideas, narrow-minded Matthew Harrison Brady (the fictional double of popular political figure William Jennings Bryan) offers to prosecute Cates.

While *Inherit the Wind* makes for fine storytelling, it makes for atrocious history. Cates, because of his stand for evolution, is portrayed as in danger of being imprisoned and losing all that's dear to him (especially the woman he loves). The real Scopes was never in such danger. Cates is portrayed as a valiant defender of truth and reason against fundamentalism and bigotry. The real Scopes had less lofty motives for defending evolution.

## 9.2 THE ACTUAL SCOPES TRIAL

John Scopes agreed to take part in the trial because local boosters put him up to it. They thought an "evolution-monkey trial" would put the town of Dayton Tennessee on the map—which it did! Scopes was a physical education teacher who taught biology part-time. Local prosecutors agreed to go along with the charade. Things got out of hand when Clarence Darrow offered to defend Scopes and William Jennings Bryan volunteered to speak for the prosecution.

Bryan, unlike Brady in the play, was not a reactionary or a fundamentalist. Bryan was a three-time Democratic presidential candidate and a progressive politician who sought to protect farmers and blue-collar workers from exploitation by big business. Unlike Brady, Bryan did not interpret the book of Genesis literally (he did not, for instance, hold that the earth was only a few thousand years old or that the world was created in six 24-hour days). Bryan personally rejected Darwinism because he saw the evidence for it as unconvincing. That by itself, however, was not enough to prompt his public opposition to Darwinism. Bryan organized public opposition to Darwinism because he saw it as justifying unrestrained capitalism as well as the militarism that led to World War I.

Darrow was not only a famous trial lawyer but also a nationally recognized lecturer who promoted agnosticism and argued publicly against religion on the basis of evolution. In 1924, the year before the Scopes trial, Darrow was the lead defense attorney in the notorious Leopold-Loeb murder case. At that trial, Darrow introduced Darwinian arguments into criminology. Nathan Leopold and Richard Loeb, two rich and well-educated college students at the University of Chicago, admitted to killing a 14-year old boy, Bobby Franks, for the thrill of committing the "perfect murder." They thought they were too smart to get caught. Darrow argued against the death penalty by suggesting that this "distressing and weird homicide" happened "because somewhere in the infinite [evolutionary] processes that go to the making up of the boy or the man something slipped."

Leopold and Loeb—Darrow kept calling them "children"—were really helpless pawns of their evolutionary past: "Nature is strong and she is pitiless. She works in her own mysterious way, and we are her victims. We have not much to do with it ourselves. Nature takes this job in hand, and we play our parts." Speaking of Richard Loeb, he asked,

> Is Dickey Loeb to blame because out of the infinite forces that conspired to form him, the infinite forces that were at work producing him ages before he was born, that because out of these infinite combinations he was born without it [i.e., normal emotional reactions]? If he is, then there should be a new definition for justice. Is he to blame for what he did not have and never had? Is he to blame that his machine is imperfect? Who is to blame?[1]

Machines act blindly and automatically—they are not responsible moral agents who can legitimately be blamed for their actions. For Darrow, evolution justified a biological determinism that turned humans into puppets of their evolutionary past.

In 1997, Edward Larson, a University of Georgia professor in the history of law, published a critical reassessment of the Scopes Trial. In *Summer for the Gods: The Scopes Trial and America's Continuing Debate Over Science and Religion,* Larson thoroughly deconstructed *Inherit the Wind*, showing just how badly the "Scopes Trial" stereotype misrepresents the actual Scopes Trial. The book shows that the debate over biological origins was—and is—far more complex than most Americans have been told. For his book, Larson was awarded the 1998 Pulitzer Prize in history.

In the actual Scopes trial, evolution, and the evidence for it, were never subjected to cross-examination. Scopes's lawyers presented extensive written statements from seven scientists stating that evolution is the correct explanation for the diversity of life on earth.[2] Statements of Drs. Metcalf, Nelson, Lipman, Judd, and Newman were read in court; statements of Drs. Cole and Curtis were also submitted in writing. The prosecution sought permission to cross-examine the five pro-Darwinian science experts whose statements were read in open court, but Darrow and the other Scopes lawyers objected, and the court refused to allow it.[3]

Certainly, the most dramatic aspect of the Scopes trial was Darrow's questioning of Bryan about the Bible. But this raises an obvious question: given that Darrow got to question Bryan about the Bible, why didn't Bryan get to question Darrow about evolution? In fact, Bryan agreed to be questioned by Darrow on his personal interpretation of the Bible *only if* Darrow agreed to be questioned on the evidence for evolution. Moreover, the court agreed that Bryan could question Darrow after Darrow questioned Bryan.[4]

But, at the conclusion of his famous examination of Bryan, Darrow unexpectedly asked the judge to instruct the jury to find his client guilty. By doing this, Darrow in effect changed the plea to guilty. By not entering an actual plea of guilty, Darrow took advantage of a technical procedural rule that preserved the right to appeal the judge's rulings—a straight guilty plea would have foreclosed the right of appeal. The upshot is that in demanding a directed verdict, Darrow closed the evidence and made it impossible for Bryan to call Darrow to the stand and question him on evolution.[5]

Darrow could easily have demanded a directed verdict against his own client, a verdict of guilty, before his examination of Bryan (in which case Bryan's defense of the Bible would never have made it into the trial transcript) or after Bryan examined him (in which case Darrow's defense of evolution would also have made it into the trial transcript). But, by demanding that his client be found guilty *right after* he examined Bryan and despite agreeing that Bryan could examine him next about evolution, Darrow made clear that his intention all along was to question Bryan and then escape questioning himself. Bryan immediately recognized this and remarked: "It is hardly fair for them [Darrow and the defense team] to bring into the limelight my views on religion and stand behind a dark lantern that throws light on other people, but conceals themselves."[6]

Because of Darrow's shrewd legal maneuvering, scientists in the Scopes trial were able to present their case for evolution without any challenge. Evolutionary theory has a long history of evading critical scrutiny and escaping proper cross-examination. The late Fred Hoyle, founder of the Institute for Astronomy at Cambridge, did not mince words when he remarked that scientific challenges to evolution have "never had a fair hearing" because "the developing system of popular education [from Darwin's day to the present] provided an ideal opportunity . . . for awkward arguments not to be discussed and for discrepant facts to be suppressed."[7]

## 9.3 THE IMPORTANCE OF KEEPING SCIENCE HONEST

Evolution, as taught in 1925, was eminently deserving of critical scrutiny and cross-examination. Back then, Darrow denounced opponents of Darwinian evolution as "bigots and ignoramuses" trying to "control the education of the United States."[8] Stereotypes like this, however, cut both ways. According to Harvard law professor Alan Dershowitz, those in 1925 who advocated for evolution included "racists, militarists, and nationalists" who used evolution "to push some pretty horrible programs" including the forced "sterilization of 'unfit' and 'inferior'" people; "the anti-immigration movement" that wanted to bar immigration of people of "inferior racial stock"; and "Jim Crow" laws that evolutionists "rationalized on grounds of the racial inferiority of blacks."[9]

Dershowitz goes on to note that the very textbook Scopes taught to high school students, Hunter's *Civic Biology*, divided humanity into five races and ranked them in terms of superiority, concluding with "the highest type of all, the Caucasians, represented by the civilized white inhabitants of Europe and America." *Civic Biology* also advocated that crime and immorality were inherited and ran in families, and that "these families have become parasitic on society. . . . If such people were lower animals, we would probably kill them off. . . . [W]e do have the remedy of separating the sexes in asylums or other places and in various ways preventing intermarriage and the possibilities of perpetuating such a low and degenerate race."[10] The lab book for Hunter's text, at Problem 160, asks students to use inheritance charts "[t]o determine some means of bettering, physically and mentally, the human race." What's more, a "note to teachers" says that "[t]he child is at the receptive age and is emotionally open to the serious lessons here involved."[11]

Of course, the scientific community today denounces all such biological racism. Nonetheless, some prominent contemporary Darwinists, like Daniel Dennett, are so assured of the truth of Darwinism that they now embrace a cultural elitism in which anyone who dissents from Darwinian orthodoxy is regarded as culturally substandard and in need of being segregated from the culturally acceptable people who embrace Darwinism. Dennett, for instance, advocates that children be forced to learn that they are "the product of evolution by natural selection" because "our future well-being depends on the education of our descendants."[12] Moreover, he advises that parents who stand in the way of such enforced education be quarantined: "Those whose visions dictate that they cannot peacefully coexist with the rest of us we will have to quarantine."[13]

But consider, the very textbook from which Scopes taught—the very book that today's scientific community insists Scopes had the absolute right to teach public school students—includes material that today's scientific community passionately rejects. Imagine a hypothetical 1925 state law—a law that permitted the teaching of eugenics as the scientific community of the time demanded, but also required that challenges to that theory be taught. Would not everyone today applaud the foresight of any state that had enacted such a law? Hear, hear! Let the science of the day have its say, but then teach its weaknesses, criticisms, and alternatives.

This hypothetical example of a state law that mandates the critical examination of the "science" of eugenics demonstrates that it is appropriate for those who oversee our school science curricula not to be slavishly bound to whatever the scientific community espouses at the moment. The population at large—who are free from the institutional incentives and professional biases that often impair the scientific community—are entirely in their rights to question a scientific theory regardless of how confidently the scientific community espouses it.

Indeed, if the history of science is any indicator, every scientific theory has faults and is eventually abandoned in favor of a better, more accurate theory. Why should we expect any different from evolutionary theory? A scientist's confidence in a theory is no guarantee that it is true. As Nobel prize winning biologist Peter Medawar put it, "I cannot give any scientist of any age better advice than this: the intensity of the conviction that a hypothesis is true has no bearing on whether it is true or not. The importance of the strength of our conviction is only to provide a proportionally strong incentive to find out if the hypothesis will stand up to critical examination."[14]

To discredit those who opposed the teaching of Hunter's *Civic Biology* in 1925, mainstream scientists and media figures insisted that religious convictions were the only motive for opposing that textbook. Dershowitz notes that even the U.S. Supreme Court agreed with the evolution-inspired eugenics program, upholding a mandatory sterilization law on the view that "three generations of imbeciles are enough."[15] But fortunately for civil rights in America, intelligent, inquiring people of good will (not "religious fanatics" or "opponents of science") questioned the reprehensible teachings of Hunter's *Civic Biology*. And fortunately, too, enough people were willing to consider both the official position of science and—to borrow a phrase from another and more recent Hollywood film—the "minority report."

So too, in our own day, intelligent, inquiring people of good will (not opponents of science and not Daniel Dennett's cultural inferiors) can question the teaching of Darwinian and other materialistic forms of evolution. It is entirely legitimate, both intellectually and scientifically, to question whether evolution operates exclusively by means of unintelligent, purely mechanistic processes like natural selection. Far from repeating the one-sidedness of the Scopes Monkey Trial, the approach embodied in this book remedies it. It does so by providing the kind of cross-examination that the Scopes science experts and lawyers should have had to face, but conveniently avoided.

## 9.4 THE SANTORUM AMENDMENT

The United States Senate has itself recognized the need for such cross-examination. In 2001, ninety-one United States Senators voted to make rational, science-based questioning of Darwinism the law of the United States by voting in favor of an amendment (offered by Senator Rick Santorum) to an education bill. The language, known as the "Santorum Amendment," mandated that "good science education should prepare students to distinguish the data or testable theories of science from philosophical or religious claims that are made in the name of science" and that "where biological evolution is taught, the curriculum should help students to understand why this subject generates so much continuing controversy."

A joint Senate-House Conference Committee eventually moved the language to its published Conference Report. Conference reports authoritatively interpret the bills they accompany. In words virtually identical to the Santorum Amendment, the senators and representatives on that committee declared in the Conference Report that "a quality science education should prepare students to distinguish the data or testable theories of science from philosophical or religious claims that are made in the name of science" and "where topics are taught that may generate controversy (such as biological evolution), the curriculum should help students understand the full range of scientific views that exist."[16]

Ninety-one Unites States Senators, along with House and Senate members of the Conference Committee, are on record favoring that the full range of scientific views about biological evolution be taught. Intelligent design is one of those views. Proponents of intelligent design do not argue that evolution and the evidence for it must be suppressed because of some alleged conflict with the Bible. Instead, they argue that evolution—specifically, the theory that evolution occurs exclusively by means of undirected mechanistic processes such as natural selection and random variation—may legitimately be questioned because the scientific evidence used to support it is weak. Noted neo-Darwinist Theodosius Dobzhansky famously asserted, "Nothing in biology makes sense except in the light of evolution."[17] In fact, nothing in biology makes sense except in the light of *evidence*.

Where does the evidence of biology lead, to unguided evolution or to intelligent design? This book, in presenting the evidence and arguments for intelligent design, provides students with the information they need to answer this question. Providing this information is not just pedagogically sound but also legally permissible. In 1987, the U.S. Supreme Court ruled in *Edwards v. Aguillard* that "teaching a variety of scientific theories about the origins of humankind to school children might be validly done with the clear secular intent of enhancing the effectiveness of science instruction."[18] By telling about the evidence and arguments for intelligent design, science educators help fulfill that Supreme Court mandate. But they do more. They also foster the true spirit of scientific inquiry.

## 9.5 POSTSCRIPT: *KITZMILLER V. DOVER*

On December 20, 2005, as this book was undergoing its final revisions, Judge John E. Jones III rendered his verdict in the first court case over intelligent design. In *Kitzmiller v. Dover*, also billed as "Scopes II," Judge Jones not only struck down a Dover school board policy advocating intelligent design but also identified intelligent design as nonscientific and fundamentally religious. Accordingly, he concluded that the teaching of intelligent design in public school science curricula violates the Establishment Clause and therefore is unconstitutional.

It is hard to imagine how a court decision could have been formulated more negatively against intelligent design.[19] Nevertheless, it would be a mistake to view this case as a decisive blow against intelligent design. True, Judge Jones's decision will put a damper on some school boards that would otherwise have been interested in advancing intelligent design. But this is not a Supreme Court decision. Nor will it be appealed to the Supreme Court since the Dover school board that instituted the controversial policy supporting intelligent design was voted out and replaced in November 2005 with a new board that campaigned on the promise of rescinding the policy. This they have now done.

Without an explicit Supreme Court decision against intelligent design, grass roots pressure to open up discussion about intelligent design in the public schools and to critically analyze its evolutionary alternatives will only increase. Because of *Kitzmiller v. Dover*, school boards and state legislators may tread more cautiously. But no court case can make the controversy over evolution disappear. In our culture, that controversy possesses an unquenchable vitality.

It is therefore naive to think that this case threatens to derail intelligent design. Even if the courts censor intelligent design at the grade and high school levels (and with the Internet censorship means nothing to the enterprising student), they remain powerless to censor intelligent design at the college and university levels. Intelligent design is quickly gaining momentum among college and graduate students. Three years ago, there was one IDEA Center at the University of California at San Diego (IDEA = Intelligent Design and Evolution Awareness—see www.ideacenter.org). Now there are thirty such centers at American colleges and universities, including the University of California at Berkeley and Cornell University. These centers are vigorously pro-intelligent design.

The significance of a court case like *Kitzmiller v. Dover* depends not on a judge's decision but on the cultural forces that form the backdrop against which the decision is made. Take the Scopes Trial. In many people's minds, it represents a decisive victory for

Darwinian evolution. Yet, in the actual trial, the decision went against evolution. Indeed, John Scopes was convicted of violating a Tennessee statute that forbade the teaching of evolution.

Judge Jones's decision may make life in the short term less pleasant for ID proponents. But the work of intelligent design will continue. In fact, it is likely to continue more effectively than if the judge had ruled in favor of intelligent design, which might have encouraged complacency, suggesting that intelligent design had already won the day when in fact intelligent design still needs to continue developing its scientific and intellectual program. In the end, that program, and not any court rulings or public policies or Hollywood films, will decide the merit of intelligent design.[20]

# Endnotes

## PREFACE

[1]James Glanz, "Biologists Face a New Theory of Life's Origin," *New York Times* (Sunday, 8 April 2001), 1. Kenneth Chang, "In Explaining Life's Complexity, Darwinists and Doubters Clash," *New York Times* (Monday, 22 August 2005), 1.

[2]The cover for the August 15, 2005, issue of *Time* was titled "Evolution Wars" and explicitly refers to "intelligent design" in the subtitle. The cover for the November 28, 2005, issue of *Newsweek* was titled "The Real Darwin" and likewise refers explicitly to "intelligent design" in the subtitle.

[3]See, for instance, the episode titled "From Whence We Came" (airdate January 16, 2005) of ABC's *Boston Legal.*

[4]On May 9, 2005, William Dembski debated the topic of intelligent design with Michael Ruse on ABC's *Nightline* (the host that night was George Stephanopoulos). On September 14, 2005, William Dembski debated this topic again with Edward Larson on Jon Stewart's *The Daily Show.*

[5]See respectively http://www.biologicinstitute.org and http://www.EvoInfo.org (last accessed September 26, 2007).

[6] There are at the time of this writing thirty-five such clubs worldwide. See http://www.ideacenter.org/clubs/locations.php (last accessed August 10, 2007).

[7]Richard Dawkins, *The Blind Watchmaker: Why the Evidence of Evolution Reveals a Universe Without Design* (New York: Norton, 1987), 1.

[8]Francis Crick, *What Mad Pursuit* (New York: Basic Books, 1988), 138.

[9]See, for instance, David J. Depew and Bruce H. Weber, *Darwinism Evolving: Systems Dynamics and the Genealogy of Natural Selection* (Cambridge, Mass.: MIT Press, 1995); Stuart Kauffman, *Investigations* (New York: Oxford University Press, 2000; and Franklin Harold, *The Way of the Cell: Molecules, Organisms and the Order of Life.* (New York: Oxford University Press, 2001): Lynn Helena Caporale, *Darwin in the Genome: Molecular Strategies in Biological Evolution* (New York: McGraw-Hill, 2003); Gerd B. Müller and Stuart A. Newman, eds., *Origination of Organismal Form: Beyond the Gene in Developmental and Evolutionary Biology* (Cambridge, Mass.: MIT Press, 2003).

[10]For a popular exposition of Wells's views on design in embryological development, see his article "Making Sense of Biology: The Evidence for Development by Design" in *Signs of Intelligence,* eds. William A.

Dembski and James M. Kushiner (Grand Rapids, Mich.: Brazos, 2001), 118–127. For his use of design as a research tool for investigating molecular structures inside the cell, see his article "Do Centrioles Generate a Polar Ejection Force?" *Rivista di Biologia/Biology Forum* 98 (2005): 37–62.

[11]Thaxton coauthored what to date remains the best-selling academic monograph on the origin of life: Charles Thaxton, Walter Bradley, and Roger Olsen, *The Mystery of Life's Origin: Reassessing Current Theories* (New York: Philosophical Library, 1984). Many now regard this book as the first book published by the intelligent design movement. See Angus Menuge, "Who's Afraid of ID?" in William A. Dembski and Michael Ruse, eds., *Debating Design: From Darwin to DNA* (Cambridge: Cambridge University Press, 2004), 36–37.

[12]Claude A. Villee, Eldra Pearl Solomon, and P. William Davis, *Biology*, 2nd ed. (Philadelphia: W. B. Saunders, 1989). Davis and Solomon also published a textbook on anatomy and physiology: *Understanding Human Anatomy and Physiology* (New York: McGraw-Hill, 1978).

[13]Dean H. Kenyon and Gary Steinman, *Biochemical Predestination* (New York: McGraw-Hill, 1969). Dean H. Kenyon, "Prefigured Ordering and Protoselection in the Origin of Life," in *The Origin of Life and Evolutionary Biochemistry* (Festschrift commemorating the fiftieth anniversary of the publication of *Proiskhozhdenie Zhizni* and the eightieth birthday of Alexander I. Oparin), eds. K. Dose, S. W. Fox, G. A. Deborin, and T. E. Pavlovskaya (New York: Plenum Press, 1974), 207–220. Dean H. Kenyon, "A Comparison of Proteinoid and Aldocyanoin Microsystems as Models for the Primordial Protocell," in *Molecular Evolution and Protobiology* (Festschrift commemorating the twenty-fifth anniversary of the pioneering thermal heteropolycondensation of amino acids and as a dedication to Sidney W. Fox on the occasion of his seventieth birthday), eds. K. Matsuno, K. Dose, K. Harada, and D. L. Rohlfing (New York: Plenum Press, 1984), 163–188.

[14]To see that fundamental debate over biological origins is indeed ongoing, consider the "Dissent from Darwin" list at www.dissentfromdarwin.org (last accessed June 14, 2007), to which over 700 scientists have signed their names. To be a signatory, a scientist must accept the following statement: "We are skeptical of claims for the ability of random mutation and natural selection to account for the complexity of life. Careful examination of the evidence for Darwinian theory should be encouraged."

[15]For entry into this literature, see William A. Dembski and Michael Ruse, eds., *Debating Design: From Darwin to DNA* (Cambridge: Cambridge University Press, 2004), especially parts I and IV.

## CHAPTER ONE Human Origins

[1]For more about Sidis, see Jim Morton, "Peridromophilia Unbound," available through http://www.sidis.net/WJSJourLinks.htm, and Grady M. Towers, "The Outsiders," http://www.prometheussociety.org/articles/Outsiders.html. Both websites last accessed June 4, 2004.

[2]Charles Darwin, *The Descent of Man and Selection in Relation to Sex*, 2nd ed. (London: John Murray, 1882), 126. Shortly after this passage, Darwin adds: "If it could be proved that certain high mental powers, such as the formation of general concepts, self-consciousness, etc., were absolutely peculiar to man, which seems extremely doubtful, it is not improbable that these qualities are merely the incidental results of other highly-advanced intellectual faculties; and these again are mainly the result of the continued use of a perfect language."

[3]See Harold J. Morowitz, *The Emergence of Everything: How the World Became Complex* (Oxford: Oxford University Press, 2002), chs. 28–32.

[4]For the distinction between a difference in kind versus a difference in degree, see Mortimer Adler, *The Difference of Man and the Difference It Makes* (New York: Fordham University Press, 1993).

[5]Francisco J. Ayala, "Darwin's Revolution," in *Creative Evolution?!*, eds. J. H. Campbell and J. W. Schopf (Boston: Jones and Bartlett, 1994), 4. The subsection from which this quote is taken is titled "Darwin's Discovery: Design without Designer."

[6]David Hull, *Darwin and His Critics: The Reception of Darwin's Theory of Evolution by the Scientific Community* (Cambridge, Mass.: Harvard University Press, 1973), 26.

[7]Francis Crick and Leslie E. Orgel, "Directed Panspermia," *Icarus* 19 (1973): 341–346.

[8]Eliot Marshall, "Medline Searches Turn Up Cases of Suspected Plagiarism," Science 279 (1998): 473–474. Lila Guterman, "Sense of Injustice Can Lead Scientists to Act Unethically, Study Finds," Chronicle of Higher Education (April 7, 2006): available online at http://chronicle.com/daily/2006/04/2006040704n.htm (last accessed August 30, 2006).

[9]Lila Guterman, "Sense of Injustice Can Lead Scientists to Act Unethically, Study Finds," Chronicle of Higher Education (April 7, 2006): available online at http://ethics.tamucc.edu/article.pl?sid=06/04/10/1512249 (last accessed June 8, 2006).

[10]C. Pellicciari, D. Formenti, C. A. Redi, and M. G. Manfredi Romanini, "DNA Content Variability in Primates," *Journal of Human Evolution* 11 (1982): 131–141.

[11]See respectively http://genomebiology.com/researchnews/default.asp?arx_id=gb-spotlight-20031215-01 and http://www.nature.ca/genome/03/a/03a_11a_e.cfm (both websites last accessed August 18, 2004).

[12]Charles G. Sibley and Jon E. Ahlquist, "DNA Hybridization Evidence of Hominid Phylogeny: Results from an Expanded Data Set," *Journal of Molecular Evolution* 26 (1987): 99–121.

[13]Jonathan Marks, "98% Alike? (What Our Similarity to Apes Tells Us About Our Understanding of Genetics)," *The Chronicle of Higher Education* (May 12, 2000): B7. See also Jonathan Marks, *What It Means to Be 98% Chimpanzee: Apes, People, and Their Genes* (Berkeley, Calif.: University of California Press, 2002).

[14]Taken and abridged from Geoffrey Simmons, *What Darwin Didn't Know* (Eugene, Oregon: Harvest House, 2004), 274-278.

[15]Available online at http://www.nature.com/nsu/040322/040322-9.html (published March 25, 2004; last accessed June 17, 2004). For the research article cited in this report, see H. H. Stedman et al., "Myosin Gene Mutation Correlates with Anatomical Changes in the Human Lineage," *Nature* 428 (2004): 415–418.

[16]Isaac Asimov, "In the Game of Energy and Thermodynamics You Can't Even Break Even," *Smithsonian* (August 1970): 10.

[17]Isaac Asimov, *Science Past—Science Future*, (New York, NY: Doubleday, 1975), 291.

[18]In all such discussions relating brain size to cognitive capacities, it is important to consider brain size not merely in absolute terms (e.g., weight or volume of brain) but also in relation to body size. Elephants, for instance, have bigger brains than humans.

[19]As stated August 18, 2004 at http://www.alexfoundation.org. This is the website of the Alex Foundation, Irene Pepperberg's research group dedicated to "psittacine intelligence and communication research." Alex's supposed ability to read was in fact never confirmed; to say that he could read was a case of linguistic anthropomorphism (see note 26). Alex died September 10, 2007 at the age of 31.

[20]Roger Lewin, "Is Your Brain Really Necessary?" *Science*, 210 (12 December 1980): 1232.

[21]Stanley L. Jaki, Brain, *Mind and Computers* (South Bend, Ind.: Gateway Editions, 1969), 115–116.

[22]Reported by Ray Kurzweil in Jay W. Richards, ed., *Are We Spiritual Machines: Ray Kurzweil vs. the Critics of Strong A.I.* (Seattle: Discovery Institute, 2002), 193.

[23]David Chalmers, *The Conscious Mind: In Search of a Fundamental Theory* (Oxford: Oxford University Press, 1996). Jeffrey Schwartz and Sharon Begley, *The Mind and the Brain: Neuroplasticity and the Power of Mental Force* (New York: HarperCollins, 2002).

[24]Barbara J. King, *Roots of Human Behavior,* 24-part audio course (Chantilly, Va.: The Teaching Company, 2001).

[25]According to her colleague Hans Christian von Baeyer, "Barbara King has even suggested that the human ability to exchange information through speech and gesture is not unique. It evolved, she believes, along with other traits we inherited from primates and should be seen as part of a continuum that extends from an amoeba's ability to extract information from its environment, through the dance of the honeybee and the song of a bird, to our modern methods of communication." Quoted from von Baeyer, *Information: The New Language of Science* (Cambridge, Mass.: Harvard University Press, 2004), 9.

[26]As an anonymous reviewer has remarked, "This already gives the ape too much credit. The linguistic complexity involved in the expression 'that bubbly yellow liquid that tastes good' is significantly greater than the symbolic communication achieved by the ape. The expression includes ostension, quality ascriptions (bubbly, yellow), value ascriptions (good), etc. These are only intelligible as parts of a larger set of established linguistic practices. That is, the ape lacks the supporting linguistic structures that the expression presupposes. To interpret the symbolic communication of the ape by an English language expression commits the very mistake being criticized—linguistic anthropomorphism." Email sent to William Dembski, June 23, 2004.

[27]Noam Chomsky, "Form and Meaning in Natural Languages," in *Language and Mind*, enlarged edition (New York: Harcourt, Brace, Jovanovich, 1972), 100.

[28]Marks, *What It Means to Be 98% Chimpanzee*, 182.

[29]Charles Darwin, Letter to W. Graham, 1881. In F. Darwin, ed., *The Life and Letters of Charles Darwin* (New York: D. Appleton & Co., 1905), 1:285. Available online at http://pages.britishlibrary.net/charles.darwin/texts/letters/letters1_08.html (last accessed 4 August 2004). For a full-scale philosophical treatment of Darwin's worry and the skepticism it forces on human knowing, see Victor Reppert, *C. S. Lewis's Dangerous Idea: In Defense of the Argument from Reason* (Downers Grove, Ill.: InterVarsity, 2003)

[30]Joan B. Silk, Sarah F. Brosnan, Jennifer Vonk, Joseph Henrich, Daniel J. Povinelli, Amanda S. Richardson, Susan P. Lambeth, Jenny Mascaro and Steven J. Schapiro, "Chimpanzees Are Indifferent to the Welfare of Unrelated Group Members," *Nature* 437, (27 October 2005): 1357–1359.

[31]Michael Ruse, "Evolutionary Ethics: A Defense," in *Biology, Ethics, and the Origins of Life*, Holmes Rolston III, ed., pp. 89–112 (Boston: Jones & Bartlett Publishers, 1995), 93.

[32]Michael Ruse and E. O. Wilson, "The Evolution of Ethics," in *Religion and the Natural Sciences: The Range of Engagement*, ed. J. E. Hutchingson (Orlando, Fl.: Harcourt and Brace, 1991), 310.

[33]Ruse, "Evolutionary Ethics: A Defense," 101.

[34]In a widely reported anecdote, the geneticist J. B. S. Haldane, when asked whether he would risk death to save a drowning brother, famously replied "No, but I would to save two brothers or eight cousins." This calculus of "inclusive fitness" is said to underlie kin selection: an individual shares half of one's brother's genes and an eighth of one's cousin's genes.

[35]Jeffrey Schloss, "Evolutionary Accounts of Altruism and the Problem of Goodness by Design," in *Mere Creation*, ed. W. A. Dembski (Downers Grove, Ill.: InterVarsity, 1998), 251.

[36]Edward O. Wilson, *On Human Nature* (Cambridge, Mass.: Harvard University Press, 1978), 155–156.

[37]Ibid., 165.

[38]Stephen Fraser, "Newly Released Letters Tell of Jesus Calling Mother Teresa 'My Little Wife'", *Scotland on Sunday* (December 8, 2002), available online at http://news.scotsman.com/international.cfm?id=1367572002 (last accessed June 8, 2006).

[39]James Rachels, *Created from Animals: The Moral Implications of Darwinism* (New York: Oxford University Press, 1990); Robert Wright, *The Moral Animal: Evolutionary Psychology in Everyday Life* (New York: Vintage Books, 1994); Leonard D. Katz, ed., *Evolutionary Origins of Morality* (New York: Norton, 1998). Compare Benjamin Wiker, *Moral Darwinism: How We Became Hedonists* (Downers Grove, Ill.: InterVarsity, 2002).

[40]For a primer on natural law, see J. Budziszewski, *What We Can't Not Know* (Dallas: Spence, 2004).

[41]Huxley's letter to Dyster, January 30, 1859. Available online at http://aleph0.clarku.edu/huxley/letters/59.html (last accessed June 18, 2004).

## CHAPTER TWO Genetics and Macroevolution

[1]In 1995 the National Association of Biology Teachers (NABT) issued the following statement on evolution: "The diversity of life on earth is the outcome of evolution: an *unsupervised, impersonal, unpredictable and natural process* of temporal descent with genetic modification that is affected by natural selection, chance, historical contingencies and changing environments." (Emphasis added.) See "NABT Unveils New Statement on Teaching Evolution," *American Biology Teacher* 58 (January 1996): 61–62. Two years later the NABT deleted the words "unsupervised" and "impersonal" to placate religious believers. But the substance of the definition was unchanged. As Eugenie Scott reports on the NABT board meeting where those words were deleted: "Evolution is still described as a 'natural process' (the only phenomena science can study), and a later bullet [point] states that natural selection 'has no specific direction or goal, including survival of a species.'" Quoted from Eugenie C. Scott, "NABT Statement on Evolution Evolves," special report of the National Center for Science Education, at http://www.ncseweb.org/resources/articles/8954_nabt_statement_on_evolution_ev_5_21_1998.asp. Thus, for the NABT, evolution remains to this day a process that operates without plan or purpose.

[2]Charles Darwin, *On the Origin of Species*, facsimile 1st ed. (1859; reprinted Cambridge, Mass.: Harvard University Press, 1964), 84.

[3]Ibid.

[4]Richard Dawkins, *The Blind Watchmaker* (New York: Norton, 1986).

[5]Steven Pinker, *How the Mind Works* (New York: Norton, 1997).

[6]Natural selection is the majority position within the biological community as to how life evolved: "Biologists now tend to believe profoundly that natural selection is the invisible hand that crafts well-

wrought forms. It may be an overstatement to claim that biologists view selection as the sole source of order in biology, but not by much. If current biology has a central canon, you have now heard it." Stuart Kauffman, *At Home in the Universe: The Search for the Laws of Self-Organization and Complexity* (New York: Oxford University Press, 1995), 150. Kauffman himself dissents from this majority view.

[7]See Stuart Kauffman, *Investigations* (New York: Oxford University Press, 2000); Brian Goodwin, *How the Leopard Changed Its Spots: The Evolution of Complexity* (New York: Scribner's, 1994); and Robert B. Laughlin, *A Different Universe* (New York: Basic Books, 2005), 168–169.

[8]Darwin, *Origin of Species,* 82.

[9]Jean Baptiste de Lamarck, *Zoological Philosophy,* trans. H. Elliot (1809; translated and reprinted London: Macmillan, 1914), 122.

[10]E. J. Ambrose, *The Nature and Origin of the Biological World* (New York: Wiley Halsted, 1982), 26.

[11]Simon Conway Morris, *Life's Solution: Inevitable Humans in a Lonely Universe* (Cambridge: Cambridge University Press, 2003), 19–20.

[12]Note that because the genetic code is degenerate (i.e., multiple codons can map onto the same amino acid—see the expository notes to this chapter), the amino acids coded for by the mutated genes may be unchanged—different triplets can map to the same amino acid.

[13]Theodosius Dobzhansky, *Genetics and the Origin of Species* (New York: Columbia University Press, 1951), 59.

[14]For a technical discussion of the difficulties here, see Michael J. Behe and David W. Snoke, "Simulating Evolution by Gene Duplication of Protein Features that Require Multiple Amino Acid Residues," *Protein Science* 13 (2004): 1–14.

[15]Darwin, *Origin of Species,* 6th edition, ch. 7, available online at http://www.literature.org/authors/darwin-charles/the-origin-of-species-6th-edition/chapter-07.html (last accessed October 10, 2003).

[16]As we shall see in Chapter 6, this same problem (where the pieces of an adaptational package confer no advantage until all the pieces are in place) recurs at the molecular level. There the adaptational packages take the form of irreducibly complex molecular machines.

[17]Ambrose, *Nature and Origin of the Biological World,* 140–41.

[18]Norbert Wiener, *Cybernetics: or Control and Communication in the Animal and the Machine,* 2nd ed. (Cambridge, Mass.; MIT Press, 1961), 132.

[19]Ambrose, *Nature and Origin of the Biological World,* 120.

[20]E. B. Ford, *Ecological Genetics* (London: Chapman and Hill, 1971).

[21]Ambrose, *Nature and Origin of the Biological World,* 123.

[22]Ibid., 143.

[23]See, for instance, John R. True and Sean B. Carroll, "Gene Co-option in Physiological and Morphological Evolution," *Annual Review of Cell and Developmental Biology* 18 (November 2002): 53–80.

[24]Quoted from Elizabeth Pennisi, "Evo-Devo Enthusiasts Get Down to Details," *Science* 298 (1 November 2002): 953. The details that evo-devo enthusiasts are getting down to are microevolutionary changes. This insightful article makes clear that macroevolution has eluded evo-devo.

[25]Ibid.

[26]William Bateson, *Materials for the Study of Variation Treated with Especial Regard to Discontinuity in the Origin of Species* (London: MacMillan, 1894).

[27]The point of logic at issue here is a principle of scientific reasoning described by John Stuart Mill, a contemporary of Darwin's. In his *System of Logic*, Mill put forward his well-known method of difference. According to the method of difference, to explain a difference in effects, one must identify a difference in causes. Alternatively, common causes cannot explain differences in effects. See John Stuart Mill, *System of Logic*, 8th ed. (London: Longmans, 1872).

# CHAPTER THREE The Fossil Record

[1]Theodosius Dobzhansky, "On Methods of Evolutionary Biology and Anthropology," *American Scientist* 46 (December 1957): 388.

[2]Ernst Mayr, *What Makes Biology Unique? Considerations on the Autonomy of a Scientific Discipline* (Cambridge: Cambridge University Press, 2004), 24–25. Emphasis in the original.

[3]Charles Darwin, *On the Origin of Species*, facsimile 1st ed. (1859; reprinted Cambridge, Mass.: Harvard University Press, 1964), 189.

[4]Ibid., 281–282.

[5]Ibid., 280.

[6]Ibid.

[7]Ibid.

[8]Quoted in R. A. Raff and E. C. Raff, eds., *Development as an Evolutionary Process* (New York: A. R Liss, 1987), 84.

[9]Stephen Jay Gould, *Wonderful Life* (New York: W. W. Norton, 1989), 64.

[10]D. Erwin, J. W. Valentine, and J. J. Sepkoski, Jr. "A Comparative Study of Diversification Events: The Early Paleozoic Versus the Mesozoic," *Evolution* 41 (1987): 1177–1186.

[11]David Raup, "Conflicts between Darwin and Paleontology," *Field Museum of Natural History Bulletin* 30(1) (1979): 25.

[12]Stephen Jay Gould, "Evolution's Erratic Pace," *Natural History* 86(5) (May 1977): 12–16.

[13]Ibid.

[14]Stephen M. Stanley, *Macroevolution: Pattern and Process* (San Francisco: W. H. Freeman, 1979), 82.

[15]Harold C. Bold, *Morphology of Plants* (New York: Harper and Row, 1967), 515.

[16]Darwin, *Origin of Species*, 280.

[17]The numbers in the last two paragraphs are taken from Michael Denton, *Evolution: A Theory in Crisis* (Bethesda, Md.: Adler & Adler, 1985), 189–90.

[18]Ibid., 187.

[19]Ibid., 191–92.

[20]Stephen Jay Gould, *The Structure of Evolutionary Theory* (Cambridge, Mass.: Harvard University Press, 2002).

[21]Ibid.

[22]Ibid., 886. Note that the reference to Dawkins (1986) is Richard Dawkins, *The Blind Watchmaker* (New York: Norton, 1986).

[23]Kim Sterelny, *Dawkins vs. Gould: Survival of the Fittest* (Cambridge, UK: Icon Books, 2001).

[24]Anne Dambricourt Malassé, Andre Debenath, and J. Pelegrin, "On New Models for the Neanderthal Debate," *Current Anthropology* 33(1) (1992): 49–54.

[25]Lynn Margulis and Dorion Sagan, *Acquiring Genomes: A Theory of the Origins of Species* (New York: Basic Books, 2002).

[26]Douglas J. Futuyma, *Evolutionary Biology*, 3rd ed. (Sunderland, Mass.: Sinauer, 1998), 146.

[27]James A. Hopson, "The Mammal-Like Reptiles: A Study of Transitional Fossils," *American Biology Teacher* 49(1), (1987): 16–26.

[28]Ibid., 19.

[29]Douglas Futuyma, *Science on Trial: The Case for Evolution* (New York: Pantheon, 1982), 85.

[30]For intricate devices such as the mammalian ear, we might speculate that the information required to build them may have existed unexpressed in the genome of the earliest therapsids before they diversified. Then this information could have been passed on, as part of the genetic blueprint, to the several therapsid variants that arose later. But there's a problem here. For natural selection to produce this information, the process of crafting the mammalian ear could not have been hidden in unexpressed genetic material. After all, natural selection only acts on expressed traits or structures. Such concealed genetic information would therefore indicate an underlying blueprint. And such blueprints are more readily reconciled with intelligent design than with Darwinian evolution.

[31]Michael J. Benton, *Vertebrate Palaeontology*, 2nd ed. (London: Chapman & Hall, 1997). J. G. M. Thewissen and E. M. Williams, "The Early Radiations of Cetacea (Mammalia): Evolutionary Pattern and Developmental Correlations," *Annual Review of Ecology and Systematics* 33 (2002): 73–90.

[32]J. G. M. Thewissen, S. T. Hussain, and M. Arif, "Fossil Evidence for the Origin of Aquatic Locomotion in Archaeocete Whales," *Science* 263 (1994): 210–212. Annalisa Berta, "What Is a Whale?" *Science* 263 (1994): 180–181.

[33]Philip D. Gingerich, S. Mahmood Raza, Muhammad Arif, Mohammad Anwar, and Xiaoyuan Zhou, "New Whale from the Eocene of Pakistan and the Origin of Cetacean Swimming," *Nature* 368 (1994): 844–847.

[34]Stephen Jay Gould, "Hooking Leviathan by Its Past," *Natural History* 103 (May 1994): 8–14.

[35]Kevin Padian, "The Tale of the Whale," National Center for Science Education Resources, http://www.ncseweb.org/resources/rncse_content/vol17/2010_the_tale_of_the _whale_12_30_1899.asp (last accessed January 3, 2007).]

[36]Leigh Van Valen, "Deltatheridia, a New Order of Mammals," *Bulletin of the American Museum of Natural History* 132 (1966): 1–126. Leigh Van Valen, "Monophyly or Diphyly in the Origin of Whales," *Evolution* 22 (1968): 37–41. Dennis Normile, "New Views of the Origins of Mammals," *Science* 281 (1998): 774–775. Richard Monastersky, "The Whale's Tale: Research on Whale Evolution," *Science News* (November 6, 1999), http://www.findarticles.com/p/articles/mi_m1200/is_19_156/ai_57828404 (last accessed January 3, 2007).

[37]John E. Heyning, "Whale Origins—Conquering the Seas," *Science* 283 (1999): 943, 1642–1643. Zhexi Luo, "In search of the whales' sisters," Nature 404 (2000): 235–239. John Gatesy and Maureen A. O'Leary, "Decipering whale origins with molecules and fossils," *Trends in Ecology and Evolution* 16 (2001): 562–570. Kenneth D. Rose, "The Ancestry of Whales," *Science* 293 (2001): 2216–2217.

[38]William Feller, *An Introduction to Probability Theory and Its Applications*, 3rd ed., vol. 1 (New York: Wiley, 1968), 33.

[39]S. Iyengar and J. Greenhouse, "Selection Models and the File Drawer Problem (with Discussion)," *Statistical Science* 3 (1988): 109–135.

[40]Tim Berra, *Evolution and the Myth of Creationism* (Stanford, Calif.: Stanford University Press, 1990), 117–119.

[41]From a presentation by Gareth Nelson in 1969 to the American Museum of Natural History, quoted in David M. Williams and Malte C. Ebach, "The Reform of Palaeontology and the Rise of Biogeography—25 Years after 'Ontogeny, Phylogeny, Palaeontology and the Biogenetic Law' (Nelson, 1978)," *Journal of Biogeography* 31 (2004): 709.

[42]Henry Gee, *In Search of Deep Time* (New York: Free Press, 1999), 23, 32, 116–117.

## CHAPTER FOUR The Origin of Species

[1]Ernst Mayr, *The Growth of Biological Thought* (Cambridge, Mass.: Harvard University Press, 1982), 403.

[2]Charles Darwin, *On the Origin of Species*, facsimile 1st ed. (1859; reprinted Cambridge, Mass.: Harvard University Press, 1964), 1.

[3]Ernst Mayr, *Animal Species and Evolution* (Cambridge, Mass.: Harvard University Press, 1963), 12.

[4]Keith Stewart Thomson, "Natural Selection and Evolution's Smoking Gun," *American Scientist* 85 (1997): 516–518.

[5]Jerry A. Coyne and H. Allen Orr, *Speciation* (Sunderland, Mass.: Sinauer Associates, 2004), 25–26.

[6]Ibid., 27, 30, and 39 respectively.

[7]Charles E. Linn, Jr., Hattie R. Dambroski, Jeffrey L. Feder, Stewart H. Berlocher, Satoshi Nojima, and Wendell L. Roelofs, "Postzygotic Isolating Factor in Sympatric Speciation in *Rhagoletis* Flies: Reduced Response of Hybrids to Parental Host-Fruit Odors," *Proceedings of the National Academy of Sciences USA* 101 (2004): 17753–17758. Available online at http://www.pnas.org/cgi/reprint/101/51/17753 (last accessed January 7, 2007).

[8]See Catherine A. Callaghan, "Instances of Observed Speciation," *The American Biology Teacher* 49 (1987): 34–36; Joseph Boxhorn, "Observed Instances of Speciation," *The Talk.Origins Archive*, September 1, 1995, available online at http://www.talkorigins.org/faqs/faq-speciation.html (last accessed January 9, 2007); Chris Stassen, James Meritt, Annelise Lilje, and L. Drew Davis, "Some More Observed Speciation Events," *The Talk.Origins Archive*, 1997, available online at http://www.talkorigins.org/faqs/speciation.html (last accessed January 9, 2007).

[9]See Justin Ramsey and Douglas W. Schemske, "Neopolyploidy in Flowering Plants," *Annual Review of Ecology and Systematics* 33 (2002): 589–639; D. M. Rosenthal, L. H. Rieseberg, and L. A. Donovan, "Re-creating Ancient Hybrid Species' Complex Phenotypes from Early-Generation Synthetic Hybrids: Three Examples Using Wild Sunflowers," *The American Naturalist* 166(1) (2005): 26–41.

[10]Douglas J. Futuyma, *Evolution* (Sunderland, Mass.: Sinauer Associates, 2005), 398.

[11]J. M. Thoday and J. B. Gibson, "Isolation by Disruptive Selection," *Nature* 193 (1962): 1164–1166. J. M. Thoday and J. B. Gibson, "The Probability of Isolation by Disruptive Selection," *The American Naturalist* 104 (1970): 219–230. Coyne and Orr, *Speciation*, 138.

[12]Theodosius Dobzhansky and Olga Pavlovsky, "Spontaneous Origin of an Incipient Species in the Drosophila Paulistorum Complex," *Proceedings of the National Academy of Sciences* 55 (1966): 727–733.

[13]Coyne and Orr, *Speciation*, 138.

[14]James R. Weinberg, Victoria R. Starczak, and Daniele Jörg, "Evidence for Rapid Speciation Following a Founder Event in the Laboratory," *Evolution* 46 (1992): 1214–1220.

[15]Francisco Rodriquez-Trelles, James R. Weinberg, and Francisco J. Ayala, "Presumptive Rapid Speciation After a Founder Event in a Laboratory Population of Nereis: Allozyme Electrophoretic Evidence Does Not Support the Hypothesis," *Evolution* 50 (1996): 457–461.

[16]E. Paterniani, "Selection for Reproductive Isolation Between Two Populations of Maize, *Zea mays* L.," *Evolution* 23 (1969): 534–547.

[17]William R. Rice and George W. Salt, "Speciation via Disruptive Selection on Habitat Preference: Experimental Evidence," *The American Naturalist* 131 (1988): 911–917. See also Coyne and Orr, *Speciation*, 138–141.

[18]Darwin, *Origin of Species*, 111.

[19]Alan Linton, "Scant Search for the Maker," *The Times Higher Education Supplement* (April 20, 2001), Book Section, 29, available online with registration at http://www.thes.co.uk/search/story.aspx?story_id=72809 (last accessed January 9, 2007).

[20]Lynn Margulis and Dorion Sagan, *Acquiring Genomes: A Theory of the Origins of Species* (New York: Basic Books, 2002), 32.

[21]Futuyma, *Evolution*, 401.

[22]Theodosius Dobzhansky, *Genetics and the Origin of Species,* (1937; reprinted New York: Columbia University Press, 1982), 12.

[23]Kenneth R. Miller, "Statement to the Kansas State Board of Education," 2005, available online at http://www.ksde.org/outcomes/sciencereviewmiller.pdf (last accessed January 16, 2007).

[24]Charles E. Linn, Jr., Hattie R. Dambroski, Jeffrey L. Feder, Stewart H. Berlocher, Satoshi Nojima, and Wendell L. Roelofs, "Postzygotic Isolating Factor in Sympatric Speciation in Rhagoletis Flies: Reduced Response of Hybrids to Parental Host-Fruit Odors," *Proceedings of the National Academy of Sciences USA* 101 (2004): 17753–17758. Available online at http://www.pnas.org/cgi/reprint/101/51/17753 (last accessed January 7, 2007).

[25]Gary Hurd, "To the Committee," review of proposed changes to Kansas State Science Standards, 2005, available online at http://www.ksde.org/outcomes/sciencereviewhurd.pdf (last accessed January 16, 2007).

[26]Richard Goldschmidt, *The Material Basis of Evolution* (New Haven: Yale University Press, 1940), 8.

[27]Ibid., 396.

[28]Scott F. Gilbert, John M. Opitz, and Rudolf A. Raff, "Resynthesizing Evolutionary and Developmental Biology," *Developmental Biology* 173 (1996): 357–372.

[29]Sean B. Carroll, "The Big Picture," *Nature* 409 (2001): 669.

[30]Christiane Nüsslein-Volhard and Eric Wieschaus, "Mutations Affecting Segment Number and Polarity in *Drosophila*," *Nature* 287 (1980): 795–801.

[31]"First Genetic Evidence Uncovered of How Major Changes in Body Shapes Occurred During Early Animal Evolution," UCSD News Release, February 6, 2002. Available online at http://ucsdnews.ucsd.edu/newsrel/science/mchox.htm (last accessed January 9, 2007).

[32]Matthew Ronshaugen, Nadine McGinnis, and William McGinnis, "Hox Protein Mutation and Macroevolution of the Insect Body Plan," *Nature* 415 (2002): 914–917.

[33]For further analysis, see Jonathan Wells, "Mutant Shrimp?—A Correction," February 11, 2002. Available online at http://www.discovery.org/scripts/viewDB/index.php?command=view&id=1118 (last accessed January 9, 2007).

[34]Arhat Abzhanov, Meredith Protas, B. Rosemary Grant, Peter R. Grant, and Clifford J. Tabin, "Bmp4 and Morphological Variation of Beaks in Darwin's Finches," *Science* 305 (2004): 1462–1465.

[35]H. Lisle Gibbs and Peter R. Grant, "Oscillating Selection on Darwin's Finches," *Nature* 327 (1987): 511–513. Jonathan Weiner, *The Beak of the Finch* (New York: Vintage Books, 1994), 104–105, 176.

[36]Marc W. Kirschner and John C. Gerhart, *The Plausibility of Life: Resolving Darwin's Dilemma* (New Haven, Conn.: Yale University Press, 2005), preface.

[37]Ibid., 237–238. See also Jonathan Wells, "What's New? A Review of The Plausibility of Life," *Books & Culture* (September/October 2006), 45–46.

[38]Hugo de Vries, *Species and Varieties: Their Origin by Mutation* (1904; reprinted New York: Garland, 1988), 825–826.

[39]Douglas Robertson, though not a proponent of intelligent design, defines intelligence as that which creates information. See Douglas Robertson, "Algorithmic Information Theory, Free Will and the Turing Test," *Complexity* 4(3) (1999): 25–34.

## CHAPTER FIVE Similar Features

[1]D. D. Davis, *The Giant Panda: A Morphological Study of Evolutionary Mechanisms* (Chicago: Field Museum of Natural History, 1964).

[2]Charles Darwin, *On the Origin of Species*, 6th ed. (London: John Murray, 1872), 383, 420.

[3]Ibid., 403.

[4]Ernst Mayr, *The Growth of Biological Thought* (Cambridge, Mass.: Harvard University Press, 1932), 232, 465.

[5]J. H. Woodger, "On Biological Transformations," in W. E. Le Gros Clark and P. B. Medawar (eds.), *Essays on Growth and Form Presented to D'Arcy Wentworth Thompson* (Oxford: Clarendon Press, 1945), 109.

[6]Alan Boyden, "Homology and Analogy," *American Midland Naturalist* 37 (1947): 648–669. Emphasis in original.

[7]Robert R. Sokal and Peter H. A. Sneath, *Principles of Numerical Taxonomy* (San Francisco: Freeman, 1963), 21.

[8]Michael T. Ghiselin, "An Application of the Theory of Definitions to Systematic Principles," *Systematic Zoology* 15 (1966): 127–130.

[9]David L. Hull, "Certainty and Circularity in Evolutionary Taxonomy," *Evolution* 21 (1967): 174–189.

[10]See Donald H. Colless, "The Phylogenetic Fallacy," *Systematic Zoology* 16 (1967): 289–295.

[11]Ronald H. Brady, "On the Independence of Systematics," *Cladistics* 1 (1985): 113–126.

[12]David B. Wake, "Homoplasy, Homology and the Problem of 'Sameness' in Biology," *Homology*, Novartis Symposium 222 (Chichester, UK: Wiley, 1999), 45, 27.

[13]Emile Zuckerkandl and Linus Pauling, "Molecular Disease, Evolution, and Genetic Heterogeneity," 189–225, in M. Kasha and B. Pullman (eds.), *Horizons in Biochemistry* (New York: Academic Press, 1962), 200–201.

[14]Bruce Runnegar, "A Molecular Clock Date for the Origin of the Animal Phyla," *Lethaia* 15 (1982): 199–205.

[15]Russell F. Doolittle, Da-Fei Feng, Simon Tsang, Glen Cho, and Elizabeth Little, "Determining Divergence Times of the Major Kingdoms of Living Organisms with a Protein Clock," *Science* 271 (1996): 470–477.

[16]Gregory A. Wray, Jeffrey S. Levinton, and Leo H. Shapiro, "Molecular Evidence for Deep Precambrian Divergences among Metazoan Phyla," *Science* 274 (1996): 568–573.

[17]See Richard A. Fortey, Derek E. G. Briggs, and Matthew A. Wills, "The Cambrian Evolutionary 'Explosion' Recalibrated," *BioEssays* 19 (1997): 429–434 and Francisco José Ayala, Andrey Rzhetsky, and Francisco J. Ayala, "Origin of the Metazoan Phyla: Molecular Clocks Confirm Paleontological Estimates," *Proceedings of the National Academy of Sciences USA* 95 (1998): 606–611.

[18]Kenneth M. Halanych, "Considerations for Reconstructing Metazoan History: Signal, Resolution, and Hypothesis Testing," *American Zoologist* 38 (1998): 929–941. See also Simon Conway Morris, "Evolution: Bringing Molecules into the Fold," *Cell* 100 (2000): 1–11.

[19]James W. Valentine, David Jablonski, And Douglas H. Erwin, "Fossils, Molecules and Embryos: New Perspectives on the Cambrian Explosion," *Development* 126 (1999): 851–859.

[20]Anna Marie A. Aguinaldo, James M. Turbeville, Lawrence S. Linford, Maria C. Rivera, James R. Garey, Rudolf A. Raff, and James A. Lake, "Evidence for a Clade of Nematodes, Arthropods and Other Molting Animals," *Nature* 387 (1997): 489–493.

[21]Michael Lynch, "The Age and Relationships of the Major Animal Phyla," *Evolution* 53 (1999): 319–325.

[22]André Adoutte, Guillaume Balavoine, Nicolas Lartillot, Olivier Lespinet, Benjamin Purd'homme, and Renaud de Rosa, "The New Animal Phylogeny: Reliability and Implications," *Proceedings of the National Academy of Sciences USA* 97 (2000): 4453–4456. Available online at http://www.pnas.org/cgi/reprint/97/9/4453 (last accessed January 10, 2007).

[23]Jaime E. Blair, Kazuho Ikeo, Takashi Gojobori, and S. Blair Hedges, "The Evolutionary Position of Nematodes," *Biomed Central Evolutionary Biology* 2 (2002): 7. Available online at http://www.biomedcentral.com/content/pdf/1471-2148-2-7.pdf (last accessed January 10, 2007).

[24]Yuri I. Wolf, Igor B. Rogozin, and Eugene V. Koonin, "Coelomata and Not Ecdysozoa: Evidence From Genome-Wide Phylogenetic Analysis," *Genome Research* 14 (2004): 29–36. Available online at http://www.genome.org/cgi/reprint/14/1/29.pdf (last accessed January 10, 2007).

[25]Hervé Philippe, Nicolas Lartillot, and Henner Brinkmann, "Multigene Analysis of Bilaterian Animals Corroborate the Monophyly of Ecdysozoa, Lophotrochozoa, and Protostomia," *Molecular Biology and Evolution* 22 (2005): 1246–1253.

[26]Martin Jones and Mark Blaxter, "Animal Roots and Shoots," *Nature* 434 (2005): 1076–1077.

[27]Antonis Rokas, Dirk Krüger, and Sean B. Carroll, "Animal Evolution and the Molecular Signature of Radiations Compressed in Time," *Science* 310 (2005): 1933–1938.

[28]Darwin, *Origin of Species*, 484.

[29]James A. Lake, Ravi Jain and Maria C. Rivera, "Mix and Match in the Tree of Life," *Science* 283 (1999): 2027–2028.

[30]Hervé Philippe and Patrick Forterre, "The Rooting of the Universal Tree of Life Is Not Reliable," *Journal of Molecular Evolution* 49 (1999): 509–523.

[31]Carl Woese, "The Universal Ancestor," *Proceedings of the National Academy of Sciences USA* 95 (1998): 6854–6859.

[32]W. Ford Doolittle, "Phylogenetic Classification and the Universal Tree," *Science* 284 (1999): 2124–2128.

[33]See W. Ford Doolittle, "Lateral Genomics," *Trends in Biochemical Sciences* 24 (1999): M5–M8 and W. Ford Doolittle, "Uprooting the Tree of Life," *Scientific American* 282 (February, 2000): 90–95.

[34]Patrick Forterre and Hervé Philippe, "Where Is the Root of the Universal Tree of Life," *BioEssays* 21 (1999): 871–879. Patrick Forterre and Hervé Philippe, "The Last Universal Common Ancestor (LUCA), Simple or Complex?" *Biological Bulletin* 196 (1999): 373–377.

[35]Carl Woese, "On the Evolution of Cells," *Proceedings of the National Academy of Sciences USA* 99 (2002): 8742–8747.

[36]Carl R. Woese, "A New Biology for a New Century," *Microbiology and Molecular Biology Reviews* 68 (2004): 173–186.

[37]W. Ford Doolittle, "If the Tree of Life Fell, Would We Recognize the Sound?" 119–133 in Jan Sapp, ed., *Microbial Phylogeny and Evolution: Concepts and Controversies* (New York: Oxford University Press, 2005), 131.

[38]E. Bapteste, E. Susko, J. Leigh, D. MacLeod, R. L. Charlebois, and W. F. Doolittle, "Do Orthologous Gene Phylogenies Really Support Tree-Thinking?" *BioMed Central Evolutionary Biology* 5 (2005): 33, available online at http://www.biomedcentral.com /content/pdf/1471-2148-5-33.pdf (last accessed January 10, 2007).

[39]For further efforts to root the tree of life in complex cells with nuclei, see S. L. Baldauf, "The Deep Roots of Eukaryotes," *Science* 300 (2003): 1703–1706. Or consider the model proposed in 2004 by Maria Rivera and James Lake. On the assumption that cells with nuclei originated when cells without nuclei fused together, Rivera and Lake inferred that "at the deepest levels . . . the tree of life is actually a ring of life." See Maria C. Rivera and James A. Lake, "The Ring of Life Provides Evidence for a Genome Fusion Origin of Eukaryotes," *Nature* 431 (2004): 152–155. In an accompanying commentary, biologists

William Martin and T. Martin Embley note that this "ring of life" is "at odds with the view of . . . simple Darwinian divergence." See William Martin and T. Martin Embley, "Early Evolution Comes Full Circle," *Nature* 431 (2004): 134–137.

[40]See M. Nishikimi, T. Kawai, and K. Yagi, "Guinea Pigs Possess a Highly Mutated Gene for L-Gulono-Gamma-Lactone Oxidase, the Key Enzyme for L-Ascorbic Acid Biosynthesis Missing in this Species," Journal of Biological Chemistry 267 (1992): 21967–21972 and M. Nishikimi, R. Fukuyama, S. Minoshima, N. Shimizu, and K. Yagi, "Cloning and Chromosomal Mapping of the Human Nonfunctional Gene for L-Gulono-Gamma-Lactone Oxidase, the Enzyme for L-Ascorbic Acid Biosynthesis Missing in Man," *Journal of Biological Chemistry* 269 (1994): 13685–13688.

[41]Y. Inai, Y. Ohta, and M. Nishikimi, "The Whole Structure of the Human Non-functional L-gulono-y-lactone Oxidase Gene—the Gene Responsible for Scurvy—and the Evolution of Repetitive Sequences Thereon," *Journal of Nutritional Science and Vitaminology* (Tokyo) 49 (2003): 315–319. They write: "When the human and guinea pig sequences (647 nucleotides in total) of the regions of exons 4, 7, 9, 10, and 12 were compared, we found 129 and 96 substitutions in humans and guinea pigs, respectively, when compared with the rat sequences [note that rats have a functional GULO gene]. . . . The same substitutions from rats to both humans and guinea pigs occurred at 47 nucleotide positions among the 129 positions where substitutions occurred in the human sequences." (319)

[42]Inai *et al.* write: "Assuming an equal chance of substitution throughout the sequences, the probability of the same substitutions in both humans and guinea pigs occurring at the observed number of positions and more was calculated to be $1.84 \times 10^{-12}$. This extremely small probability indicates the presence of many mutational hot spots in the sequences." (317)

[43]Helen Pearson, "Silent Mutations Speak Up: Overlooked Genetic Changes Could Impact on Disease," *Nature* (December 21, 2006): news item, published online at http://www.nature.com/news/2006/061218/full/061218-12.html (last accessed 11 January 2007). This was big news at the end of 2006. *Science*NOW reported on it as follows: "Another dogma in cell biology seems about to be toppled: If a mutation in a gene doesn't change the basic sequence of building blocks, then it has no effect. Chava Kimchi-Sarfaty of the U.S. Food and Drug Administration in Bethesda, Maryland, and colleagues report online this week in *Science* that such 'silent mutations' can, under certain circumstances, determine how well a final protein performs—an 'extremely provocative' result, says cell biologist William Skach of Oregon Health & Science University in Portland." Quoted from Mary Beckman, "The Sound of a Silent Mutation," *Science*NOW Daily News (December 22, 2006): published online at http://sciencenow.sciencemag.org/cgi/content /full/2006/1222/2 (last accessed January 11, 2007).

[44]Chava Kimchi-Sarfaty, Jung Mi Oh, In-Wha Kim, Zuben E. Sauna, Anna Maria Calcagno, Suresh V. Ambudkar, and Michael M. Gottesman, "A 'Silent' Polymorphism in the MDR1 Gene Changes Substrate Specificity," *Science* (December 21, 2006): published online at http://www.sciencemag.org/cgi/content/abstract/1135308 (last accessed January 11, 2007).

[45]How is it possible for different DNA sequences that map onto the same amino acid sequence to induce different proteins? Computer engineer David Springer conjectures that "ribosomes process codons at different rates when the codons differ only by a redundant nucleotide replacement." He offers the following analogy for the effect this has on protein folding: "Think of the ribosome like a caulk gun producing a bead consisting of amino acid polymers that fold as they come out of the gun. If the rate at which the bead comes out changes, then the shape it folds into changes as well." He also considers the possibility that "RNA molecules dependent on specific gene sequences alter the way the protein is processed after the ribosome finishes producing it." See David Springer, "The Sound of the Neutral Theory Exploding," *Uncommon Descent* (December 23, 2006): published online at http://www.uncommondescent.com /archives/1901 (last accessed January 11, 2007).

[46]Darwin, *Origin of Species*, 396.

[47]Ibid., 387.

[48]Ibid., 395–396.

[49]Charles Darwin, letter to Asa Gray, Sept. 10, 1860, in Francis Darwin, ed., *The Life and Letters of Charles Darwin*, Vol. II (New York: D. Appleton and Company, 1896), 131.

[50]Darwin, *Origin of Species*, 395.

[51]Charles Darwin, *The Descent of Man*, Modern Library Reprint Edition (New York: Random House, 1936), 398.

[52]Ibid., 411.

[53]Jane M. Oppenheimer, "Haeckel's Variations on Darwin," in H. M. Hoenigswald and L. F. Wiener, eds., *Biological Metaphor and Cladistic Classification* (Philadelphia: University of Pennsylvania Press, 1987), 134.

[54]Stephen Jay Gould, "Abscheulich! Atrocious!" *Natural History* (March 2000): 42–49.

[55]Elizabeth Pennisi, "Haeckel's Embryos: Fraud Rediscovered," *Science* 277 (1997): 1435.

[56]Walter Garstang, "The Theory of Recapitulation: A Critical Restatement of the Biogenetic Law," *Journal of the Linnean Society* (Zoology) 35 (1922): 81–101.

[57]Ibid.

[58]Gavin de Beer, *Embryos and Ancestors*, 3rd ed. (Oxford: Clarendon Press, 1958), 10.

[59]Ibid., 164.

[60]Ibid., 172.

[61]Michael Behe, a leading proponent of intelligent design writes: "I find the idea of common descent (that all organisms share a common ancestor) fairly convincing, and have no particular reason to doubt it. See his book *Darwin's Black Box* (New York: Free Press, 1996), 5.

## CHAPTER SIX Irreducible Complexity

[1]Bruce Alberts, "The Cell as a Collection of Protein Machines: Preparing the Next Generation of Molecular Biologists," *Cell* 92 (8 February 1998): 291.

[2]Adam Wilkins, "A Special Issue on Molecular Machines," *BioEssays* 25(12) (December 2003): 1146.

[3]This definition of irreducible complexity is William Dembski's refinement and generalization of Michael Behe's original definition. For Behe's original definition see *Darwin's Black Box: The Biochemical Challenge to Evolution* (New York: Free Press, 1996), 39. For Dembski's refinement and generalization, see *No Free Lunch: Why Specified Complexity Cannot Be Purchased without Intelligence* (Lanham, Md.: Rowman and Littlefield, 2002), sec. 5.9. Behe himself has adopted this refinement of his original definition—see his essay "Irreducible Complexity: Obstacle to Darwinian Evolution," 359, in W. Dembski and M. Ruse, eds., *Debating Design: From Darwin to DNA* (Cambridge: Cambridge University Press, 2004), 352–370.

[4]For a jaw-dropping look at the wealth of irreducibly complex systems inside the cell, see the XVIVO animation titled *The Inner Life of the Cell.* An abridged version of this video can be viewed at http://www.studiodaily.com/main/technique/tprojects/6850.html (last accessed January 25, 2007). The full-length version can be viewed at http://multimedia.mcb.harvard.edu/media.html (last accessed January 25, 2007).

[5]See Howard C. Berg, *Random Walks in Biology*, exp. ed. (Princeton: Princeton University Press, 1993), 134. Berg writes: "*E. coli* has receptors for oxygen and other electron acceptors, sugars, amino acids, and dipeptides. It monitors the occupancy of these receptors as a function of time. The probability that a cell will run (rotate its flagella counterclockwise) rather than tumble (rotate its flagella clockwise) depends on the time rate of change of receptor occupancy. We know from responses of cells to short pulses of chemicals delivered by micropipettes that this measurement spans some 4 sec. A cell compares the occupancy of a given receptor measured over the past second—the aspartate receptor is the only receptor that has been studied in detail—with that measured over the previous 3 sec and responds to the difference. Now given rotational diffusion, *E. coli* wanders off course about 60 degrees in 4 sec. If measurements of differences in concentration took much longer than this, they would not be relevant, because cells would change course before the results could be applied. On the other hand, if these measurements were made on a much shorter time scale, their precision would not be adequate. E. coli counts molecules as they diffuse to its receptors, and this takes time. The relative error (the standard deviation divided by the mean) decreases with the square root of the count. Thus in deciding whether life is getting better or worse, *E. coli* uses as much time as it can, given the limit set by rotational Brownian movement."

[6]Michael Behe, *Darwin's Black Box* (New York: Free Press, 1996), 69–73.

[7]John Postgate, *The Outer Reaches of Life* (Cambridge: Cambridge University Press, 1994), 160.

[8]Darwin, *Origin of Species,* 189.

[9]H. Allen Orr, "Darwin v. Intelligent Design (Again)," *Boston Review* (December/January 1996-1997): 29.

[10]Ibid., 29.

[11]See L. Nguyen, I. T. Paulsen, J. Tchieu, C. J. Hueck, M. H. Saier Jr., "Phylogenetic Analyses of the Constituents of Type III Protein Secretion Systems," *Journal of Molecular Microbiology and Biotechnology* 2(2) (2000):125–44.

[12]See Kenneth R. Miller, *Finding Darwin's God* (New York: HarperCollins, 1999), ch. 5.

[13]Harold, *Way of the Cell*, 205.

[14]Lynn Margulis and Dorion Sagan, *Acquiring Genomes: A Theory of the Origins of Species* (New York: Basic Books, 2002), 103.

[15]For reviews in the popular press see James Shreeve, "Design for Living," *New York Times*, Book Review Section (4 August 1996): 8; Paul R. Gross, "The Dissent of Man," *Wall Street Journal* (30 July 1996): A12; and Boyce Rensberger, "How Science Responds When Creationists Criticize Evolution," *Washington Post* (8 January 1997): H01. For reviews in the scientific journals see Jerry A. Coyne, "God in the Details," *Nature* 383 (19 September 1996): 227–228; Neil W. Blackstone, "Argumentum Ad Ignorantiam," *Quarterly Review of Biology* 72(4) (December 1997): 445–447; and Thomas Cavalier-Smith, "The Blind Biochemist," *Trends in Ecology and Evolution* 12 (1997): 162–163.

[16]See John Catalano's web page titled "Behe's Empty Box": http://www.world-of-dawkins.com/Catalano/box/behe.htm (last accessed January 19, 2007).

[17]Darwin, *Origin of Species*, 82.

[18]Richard Dawkins, *The Blind Watchmaker* (New York: Norton, 1987), 287.

## CHAPTER SEVEN Specified Complexity

[1]See, for instance, chapters 2 and 3 of William A. Dembski, *No Free Lunch: Why Specified Complexity Cannot Be Purchased Without Intelligence* (Lanham, Md.: Rowman and Littlefield, 2002).

[2]See William A. Dembski and Michael Ruse, eds., *Debating Design: From Darwin to DNA* (Cambridge: Cambridge University Press, 2004), part IV.

[3]Leslie Orgel, *The Origins of Life* (New York: Wiley, 1973), 189.

[4]See http://www.mdl-research.org (last accessed November 15, 2006).

[5]More likely than not, this succession of identical automobiles would be a publicity stunt by the auto manufacturer or a local auto dealership. In that case, this succession would be due to design rather than to chance.

[6]The actual mathematics underlying probabilistic resources is better illustrated with the following example: imagine a large wall with $N$ identically-sized nonoverlapping targets painted on it and $M$ arrows in your quiver. Let us say that the probability of hitting any one of these targets, taken individually, with a single arrow by chance is $p$. Then the probability of hitting any one of these $N$ targets, taken collectively, with a single arrow, by chance, is bounded by $N \times p$, and the probability of hitting any of these $N$ targets with at least one of your $M$ arrows by chance is bounded by $M \times N \times p$. In this case, the number of replicational resources corresponds to $M$ (the number of arrows in your quiver), the number of specificational resources corresponds to $N$ (the number of targets on the wall), and the total number probabilistic resources corresponds to the product $M \times N$. To the degree that the probability $p$ of a specified event is so small that the number $M \times N \times p$ itself comes close to zero, this event exhibits specified complexity and is properly attributed to design rather than to chance.

[7]See William A. Dembski, *The Design Inference: Eliminating Chance through Small Probabilities* (Cambridge: Cambridge University Press, 1998), chs. 5 and 6. See also Dembski, *No Free Lunch,* chs. 2 and 3. For the most up to date formulation of specified complexity, at the time of this writing, see William A. Dembski, "Specification: The Pattern That Signifies Intelligence," typescript, available online at http://www. design-inference.com/documents/2005.06.Specification.pdf (last accessed November 15, 2006).

[8]Quoted on the institute's homepage: http://www.seti.org (last accessed November 20, 2006).

[9]Seth Shostak, "SETI and Intelligent Design," SETI Institute, posted December 1, 2005, available online at http://www.space.com/searchforlife/seti_intelligentdesign_ 051201.html (last accessed November 17, 2006).

[10]Almost 300 years ago, the mathematician Abraham de Moivre expressed this intuition as follows: "The same Arguments which explode the Notion of Luck, may, on the other side, be useful in some Cases to establish a due comparison between Chance and Design: We may imagine Chance and Design to be, as it were, in Competition with each other, for the production of some sorts of Events, and may calculate what Probability there is, that those Events should be rather owing to one than to the other. To give a familiar Instance of this, Let us suppose that two Packs of Piquet-Cards being sent for, it should be perceived that there is, from Top to Bottom, the same Disposition of the Cards in both packs; let us likewise suppose that, some doubt arising about this Disposition of the Cards, it should be questioned whether it ought to be attributed to Chance, or to the Maker's Design: In this Case the Doctrine of Combinations decides the Question; since it may be proved by its Rules, that there are the odds of above 263130830000 Millions of Millions of Millions of Millions to One, that the Cards were designedly set in the Order in which they were found." See Abraham de Moivre, *The Doctrine of Chances* (1718; reprinted New York: Chelsea, 1967), v.

[11]See Ian Hacking, *Logic of Statistical Inference* (Cambridge: Cambridge University Press, 1965), 82.

[12]Simon Singh, *The Code Book: The Evolution of Secrecy from Mary Queen of Scots to Quantum Cryptography* (New York: Doubleday, 1999), 10–13.

[13]Here is how professional skeptic Michael Shermer states the objection: "Perceiving the world as well designed and thus the product of a designer, . . . may be the product of a brain adapted to finding patterns in nature. We are pattern-seeking as well as pattern-finding animals. . . . Finding patterns in nature may have an evolutionary explanation: There is a survival payoff for finding order instead of chaos in the world, and being able to separate threats (to fight or flee) from comforts (to embrace or eat, among other things), which enabled our ancestors to survive and reproduce. We are the descendants of the most successful pattern-seeking members of our species. In other words, we were designed by evolution to perceive design." Quoted from Michael Shermer, *Why Darwin Matters: The Case Against Intelligent Design* (New York: Times Books, 2006), 38–39.

This passage underscores an important distinction: whether all patterns we find in nature are simply imposed by us on nature on account of our evolutionary conditioning, and thus signify nothing about any underlying design; or whether there can be patterns that reliably point us to the activity of an intelligent agent, as we are claiming for specified complexity. Shermer's observation that we are pattern-seeking, pattern-finding animals, however, does nothing to draw this distinction. Clearly, with the search for extraterrestrial intelligence, even Shermer would admit that radio signals coming in from outer space could exhibit patterns that would leave no doubt about their intelligent origin and thus could not be chalked up to our brains being conditioned by evolution to see patterns that we are merely making up. We all know the difference between patterns that result from an overactive imagination and patterns that are really there. Specified complexity makes this difference explicit.

[14]This connection between engineering and molecular biology parallels the connection between mathematics and physics. Nobel laureate physicist Eugene Wigner noted the remarkable way that mathematics elucidates physical reality: "The appropriateness of the language of mathematics for the formulation of the laws of physics is a wonderful gift which we neither understand nor deserve." We are essentially making the same point about the relation between engineering and molecular biology: "The appropriateness of the language of engineering for the elucidation of molecular machines inside the cell is a wonderful gift which we neither understand nor deserve." Far from undercutting design, this connection establishes it. For the Wigner quote, see "The Unreasonable Effectiveness of Mathematics in the Physical Sciences," *Communications on Pure and Applied Mathematics* 13 (1960): 1–14.

[15]See Dawkins's foreword to Niall Shanks, *God, the Devil, and Darwin: A Critique of Intelligent Design Theory* (Oxford: Oxford University Press, 2004).

[16]Richard Dawkins, *The Blind Watchmaker* (New York: Norton, 1987), 9.

[17]Ibid., 139, 145–146.

[18]Richard Dawkins, *Climbing Mount Improbable* (New York: Norton, 1996).

[19]George Wald, "The Origin of Life," in *The Physics and Chemistry of Life*, edited by the editors of Scientific American Books (New York: Simon & Schuster, 1955), 12.

[20]Émile Borel, "*Mécanique Statistique et Irréversibilité*," J. Phys. 5e série 3 (1913): 189–196.

[21]See M. P. Frank, "The Physical Limits of Computing," *Computing in Science and Engineering* 4(3) (2002): 16–26.

[22]See Seth Lloyd, "Computational Capacity of the Universe," *Physical Review Letters* 88(23) (2002): 7901–7904. For a popularization of this work, see Seth Lloyd, *Programming the Universe: A Quantum Computer Scientist Takes on the Cosmos* (New York: Knopf, 2006). Lloyd shows that if the total energy of the known physical universe were, over its multibillion year history, dedicated to performing computations,

it could perform at most $10^{120}$ bit operations. Since that number of operations can at best exhaustively search a space of $10^{120}$ possibilities, and since a complete catalog of such a space requires, at a minimum, 400 bits of information per possibility, it follows that chance, at the level of the entire known physical universe, can effectively search for messages whose character strings contain at most 400 bits of information. With an alphabet of 30 characters (i.e., Roman capital letters plus space plus three punctuation symbols), 400 bits corresponds to messages 82 characters in length.

[23]If the first 82 letters of Hamlet's soliloquy is the most that the entire universe could ever produce by chance, you might wonder what's the longest portion of Shakespeare that humans have ever been able to produce by using chance-based processes. On July 1, 2003, a Monkey Shakespeare Simulator was posted on the Internet. The simulator did not use real monkeys but rather a computerized random letter generator in which each "monkey" types one letter per second and the number of monkeys continually increases. The simulator compares its output with the entire works of Shakespeare and reports matches. After a year and a half, the longest match produced by the simulator was 24 letters from a line in the second part of *King Henry IV*. To accomplish this match took the equivalent of 2,738 trillion trillion trillion monkey-years to produce. A year later, the record was extended to over 30 letters, which took many more trillions of monkey-years to produce. Information about this experiment is widely available on the web. The site that originally posted this experiment was available at the following link: http://user.tninet.se/~ecf599g/aardasnails/java/Monkey/webpages/index.html. This link was active still in 2006 but no longer active in 2007. To see what was at this link, use the "Way Back Machine" at http://www.archive.org, which keeps a complete record of websites both current and defunct.

[24]What would happen if we actually did place monkeys at a typewriter? In 2003, lecturers and students from Plymouth University in southwest England decided to find out. They took a computer to the Paignton Zoo, about sixty miles away, and put it in a monkey enclosure housing six crested macaques. At first, a male monkey started bashing on the computer with a rock. "Another thing they were interested in was in defecating and urinating all over the keyboard," remarked Mike Phillips, who directs the university's Institute of Digital Arts and Technologies. After a month, the monkeys had produced the equivalent of five typed pages consisting almost entirely of the letter "S." They failed to produce anything remotely resembling a word. See Brian Bernbaum, "The Odd Truth: Monkey Theory Proven Wrong," *CBS News* (May 9, 2003), available online at http://www.cbsnews.com/stories/2003/05/12/national/main553500.shtml (last accessed February 5, 2007).

[25]Quoted from "Darwinism under the Microscope," PBS television interview of William A. Dembski and Eugenie Scott by Peter Robinson for the program *Uncommon Knowledge*, filmed December 7, 2001, transcript available online at http://www.hoover.org/publications/uk/3004521.html (last accessed February 5, 2007). Richard Dawkins makes essentially the same point in *The Blind Watchmaker*, 46, 141–142.

[26]Dawkins, *The Blind Watchmaker*, 1.

[27]Francis Crick, *What Mad Pursuit* (New York: Basic Books, 1988), 138.

[28]Fred Hoyle, *The Intelligent Universe* (New York: Holt, Reinhart, and Winston, 1984), 19.

[29]See, for instance, Dembski, *No Free Lunch*, sec. 5.10.

[30]Ibid. See also Angus Menuge, *Agents Under Fire: Materialism and the Rationality of Science* (Lanham, Md.: Rowman and Littlefield, 2004), ch. 4.

[31]See Hubert P. Yockey, *Information Theory and Molecular Biology* (Cambridge: Cambridge University Press, 1992), 220–221.

[32]Note that this inequality need not be a strict equality because it can be refined with additional terms. For instance, consider the *retention probability* $\boldsymbol{p}_{\text{reten}}$, the probability that items available at the right time and

in the right place stay at the right place long enough (i.e., are retained) for the bacterial flagellum (or whatever irreducibly complex system is in question) to be properly constructed. The retention probability is thus conditional on the availability, synchronization, and localization probabilities and could be inserted as a factor after these terms in the origination inequality.

Or consider the *proportionality probability* $p_{propor}$, the probability that items available at the right time, in the right place, and long enough occur in the right proportion for the bacterial flagellum to be properly constructed. The protein that goes into the flagellum's whip-like tale requires tens of thousands of subunits; proteins for other parts of the flagellum require only a few hundred subunits. Without the right proportion of suitable parts (subunits), no functioning flagellum can be built. The proportionality probability is conditional on availability, synchronization, localization, and retention probabilities and could be inserted as a factor after these terms in the origination inequality.

[33]Michael J. Behe and David W. Snoke, "Simulating Evolution by Gene Duplication of Protein Features that Require Multiple Amino Acid Residues," *Protein Science* 13 (2004): 1–14.

[34]Ibid., 11.

[35]For reasoning probabilistically with waiting times, see William Feller, *An Introduction to Probability Theory and Its Applications*, vol. 1, 3rd ed. (New York: Wiley, 1968), secs. 2.7, 11.3, and 17.2.

[36]In fact, a strict Darwinian explanation for antibiotic resistance seems to be more the exception than the rule. In times of environmental stress, bacteria go into a programmed defense that constitutes a targeted search for gene combinations that will enable at least a few of the bacteria's descendants to survive (the genetic changes here are therefore not random mutations as understood within neo-Darwinism). To see this, consider the following abstract from an article in *Cell*: "According to classical evolutionary theory, phenotypic variation originates from random mutations that are independent of selective pressure. However, recent findings suggest that organisms have evolved mechanisms to influence the timing or genomic location of heritable variability. Hypervariable contingency loci and epigenetic switches increase the variability of specific phenotypes; error-prone DNA replicases produce bursts of variability in times of stress. Interestingly, these mechanisms seem to tune the variability of a given phenotype to match the variability of the acting selective pressure. Although these observations do not undermine Darwin's theory, they suggest that selection and variability are less independent than once thought." Quoted from O. J. Rando and K. J. Verstrepen, "Timescales of Genetic and Epigenetic Inheritance" (review) *Cell* 128 (2007): 655–668.

[37]Charles Darwin, *On the Origin of Species*, facsimile 1st ed. (1859; reprinted Cambridge, Mass.: Harvard University Press, 1964), 189.

[38]Robert Koons, "The Check Is in the Mail: Why Darwinism Fails to Inspire Confidence," in W. A. Dembski, ed., *Uncommon Dissent: Intellectuals Who Find Darwinism Unconvincing*, 3–22 (Wilmington, Del.: ISI Books, 2004), 13–14.

[39]"Even if there were no actual evidence in favour of the Darwinian theory," writes Richard Dawkins, one "should still be justified in preferring it over all rival theories." Quoted from Dawkins, *The Blind Watchmaker*, 287.

[40]Darwin, *Origin of Species*, 2.

[41]See Carl Zimmer's *Evolution: The Triumph of an Idea* (New York: HarperCollins, 2001). The cover of this book depicts different eyes of varying complexity. What neither the cover nor the book itself nor the Darwinian community as a whole has accomplished is to provide a detailed, testable explanation of how even one of these eyes could have evolved by Darwinian means from simpler precursors.

[42]Note that merely pointing out that proteins in the flagellum are similar to other proteins serving other functions does not render flagellar evolution more plausible, much less solve the availability problem. For an example of this type of faulty reasoning, see M. J. Pallen and N. J. Matzke, "From *The Origin of Species* to the Origin of Bacterial Flagella," *Nature Reviews Microbiology* 4(10) (2006): 784–790. The problem is how classes of similar proteins became available in the first place. In the evolution of the flagellum, there was a point early on at which none of the proteins that appear in the flagellum, whether identical or similar, existed at all. How did these classes of proteins that appear in the flagellum come to exist in the first place? That's what availability asks, and that's what conventional accounts of biological evolution leave unanswered.

[43]In fact, if all evolution were doing was preserving meaning/function, we would be talking about neutral rather than Darwinian evolution. Strictly speaking, Darwinian evolution should be continually enhancing meaning/function and not merely maintaining its status quo. Even so, a necessary condition for enhancing meaning/function is preserving it. For neutral evolution, see Motoo Kimura, *The Neutral Theory of Molecular Evolution* (Cambridge: Cambridge University Press, 1983).

[44]Computers remain linguistically challenged. Indeed, natural language processing shows no ability to ascertain which sentences are meaningful, to say nothing of ascertaining their meaning.

[45]As Douglas Axe notes, "Mutagenesis studies and alignments of homologous sequences have demonstrated that protein function typically is compatible with a variety of amino-acid residues at most exterior non-active-site positions." Quoted from Douglas D. Axe, "Extreme Functional Sensitivity to Conservative Amino Acid Changes on Enzyme Exteriors," *Journal of Molecular Biology* 301 (2000): 585.

[46]D. D. Axe, N. W. Foster, and A. R. Fersht, "A Search for Single Substitutions That Eliminate Enzymatic Function in a Bacterial Ribonuclease," *Biochemistry* 7(20) (1998): 7157–7166.

[47]See S. Meier, P. R. Jensen, C. N. David, J. Chapman, T. W. Holstein, S. Grzesiek, and S. Ozbek, "Continuous Molecular Evolution of Protein-Domain Structures by Single Amino Acid Changes," *Current Biology* 17(2) (2007):173–178. Meier *et al.* focused on cysteine-rich domains of cnidarian nematocyst proteins. The press release for this article is revealing. Translated, it reads: "The evolution of complex characteristics such as new protein structures by the process of the evolution is to a large extent unresolved. Whereas proponents of intelligent design theory, as for instance the U.S.-American scientist Michael J. Behe, exclude the invention of new, complex protein structures via few mutational steps, evolutionary biologists have found indications that new proteins can originate out of transitional forms that combine original and novel characteristics. Until now, however, this has only demonstrated by the accumulation of artificial mutations, which merely simulate evolutionary processes." At the very least this press release indicates that there is a genuine scientific debate over protein evolution and that intelligent design is a key player in this debate. For the press release, go to http://idw-online.de/pages/en/news?id= 193184 (last accessed March 3, 2007).

[48]The ideas in this paragraph derive from a personal correspondence between WmAD and Cornelius Hunter (February 1, 2007).

[49]Douglas D. Axe, "Estimating the Prevalence of Protein Sequences Adopting Functional Enzyme Folds," *Journal of Molecular Biology* 341 (2004): 1298.

[50]See John R. Koza, *Genetic Programming: On the Programming of Computers by Means of Natural Selection* (Cambridge, Mass.: MIT Press, 1992) as well as John R. Koza, Forrest H. Bennett III, David Andre, and Martin A. Keane, *Genetic Programming III: Darwinian Invention and Problem Solving* (San Francisco: Morgan Kaufmann, 1999).

[51]Axe, "Estimating the Prevalence," 1295. Note that the implicit assumption here of uniform probability, i.e., of treating as probabilistically equivalent all proteins with this pattern of hydrophobic interactions, is

entirely warranted because the random processes that would be transforming these proteins are, according to neo-Darwinian theory, mediated principally through genetic point mutations that operate uniformly across the genome.

[52]Borel writes: "If we turn our attention, not to the terrestrial globe, but to the portion of the universe accessible to our astronomical and physical instruments, . . . we may be led to set at $10^{-50}$ the value of negligible probabilities on the cosmic scale." See Émile Borel, *Probabilities and Life*, trans. M. Baudin (New York: Dover, 1962), 28. More stringent universal probability bounds have been proposed since (see Dembski, *The Design Inference*, chs. 1 and 6), but for practical purposes, Borel's will do.

[53]It also didn't hurt the evolvability of THE CAT SAT ON THE MAT that the words in this sentence were all very short. Try evolving with the same small letter changes that we permitted here the following meaningful sentence from mathematics: DIFFEOMORPHIC DIFFERENTIABLE MANIFOLDS CONSTITUTE HOMEOMORPHIC TOPOLOGICAL STRUCTURES. Wordox lexicons, which list words first by length and then alphabetically, suggest that sentences with long and abstruse words such as this will be unevolvable via small changes in letters. For a Wordox lexicon, see http://aaron.doosh.net/lexicon (last accessed March 12, 2007).

[54]There's a long-standing debate in the evolutionary biology community over whether natural selection functions primarily at the level of the gene, cell, organism, group, species, etc. See Samir Okasha, *Evolution and the Levels of Selection* (Oxford: Oxford University Press, 2007).

[55]Darwin, *Origin of Species*, 82.

# CHAPTER EIGHT The Origin of Life

[1]George Polya, *How to Solve It*, 1st ed. (Princeton: Princeton University Press, 1945). This quote does not occur in the second edition, which was published in 1957. For this quote and others from the out-of-print 1945 edition, see http://www-gap.dcs.st-and.ac.uk/~history/Quotations/Polya.html (last accessed November 30, 2006).

[2]Michael Denton, *Evolution: A Theory in Crisis* (Bethesda, Md.: Adler & Adler, 1985), 328–329.

[3]Harvard biologist Richard Lewontin in *The New York Review of Books* writes, "We take the side of science *in spite of* the patent absurdity of some of its constructs, *in spite of* its failure to fulfill many of its extravagant promises of health and life, *in spite of* the tolerance of the scientific community for unsubstantiated just-so stories, because we have a prior commitment, a commitment to materialism. It is not that the methods and institutions of science somehow compel us to accept a material explanation of the phenomenal world, but, on the contrary, that we are forced by our *a priori* adherence to material causes to create an apparatus of investigation and a set of concepts that produce material explanations, no matter how counterintuitive, no matter how mystifying to the uninitiated." Quoted in Richard Lewontin, "Billions and Billions of Demons" (review of Carl Sagan's *The Demon-Haunted World: Science as a Candle in the Dark*), *The New York Review of Books* (9 January 1997): 31. Lewontin is saying more here than that some scientific ideas are counterintuitive. Rather, he is saying that no evidence whatsoever can overturn science's commitment to materialism. But if that is the case, then the commitment to materialism in science is itself not scientific because scientific claims are always in contact with evidence and capable of being overturned in light of new evidence.

[4]As seen at http://www.studiodaily.com/main/casestudies/6850.html (last accessed December 30, 2006).

[5]Obligate parasites, such as viruses and bacterial endosymbionts, which cannot live independently of host organisms, do not pose a counterexample to this claim that the simplest cell requires hundreds of genes to handle its basic tasks of living. That's because these parasites require for their very existence cells that *can* live independently of host organisms, and such non-parasitic cells *do* require a full complement of genes and capabilities, which puts them well above the minimum complexity threshold described here. If you will, obligate parasites are just like other cells in requiring hundreds if not thousands of genes, only they don't keep them on hand but let other organisms maintain those genes for them. For the simplest bacterial endosymbiont at the time of this writing (i.e., *Carsonella ruddii* with a 160-kilobase genome), see the work of Nancy Moran at the University of Arizona, http://eebweb.arizona.edu/faculty/moran/research.htm (last accessed March 28, 2007).

[6]Lynn Margulis has proposed that eukaryotic cells evolved by one prokaryotic cell swallowing another, with the swallowed cell then serving as the nucleus of the newly formed eukaryotic cell. This proposal remains speculative, however. It is not clear by what mechanisms a prokaryotic cell could engulf and assimilate another prokaryotic cell and thereafter work together symbiotically (phagocytosis, by which eukaryotic cells ingest food particles, seems not to be the right model here). Moreover, there are many differences between eukaryotic and prokaryotic cells that are not explained by Margulis's model, including differences in cell division, cytoplasmic membranes, locomotor organelles, respiratory enzymes, and electron transport chains. See Lynn Margulis and Dorion Sagan, *Acquiring Genomes: A Theory of the Origins of Species* (New York: Basic 2002), ch. 3.

[7]Karl Popper, "Scientific Reduction and the Essential Incompleteness of All Science," *Studies in the Philosophy of Biology* 259 (1974): 270. Emphasis in original.

[8]See http://astrobiology.arc.nasa.gov/about/index.cfm (last accessed November 16, 2006).

[9]A. I. Oparin, *The Origin of Life on Earth*, 3rd ed (1924; reprinted, revised, and translated New York: Academic Press, 1957).

[10]When Darwin wrote his *Origin of Species* in the mid 1800s, the cell was conceived as a blob of Jell-O enclosed by a membrane. Accordingly, it was thought to be so simple that it could have easily originated spontaneously and quickly. By the 1920s, when Oparin published his hypothesis suggesting that the first cell formed gradually, scientists understood the cell to be far more complex than they had previously imagined. They therefore recognized that without intelligent design the formation of something as complex as a cell must have required extended periods of time.

[11]J. B. S Haldane, "The Origin of Life," *Rationalist Annual* 148 (1928): 3–10.

[12]J. B. Corliss, J. A. Baross, and S. E. Hoffman, "An Hypothesis concerning the Relationship between Submarine Hot Springs and the Origin of Life on Earth," *Oceanologica Acta* 4(suppl.) (1981): 59–69.

[13]C. Huber and G. Wächtershäuser, "Peptides by Activation of Amino Acids with CO on (Ni,Fe)S Surfaces: Implications for the Origin of Life," *Science* 281 (1998): 670–72.

[14]Jay A. Brandes, Nabil Z. Boctor, George D. Cody, Benjamin A. Cooper, Robert M. Hazen, and Hatten S. Yoder Jr., "Abiotic Nitrogen Reduction on the Early Earth," *Nature* 395 (24 September 1998): 365.

[15]Paul Davies makes this point as follows: "There would have been no lack of available energy sources on the early Earth to provide the work needed to forge [biologically significant chemical] bonds, but just throwing energy at the problem is no solution. The same energy sources that generate organic molecules also serve to destroy them. To work constructively, the energy has to be targeted at the specific reaction required. Uncontrolled energy input, such as simple heating, is far more likely to prove destructive than constructive." The same could be said for turbulence. Quoted from Paul Davies, *The Fifth Miracle: In Search for the Origin and Meaning of Life* (New York: Simon & Schuster, 1999), 89–90.

[16]J. H. Carver, "Prebiotic Atmospheric Oxygen Levels," *Nature* 292 (1981): 136–38; James F. Kasting, "Earth's Early Atmosphere," *Science* 259 (1993): 920–26.

[17]Philip H. Abelson, "Chemical Events on the Primitive Earth," *Proceedings of the National Academy of Sciences USA* 55 (1966): 1365–1372.

[18]Marcel Florkin, "Ideas and Experiments in the Field of Prebiological Chemical Evolution," *Comprehensive Biochemistry* 29B (1975): 241–242.

[19]Sidney W. Fox and Klaus Dose, *Molecular Evolution and the Origin of Life*, rev. ed. (New York: Marcel Dekker, 1977), 43, 74–76.

[20]Jon Cohen, "Novel Center Seeks to Add Spark to Origins of Life," *Science* 270 (1995): 1925–1926.

[21]Heinrich D. Holland, *The Chemical Evolution of the Atmosphere and Oceans.* (Princeton: Princeton University Press, 1984), 99–100.

[22]Gordon Schlesinger and Stanley L. Miller, "Prebiotic Synthesis in Atmospheres Containing CH4, CO, and CO2: I. Amino Acids," *Journal of Molecular Evolution* 19 (1983): 376.

[23]John Horgan, "In the Beginning . . . ," *Scientific American* (February 1991): 116–126.

[24]Alan Schwartz, editor of *Origins of Life and Evolution of Biospheres,* notes, "A problem which is familiar to organic chemists is the production of unwanted byproducts in synthetic reactions. For prebiotic chemistry, where the goal is often the simulation of conditions on the prebiotic Earth and the modeling of a spontaneous reaction, it is not surprising—but nevertheless frustrating—that the unwanted products may consume most of the starting material and lead to nothing more than an intractable mixture, or 'gunk'." Quoted from Alan W. Schwartz, "Intractable Mixtures and the Origin of Life," *Chemistry & Biodiversity* 4(4) (2007): 656.

[25]Charles Thaxton, Walter Bradley, and Roger Olsen, *The Mystery of Life's Origin: Reassessing Current Theories* (New York: Philosophical Library, 1984), 104–6.

[26]Robert Shapiro, "A Simpler Origin of Life," *Scientific American* (February 12, 2007): available online at http://sciam.com/print_version.cfm?articleID=B7AABF35-E7F2-99DF-309B8CEF02B5C4D7 (last accessed April 13, 2007).

[27]J. Brooks and G. Shaw, *Origin and Development of Living Systems* (New York: Academic Press, 1973), 359.

[28]Robert M. Hazen, Timothy R. Filley, and Glenn A. Goodfriend. "Selective Adsorption of L- and D-Amino Acids on Calcite: Implications for Biochemical Homochirality," *Proceedings of the National Academy of Sciences* 98 (8 May 2001): 5487–90.

[29]Jessica Gorman, "Rocks May Have Given a Hand to Life," *Science News* 159(18) (5 May 2001), available online at http://www.sciencenews.org/20010505/fob1.asp (last accessed February 28, 2007).

[30]For simplicity, we limited ourselves in this analogy to linear arrangements of letters. Yet, because the bonding between amino-acid residues allows for branching, we could have extended the analogy by not only rotating letters but also having them assume a branched pattern such as the following:

[31]J. Bowie and R. Sauer, "Identifying Determinants of Folding and Activity for a Protein of Unknown Sequences: Tolerance to Amino Acid Substitution," *Proceedings of the National Academy of Sciences* 86 (1989): 2152–56. J. Bowie, J. Reidhaar-Olson, W. Lim, and R. Sauer, "Deciphering the Message in Protein Sequences: Tolerance to Amino Acid Substitution," *Science* 247 (1990): 1306–10. J. Reidhaar-Olson and R. Sauer, "Functionally Acceptable Solutions in Two Alpha-Helical Regions of Lambda Repressor," *Proteins, Structure, Function, and Genetics* 7 (1990): 306–10. See also Michael Behe, "Experimental Support for Regarding Functional Classes of Proteins to be Highly Isolated from Each Other," in *Darwinism: Science or Philosophy?*, eds. J. Buell, and G. Hearn (Dallas: Foundation for Thought and Ethics, 1994), 60–71; and Hubert Yockey, *Information Theory and Molecular Biology* (Cambridge: Cambridge University Press, 1992), 246–58.

[32]Douglas D. Axe, "Extreme Functional Sensitivity to Conservative Amino Acid Changes on Enzyme Exteriors," *Journal of Molecular Biology* 301 (2000): 585–595.

[33]The work of Sauer and Axe avoids the "retrospective fallacy" that design-critic Kenneth Miller attributes to intelligent design. When design theorists calculate the improbability of certain proteins, here's what they do according to Miller: "'It's what statisticians call a retrospective fallacy.' It is like equating the odds of drawing two pairs in poker with the odds of drawing a particular two-pair hand—say a pair of red queens, a pair of black 10s and the ace of clubs. 'By demanding a particular outcome, as opposed to a functional outcome, you stack the odds,' Miller says. What these calculations fail to recognise is that many different protein sequences can be functional. It is not uncommon for proteins in different species to vary by 80 to 90 per cent, yet still perform the same function." By contrast, Sauer and Axe have taken pains to avoid this fallacy, calculating not the improbability of a particular amino-acid sequence but the improbability of any sequence with the same fold and function. Note that the term "retrospective fallacy" is idiosyncratic: in fact, statisticians do not use this term to describe the mistake to which Miller is referring. What Miller is calling a fallacy is simply a failure to calculate the relevant probability. The passage cited is from Bob Holmes and James Randerson, "A Skeptic's Guide to Intelligent Design," *New Scientist* 187 (July 9, 2005): 11.

[34]"The Murchison meteorite contains a complex suite of amino acids. Whereas terrestrial samples are dominated by the 20 protein amino acids, over 70 amino acids have been positively identified in this meteorite, many of which appear to be uniquely extraterrestrial." S. Pizzarello and J. R. Cronin, "Alanine Enantiomers in the Murchison Meteorite," *Nature* 394(6690) (July 16, 1998): 236.

[35]See, for instance, "Photosynthesis Analysis Shows Work of Ancient Genetic Engineering," *Science Daily* (November 22, 2002): available online at http://www.sciencedaily.com/releases/2002/11/021122074236.htm (last accessed April 16, 2007). The article summarizes an Arizona State University press release on some work of Robert Blankenship. Merely by examining DNA sequences in extant bacteria, Blankenship and colleagues claimed to obtain "clear evidence that photosynthesis did not evolve through a linear path of steady change and growing complexity but through a merging of evolutionary lines that brought together independently evolving chemical systems—the swapping of blocks of genetic material among bacterial species known as horizontal gene transfer." Other than showing that *if* photosynthesis evolved, its evolutionary history must have been complicated, Blankenship's research does nothing to describe the actual path of evolution, much less to determine whether such an evolutionary path could have proceeded on purely materialistic principles.

[36]Compare with the previous endnote the following article, also co-authored by Robert Blankenship, which indicates how much yet remains to be understood about the photosynthetic apparatus and its superb design capabilities: Gregory S. Engel, Tessa R. Calhoun, Elizabeth L. Read, Tae-Kyu Ahn, Tomāiš Mančal, Yuan-Chung Cheng, Robert E. Blankenship, and Graham R. Fleming, "Evidence for Wavelike Energy Transfer through Quantum Coherence in Photosynthetic Systems," *Nature* 446 (12 Apr 2007): 782–786.

[37]In saying that early-earth history was more conducive toward the breakdown rather than the buildup of biological complexity, we are not backhandedly invoking the second law of thermodynamics. Instead, we are simply looking at the material processes that most likely were available on the early earth and, by taking them on their own terms, inferring that they were unlikely to facilitate the buildup of biological complexity.

[38]Thaxton et al., *Mystery of Life's Origin*, 162.

[39]Ibid., 174–76.

[40]The authors of this book are in principle opposed to referring to DNA as "doing" anything at all. More accurate would be to say "the cell needs such-and-such in order to make copies of its DNA." Unfortunately, in the current state of biological discourse, it is almost impossible to say anything about these issues without begging many questions!

[41]Gerald Joyce, "RNA Evolution and the Origins of Life," *Nature* 338 (1989): 217–24.

[42]Shapiro, "A Simpler Origin of Life." To illustrate the magnitude of the problem, Shapiro considers the analogy of a golfer "who having played a golf ball through an 18-hole course, then assumed that the ball could also play itself around the course in his absence. He had demonstrated the possibility of the event; it was only necessary to presume that some combination of natural forces (earthquakes, winds, tornadoes and floods, for example) could produce the same result, given enough time. No physical law need be broken for spontaneous RNA formation to happen, but the chances against it are so immense, that the suggestion implies that the non-living world had an innate desire to generate RNA. The majority of origin-of-life scientists who still support the RNA-first theory either accept this concept (implicitly, if not explicitly) or feel that the immensely unfavorable odds were simply overcome by good luck." Two comments are in order: (1) invoking "immense good luck" does not constitute a scientific theory and (2) the implicit small probability argument here suggests that such a formation of RNA would exhibit specified complexity and therefore trigger a design inference.

[43]Joyce, "RNA Evolution and the Origins of Life."

[44]Quoted in Robert Irion, "RNA Can't Take the Heat," *Science* 279 (1998): 1303.

[45]In "A Simpler Origin of Life," Shapiro considers the possibility of a "pre-RNA world" in which "the bases, the sugar or the entire backbone of RNA have been replaced by simpler substances, more accessible to prebiotic syntheses." Nevertheless, if origin-of-life researchers want such a nonnucleotide "soup" to produce a replicator, even a simple one, they will need to overcome some daunting probabilistic hurdles. According to Shapiro, "the spontaneous appearance of any such replicator without the assistance of a chemist faces implausibilities that dwarf those involved in the preparation of a mere nucleotide soup." Citing Gerald Joyce and Leslie Orgel for regarding it a "near miracle" that RNA chains on the lifeless Earth could form spontaneously, Shapiro concludes that it would similarly be a miracle for any of "the proposed RNA substitutes" to arrange themselves into a replicator.

[46]Ian Stewart, *Life's Other Secret: The New Mathematics of the Living World* (New York: John Wiley, 1998), 48.

[47]More realistic assessments of the origin of life problem can be found in Thaxton *et al., Mystery of Life's Origin* (1984); Robert Shapiro, Origins, *A Skeptics Guide to the Creation of Life on Earth* (New York: Summit Books, 1986); Paul Davies, *The Fifth Miracle* (1999); and Hubert Yockey, I*nformation Theory, Evolution, and the Origin of Life* (Cambridge: Cambridge University Press, 2005).

[48]Harold J. Morowitz, *The Emergence of Everything: How the World Became Complex* (New York: Oxford University Press, 2002), 76. Emphasis added.

[49]"Perhaps two-thirds of scientists publishing in the origin-of-life field (as judged by a count of papers published in 2006 in the journal *Origins of Life and Evolution of the Biosphere*) still support the idea that

life began with the spontaneous formation of RNA or a related self-copying molecule." Shapiro, "A Simpler Origin of Life." Throughout this article Shapiro refers to the "daunting probabilistic hurdles" that these genetics-first scenarios must overcome.

[50]Morowitz, *The Emergence of Everything*, 76.

[51]The sheer diversity of these explanations shows that origin-of-life theorizing is unconstrained by data, offers no insight into how life actually got started, and is limited only by our imagination. Now imagination may be a good thing, but it is too thin a soup on which to nourish science.

[52]Stuart Kauffman, *At Home in the Universe: The Search for the Laws of Self-Organization and Complexity* (New York: Oxford University Press, 1995), 274.

[53]Christian de Duve, *Singularities: Landmarks on the Pathways to Life* (Cambridge: Cambridge University Press, 2004).

[54]Günter Wächtershäuser, "Evolution of the First Metabolic Cycles," *Proceedings of the National Academy of Sciences* 87 (1990): 200–204 and Günter Wächtershäuser, "Life as We Don't Know It," *Science* 289 (2000): 1307–1308.

[55]Michael Russell, "Life from the Depths," *Science Spectra* 1 (1996): 26. See also William Martin and Michael Russell, (2002). "On the Origins of Cells: A Hypothesis for the Evolutionary Transitions from Abiotic Geochemistry to Chemoautotrophic Prokaryotes, and from Prokaryotes to Nucleated Cells," *Philosophical Transactions of the Royal Society: Biological Sciences* 358 (2002): 59–85.

[56]David W. Deamer, "The First Living Systems: A Bioenergetic Perspective," *Microbiology and Molecular Biology Reviews* 61 (1997): 239–261.

[57]See Platts's website http://www.pahworld.com (last accessed April 20, 2007). For a popular exposition of Platts's views, see Robert M. Hazen, *Genesis: The Scientific Quest for Life's Origin* (Washington, D.C.: Joseph Henry Press, 2005), ch. 17.

[58]Alexander G. Cairns-Smith, *Seven Clues to the Origin of Life* (Cambridge: Cambridge University Press, 1985); and Alexander G. Cairns-Smith and Hyman Hartman, eds., *Clay Minerals and the Origin of Life* (Cambridge: Cambridge University Press, 1986).

[59]See Harold J. Morowitz, *Beginnings of Cellular Life: Metabolism Recapitulates Biogenesis* (New Haven: Yale University Press, 1992), chs. 10 and 12; Morowitz, *The Emergence of Everything*, 80.

[60]Hazen, *Genesis*, 209–210.

[61]Ibid., 151. Emphasis added.

[62]Wächtershäuser, like Morowitz, would like to reproduce such cycles under realistic prebiotic conditions. His iron-sulfur model, however, seems inadequate to the task. In "Self-Organizing Biochemical Cycles," *Proceedings of the National Academy of Sciences* 97(23) (2000): 12506, Leslie Orgel analyzes Wächtershäuser's model and argues that there is "no reason to expect that multistep cycles such as the reductive citric acid cycle will self-organize on the surface of FeS/FeS2 or some other mineral." Orgel immediately adds, "While it seems almost impossible that a cycle of reactions as complicated as the reductive citric acid cycle could self-organize on a mineral surface, Wächtershäuser's suggestion does raise an interesting and important question. How much self-organization is it reasonable to expect on a mineral surface in the absence of *evolved, informational catalysts*?" [Emphasis added.] Here is Orgel's answer: "It is not clear that any surface is likely to catalyze two or more unrelated chemical reactions." Nonspecialists may read this last statement as conceding that there is no solid empirical evidence that mineral surfaces can catalyze metabolic cycles—*in the absence of evolved, informational catalysts!*

[63]We are indebted to Stephen Meyer for many of the insights in this section.

[64]Throughout these discussions it is important to keep referring back to section 8.1, always keeping in mind that what needs to be explained is the cell as we know it and not the simplistic examples of life constantly put forward by origin-of-life researchers.

[65]This work is summarized in Julius Rebek Jr., "Synthetic Self-Replicating Molecules," *Scientific American* 271(1) (1994): 48–55.

[66]John Horgan, "In the Beginning," *Scientific American* 264(2) (Feb. 1991): 120.

[67]See Lawrence Hurst and Richard Dawkins, "Life in a Test Tube," *Nature* 357 (21 May 1992):198–99.

[68]Horgan, "In the Beginning," 120. In fact, Rebek's experiment makes a strong case for the tremendous power of a little design in overcoming random chemistry. In other words, it illustrates the power of intelligent design!

[69]Sol Spiegelman, "An *In Vitro* Analysis of a Replicating Molecule," *American Scientist* 55 (1967): 221. See also Norman R. Pace and Sol Spiegelman, "*In Vitro* Synthesis of an Infectious Mutant RNA with a Normal RNA Replicase," *Science* 153 (1966): 64–67.

[70]Brian Goodwin, *How the Leopard Changed Its Spots: The Evolution of Complexity* (New York: Scribner's, 1994), 35–36. So here we have a self-replicating system that evolved into something simpler. One might wonder whether the system also submitted a purchase order for more monomers when the first batch ran out. Now that would be true self-replication!

[71]Note that we are not saying that Darwinism requires that evolution proceed toward simplicity. Our point is simply that Darwinism, in itself, does not mandate increasing complexity and inherently favors simplicity. Thus, if we see steadily increasing complexity, something besides selection and variation must be at work. Now, Darwinists have offered rationales for why we might expect increasing complexity strictly on Darwinian grounds (e.g., the irreversibility of certain changes, the complexity cost of arms races, and the lower wall of complexity below which things are dead). But all such rationales are post hoc—in each case the opposite might well have happened and Darwinism could still be true. Thus, we can imagine (and even program on computer) Darwinian evolutionary scenarios in which reversibility has a selective advantage, in which arms races are won by simplifying, and in which the lower wall of complexity is an absorbing barrier where maximal fitness is conferred by maximal simplicity. Bottom line: Whether evolution in a Darwinian scenario tends toward increasing or decreasing complexity depends on factors other than those dictated by Darwinian theory; moreover, since increased complexity invariably incurs a fitness cost (i.e., there's more to monitor), Darwinian theory inherently favors decreasing complexity.

[72]Theodosius Dobzhansky, discussion of G. Schramm's paper in *The Origins of Prebiological Systems and of Their Molecular Matrices*, ed. S. W. Fox (New York: Academic Press, 1965), 310.

[73]Though note: proof of concept works only when one proves the concept. Origin-of-life researchers are a long way from establishing proof of concept. Indeed, it has completely eluded them. Their willingness to embrace just about any highly speculative scenario for life's origin suggests that in fact they are giving up on proof of concept and acting out of desperation, trying to shore up a materialistic explanation of life's origin when life is clearly telling us that its origin is not materialistic.

[74]Kauffman, *At Home in the Universe,* 31. This passage was written in the mid 1990s. More recently, Kauffman has gone further, denying not just that we don't know the specifics of how life originated but that we don't even have a theory for how life might have originated: "[W]e entirely lack a theory of organization of process, yet the biosphere, from the inception of life to today manifestly propagates organization of process." See Stuart Kauffman, Robert K. Logan, Robert Este, Randy Goebel, David Hobill, and Ilya Shmulevich, "Propagating Organization: An Enquiry," *Biology and Philosophy* (2007): forthcoming, available online at http://personal.systemsbiology.net /ilya/Publications/ BiolPhilosPropagatingOrganization.pdf (last accessed April 23, 2007).

[75]Note that this bracket insertion was in the original source for this quote. See the next note.

[76]Jason Socrates Bardi, "Life—What We Know, and What We Don't," *TSRI News & Views* (the online weekly of The Scripps Research Institute) 3(30) (October 11, 2004), available online at http://www.scripps.edu/newsandviews/e_20041011/ghadiri.html (last accessed April 23, 2007).

[77]George M. Whitesides, "Revolutions in Chemistry" (Priestly Medalist address), *Chemical & Engineering News* 85(13) (March 26, 2007): 12–17, available online at http://pubs.acs.org/cen/coverstory/85/8513cover1.html (last accessed April 23, 2007).

[78]Ibid., emphasis added.

[79]Davies, *The Fifth Miracle*, 19.

[80]Another famous chemist made precisely this point in the mid-20th century: "A book or any other object bearing a pattern that communicates information is essentially irreducible to physics and chemistry. . . . We must refuse to regard the pattern by which the DNA spreads information as part of its chemical properties." Michael Polanyi, "Life Transcending Physics and Chemistry," *Chemical & Engineering News* (August 21, 1967): 62.

[81]David Baltimore, "DNA Is a Reality beyond Metaphor," *Caltech and the Human Genome Project* (2000): available online at http://pr.caltech.edu:16080/events/dna/dnabalt2.html (last accessed April 23, 2007).

[82]Manfred Eigen, *Steps Towards Life: A Perspective on Evolution*, trans. Paul Woolley (Oxford: Oxford University Press, 1992), 12.

[83]Eörs Szathmáry and John Maynard Smith, "The Major Evolutionary Transitions," *Nature* 374 (1995): 227–232.

[84]Christian de Duve, *Vital Dust: Life as a Cosmic Imperative* (New York: Basic Books, 1995). This book is divided in seven parts. These are the headings for the first four parts.

[85]Ibid., 10.

[86]All quotes in this paragraph are from Francis S. Collins, *The Language of God: A Scientist Presents Evidence for Belief* (New York: Free Press, 2006), 92–93.

[87]All quotes in this paragraph are from Francis S. Collins, "Faith and the Human Genome," *Perspectives on Science and Christian Faith* 55(3) (2003): 152.

[88]All quotes in this paragraph are from Hazen, *Genesis*, 80.

[89]Compare molecular biologist James Shapiro's work on "natural genetic engineering." Shapiro is neither a Darwinist nor an ID proponent but regards the ability of cells to do their own engineering as crucial to biological innovation and therefore to evolution. For a representative sample of his work, see http://shapiro.bsd.uchicago.edu/index3.html?content=genome.html (last accessed April 27, 2007).

[90]Hubert Yockey, "Self-Organization Origin of Life Scenarios and Information Theory," *Journal of Theoretical Biology* 91 (1981): 13–16.

[91]William A. Dembski, *The Design Revolution: Answering the Toughest Questions about Intelligent Design* (Downers Grove, Ill.: InterVarsity, 2004), 151–152.

[92]For instance, it is a fallacy of composition to argue that because bricks are hard, homogeneous, rectangular solids, therefore houses composed of bricks are likewise hard, homogeneous, rectangular solids.

[93]For instance, it is a fallacy of division to argue that because houses are dwelling places for humans, therefore the bricks composing houses are likewise dwelling places for humans.

[94]See http://www.protolife.net (last accessed April 27, 2007).

[95]See http://www.protolife.net/company/profile.php (last accessed April 27, 2007). Emphasis added.

[96]Note that if the precursor exists in duplicate, one version might evolve into the later system, the other might stay put and coexist with the later system.

[97]See Cullen Schaffer, "A Conservation Law for Generalization Performance," *Machine Learning: Proceedings of the Eleventh International Conference*, eds. H. Hirsh and W. W. Cohen, 259–265 (San Francisco: Morgan Kaufmann, 1994); Thomas M. English, "Some Information Theoretic Results on Evolutionary Optimization," *Proceedings of the 1999 IEEE Congress on Evolutionary Computation* 1 (1999): 788–795; and William A. Dembski and Robert J. Marks II, "The Conservation of Active Information," 2007, typescript, http://www.EvoInfo.org (last accessed June 8, 2007). The last paper by Dembski and Marks constitutes the most powerful and general formulation of conservation of information at the time of this writing.

[98]Richard Dawkins, *The Blind Watchmaker* (New York: Norton, 1987), 50.

[99]Ibid., 13.

[100]Ibid., 316.

[101]Ibid., 141.

[102]Richard Dawkins, *Climbing Mount Improbable* (New York: Norton, 1996).

[103]For details, see Dembski and Marks, "The Conservation of Active Information."

[104]Charles S. Peirce, "How to Make Our Ideas Clear," *Popular Science Monthly* 12 (January 1878): 286–302.

[105]Compare Robert Laughlin's ideas about organizing principles in nature, which may be readily extended to teleological organizing principles. See Robert Laughlin, *A Different Universe: Reinventing Physics from the Bottom Down* (New York: Basic Books, 2005).

[106]Michael Ruse, *Can a Darwinian Be a Christian?* (Cambridge: Cambridge University Press, 2001), 64. Emphasis added.

[107]To watch a webcast of the entire debate, go to http://www.bu.edu/com/greatdebate/fall05/index.html (last accessed April 30, 2007).

[108]"It is not implausible that life emerged as a phase transition to collective autocatalysis once a chemical minestrone, held in a localized region able to sustain adequately high concentrations, became thick enough with molecular activity." Kauffman, *At Home in the Universe*, 274.

[109]See Richard Dawkins, *The God Delusion* (New York: Houghton Mifflin, 2006).

[110]Richard Dawkins, *River Out of Eden: A Darwinian View of Life* (New York: Basic Books, 1995), 98.

## CHAPTER NINE Epilogue: The "Inherit the Wind" Stereotype

[1]This and the other quotes by Darrow from *The State of Illinois v. Nathan Leopold & Richard Loeb* are drawn from his closing argument, delivered August 22, 1924 in Chicago, Illinois. It is available online at http://www.law.umkc.edu/faculty/projects/ftrials/leoploeb/darrowclosing.html (last accessed February 22, 2006).

[2]See *The Scopes Trial*, 229–280. The verbatim transcript of the Scopes Trial was published in 1925 by the National Book Company and then reprinted 1990 by Gryphon in their Notable Trials Library series. The transcript is available at http://www14.inetba.com/gryphon/item43.ctlg (last accessed 6 February 2003).

[3]Ibid., 206, 220–221, 223. The prosecution is partly responsible for the failure of evolution to undergo cross-examination in the Scopes Trial. According to the prosecution, the Tennessee statute outlawed the teaching of evolution regardless of whether evolution was true, and thus all evidence for the truth of evolution was irrelevant and should be excluded. The judge agreed, and initially kept the scientists from testifying, which meant, of course, that the prosecution lost its initial opportunity to cross-examine them. It was later that the judge allowed the scientists' written testimony to be read in open court, but then denied the prosecution any chance to challenge that testimony. Yet, once that testimony was admitted into the proceedings, the truth of evolution did become an issue, and fairness required that it be subjected to cross-examination.

[4]Ibid., 284, 288.

[5]Ibid., 306–307.

[6]Ibid., 308.

[7]Fred Hoyle, *Mathematics of Evolution* (Memphis, Tenn.: Acorn Enterprises, 1999), 106.

[8]*The Scopes Trial*, 299.

[9]*The Scopes Trial*, introduction by Alan Dershowitz.

[10]*A Civic Biology* presents studies of two families—the Jukes family and the Kallikak family. The Jukeses are said to descend from "Margaret, the mother of criminals," and to have cost New York in the period of 75 years "the care of prisons and asylums of considerably over a hundred feeble-minded, alcoholic, immoral, or criminal persons." The Kallikaks are said to descend from Martin Kallikak and "a feeble minded girl" whose 480 descendants include "33 sexually immoral, 24 confirmed drunkards, 3 epileptics, and 143 feeble-minded. The man who started this terrible line of immorality and feeble-mindedness later married a normal Quaker girl. From this couple a line of 496 descendants have come, with no cases of feeble-mindedness. The evidence and the moral speak for themselves!" These and other quotes from *A Civic Biology* are available on the Internet: http://www.eugenics-watch.com/roots/chap08.html and http://www.iit.edu/departments/humanities/impact/colloquium/ongley_2001f.html.

[11]Hunter, *Laboratory Problems in Civic Biology*, 182. Ironically, in light of this epilogue, the lab book contains nothing on evolution. Apparently, Hunter and his publishers felt it was more important for the "receptive" young students of 1925 to learn eugenics than to learn evolution.

[12]Daniel Dennett, *Darwin's Dangerous Idea* (New York: Simon & Schuster, 1995), 519.

[13]Ibid.

[14]Peter B. Medawar, *Advice to a Young Scientist* (New York: Basic Books, 1981), 39.

[15]So wrote Justice Oliver Wendell Holmes in 1927 in the case of *Buck v. Bell*, 274 U.S. 200 (1927).

[16]See Bruce Chapman and David DeWolf, "Why the Santorum Language Should Guide State Science Education Standards," available online at http://www.discovery. org/articleFiles/PDFs/santorum LanguageShouldGuide.pdf (last accessed September 14, 2004).

[17]Theodosius Dobzhansky, "Nothing in Biology Makes Sense Except in the Light of Evolution," *The American Biology Teacher* 35 (1973): 125–129.

[18]*Edwards v. Aguillard*, 482 U.S. 578 (1986), 594.

[19]For the actual decision, see http://www.pamd.uscourts.gov/kitzmiller/kitzmiller_ 342.pdf (last accessed February 22, 2006).

[20]For a thorough examination of this case, see David DeWolf, John West, Casey Luskin, and Jonathan Witt, *Traipsing into Evolution: Intelligent Design and the Kitzmiller vs. Dover Decision* (Seattle: Discovery Institute Press, 2006).

# Glossary

**abiogenesis** The formation of living organisms from nonliving materials by undirected chemical means.

**allele** An alternative form of a gene. An allele is one of a pair of genes that occupy corresponding positions on homologous chromosomes and lead to alternative expressions of a single trait.

**allopatric speciation** The origin of a new species through reproductive isolation from its ancestral population where the reproductive isolation is caused by geographic separation. *Compare* SYMPATRIC SPECIATION.

**amino acid** An organic compound that contains one or more amino groups and one or more acidic carboxyl groups. Amino acids can be polymerized to form peptides and proteins. Only 20 amino acids normally occur in living things.

**analogy (analogous structure)** A body part similar in function to that of another organism, but at best superficially similar in structure. Such functional similarities are attributed within Darwinism not to inheritance from a common ancestor, as in homology, but to environmental pressure. One example is the wing of an insect and the wing of a bird.

**apparent design** The term "design" appears widely throughout the biological literature. Many Darwinists agree that complex structures in living things appear to have been designed for a purpose. Yet, Darwinists deny that the directed organization of living things results from the purposive activity of an actual intelligence. Instead, they attribute it to blind material forces such as natural selection and random variation. By contrast, INTELLIGENT DESIGN emphasizes that the design in living things is not merely apparent but actual.

**argument-from-ignorance objection** The charge that a lack of information or evidence for some claim is being used as positive evidence for an opposing claim.

**argument from personal credulity** Concluding that a proposition is true in the absence of convincing evidence because one's presuppositions demand that it be true.

**argument from personal incredulity** Concluding that a proposition is false because one personally cannot think of a good reason for how it could be true.

**artifact hypothesis** The claim that ancestors to the Cambrian phyla did exist but that their evolutionary history has been hidden by poor preservation, insufficient sampling, or a combination of the two.

**autogeny** (1) Control of the expression of a gene by the product of that gene. (2) The claim that life arose from self-organizing chemical reactions.

**autotroph** An organism that does not depend on other organisms for its source of nutrients. *Compare* HETEROTROPH.

**biomacromolecules** High molecular-weight chemical compounds in living matter, such as nucleic acids, proteins, and complex carbohydrates.

**blind watchmaker** Richard Dawkins's metaphor for natural selection as a unintelligent, purposeless material process that produces in living things the appearance of artifactual features that might otherwise be ascribed to the skill of a designer ("watchmaker"). Natural selection, for Dawkins, stands in for a watchmaker but is blind.

**bottleneck effect** The reduction of a population's gene pool and the accompanying changes in gene frequency produced when a few members survive the widespread elimination of a species.

**Burgess Shale** A Middle Cambrian (approximately 515 million years ago) black shale in British Columbia that contains exquisitely preserved fossils, including many soft-bodied organisms.

**Cambrian explosion** A period during the Cambrian estimated to have lasted no longer than 5 or 10 million years in which most of the animal phyla appear suddenly in the fossil record with no evident precursors.

**cassette mutagenesis** A procedure in which molecular biologists systematically mutate or alter individual DNA codons to determine the effect of those alterations on protein folding or function.

**catalyst (catalyze, catalytic)** A substance that increases the rate of a chemical reaction without itself undergoing permanent chemical change.

**chaperones** Proteins that help other proteins to achieve their correct 3-dimensional shape.

**chemical evolution** The view that life originated through purely chemical transformations of nonliving matter. Starting from primordial simplicity, such chemical transformations are supposed eventually to have brought about a living cell. Also known as PREBIOTIC EVOLUTION.

**Chengjiang biota/fauna** Fossils from the lower Cambrian (approximately 530 million years ago) Maotianshan Shale of China, known for their exquisite soft-body preservation.

**chirality** The handedness of a molecule that can exist in two non-superimposable mirror images, like our left and right hands.

**chromosome** A single large molecule of DNA together with associated proteins that hold it together. Chromosomes organize the genetic material inside a cell. In prokaryotic cells, they can be circular. In eukaryotic cells, they are threadlike structures inside the nucleus.

**cilia** Short hair-like projections on cells that undulate and typically function in locomotion.

**coacervates** Organized ball-like collections of proteins, carbohydrates, or other organic substances formed in a solution by water-repellent electrostatic charges.

**codon** A triplet of nucleotides forming a basic unit of the genetic code in DNA or RNA.

**coevolution** A form of evolution in which biological structures and functions both change so that as structures evolve they acquire new functions.

**colloid** A gel-like material in which small particles that resist filtering and settling are dispersed.

**common descent** Also known as *universal common ancestry*. The claim that all currently living organisms are biological descendants of a single organism in the past.

**complexity** The degree of difficulty to solve a problem or achieve a result. The most common forms of complexity are *probabilistic* (as in the probability of obtaining some outcome) or *computational* (as in the memory or computing time needed for an algorithm to solve a problem). *Kolmogorov complexity* is a form of computational complexity that measures the length of the minimum program needed to solve a computational problem. *Descriptive complexity* is likewise a form of computational complexity, but generalizes Kolmogorov complexity by measuring the size of the minimum description needed to characterize a pattern.

**complex specified information** Information that is both complex and specified. Synonymous with SPECIFIED COMPLEXITY.

**composition, fallacy of** To argue that what is true of the parts must be true of the whole.

**computational complexity** *See* COMPLEXITY.

**convergent evolution** The evolution of biological structures or species that exhibit similar function or appearance but do not derive from a common ancestor sharing the similarity.

**creation** The view that a CREATOR brought the world into existence and ordered it. Unlike a carpenter, who takes pre-existing materials and organizes them, a creator is responsible for the existence of those materials. Creation therefore differs from design: design merely characterizes the ordering of pre-existing materials; creation also characterizes their ultimate source.

**creation science** Also known as SCIENTIFIC CREATIONISM and often referred to, in abbreviated form, simply as CREATIONISM. An approach to the origin and formation of the natural world that attempts to make sense of the data of science in light of the account of creation given in the biblical book of Genesis. Unlike the theory of INTELLIGENT DESIGN, which begins with the data of science (such as the fossil record or DNA), creation science begins with the assumption that the account of creation in Genesis is historically and scientifically accurate. Creation science includes six key tenets: (1) There was a sudden creation of the universe, energy, and life from nothing. (2) Random variation and natural selection are insufficient to bring about the development of all living kinds from a single organism. (3) Changes of the originally created kinds of plants and animals occur only within fixed limits. (4) There is a separate ancestry for humans and apes. (5) The earth's geology can be explained in terms of catastrophes, primarily by the occurrence of a worldwide flood. (6) The earth and living kinds had a relatively recent inception (on the order of ten thousand years).

**creationism** (1) The view that a CREATOR brought the world into existence and ordered it. (2) A widely used abbreviation for CREATION SCIENCE or SCIENTIFIC CREATIONISM.

**creator** In theology, a being outside the material world that brings the material world into existence and then orders it. A creator is a designer, but a designer need not be a creator. *Compare* DESIGNER.

**cytochrome c** An iron-bearing molecule utilized in the electron transport chains of mitochondria and chloroplasts. Cytochrome c is used in protein comparison studies aimed at determining evolutionary relationships.

**Darwinian pathway** *See* DIRECT DARWINIAN PATHWAY and INDIRECT DARWINIAN PATHWAY.

**Darwinism** The theory that all living things descended from a common ancestor (*see* COMMON DESCENT) through unguided processes such as NATURAL SELECTION acting on RANDOM VARIATIONs. To say that such processes are unguided is to say that they produce the complexity and diversity of life without the aid of intelligence.

**deep divergence hypothesis** The claim that divergence among many of the animal phyla occurred long before the Cambrian Explosion, typically around 1 billion years ago.

**descriptive complexity** A measure of the difficulty needed to describe a pattern. *See* COMPLEXITY.

**design (as entity)** An event, object, or structure that an intelligence brought about by matching means to ends.

**design (as process)** A four-part process by which a DESIGNER forms a designed object: (1) A designer conceives a purpose. (2) To accomplish that purpose, the designer forms a plan. (3) To execute the plan, the designer specifies building materials and assembly instructions. (4) The designer or some surrogate applies the assembly instructions to the building materials. What emerges is a designed object. The designer is successful to the degree that the object fulfills the designer's purpose.

**designer** An intelligent agent that arranges material structures to accomplish a purpose. Whether this agent is personal or impersonal, conscious or unconscious, part of nature or beyond nature, active through miraculous interventions or through ordinary physical causes are all possibilities within the theory of INTELLIGENT DESIGN. In particular, the designer need not be a CREATOR.

**design inference** A form of inference in which design or intelligent agency is attributed to an event, object, or structure because it exhibits SPECIFIED COMPLEXITY. Design inferences gauge what is within the reach of chance-based mechanisms and what is beyond their reach.

**direct Darwinian pathway** An evolutionary progression in which a biological structure changes over time while its function stays the same. In a direct Darwinian pathway, natural selection enhances or improves a given function but does not change it. *Compare* INDIRECT DARWINIAN PATHWAY.

**disparity** Major variations in morphology or body plan of organisms. For example, insects and vertebrates have completely different body architectures.

**diversity** Minor variations within a basic body plan or biological form. For example, a golden retriever and St. Bernard have essentially the same body architecture.

**division, fallacy of** To argue that what is true of the whole must be true of the parts.

**DNA** Deoxyribonucleic acid, which is the basis for the genetic code utilized by all organisms. It consists of two long, intertwined chains of nucleotide subunits that carry the pattern needed to specify the sequence of amino acids in proteins.

**dominance (dominant)** The phenotypic expression of one allele over another when both are present in a heterozygote (thus masking the effect of the second allele).

**domain** (or **protein domain**) An autonomously folding portion of a protein that serves as a functional module.

**Ediacaran fauna** Fossilized Precambrian (Vendian) multicellular organisms with no obvious evolutionary relationships to other organisms. These fossils predate the fossils of the Cambrian by 30 to 50 million years and were first found in the Ediacara Hills of Australia.

**embryo** The early developmental stage of a multicellular organism.

**embryology** The study of the development of multicellular organisms from fertilization to the point where all major organ systems have formed.

**emergence** Within SELF-ORGANIZATIONAL THEORIES, the view that novel structures exhibiting novel functions can arise spontaneously as the parts of these structures organize themselves.

**entropy** A quantitative measure of the disorder of a system, inversely related to the amount of energy the system has available to do work. The more that energy has become dispersed within a system, the less work the system can perform and the greater its entropy. Biological systems are highly organized, carry out numerous work cycles, and therefore have low entropy.

**epistemic** Relating to what we know. *Compare* ONTOLOGICAL.

**eukaryote (eukaryotic)** A cell that has membrane-bound organelles, including (most significantly) a NUCLEUS.

**evolution** Change over time, often with an emphasis on the cumulative character of the change. In this broad sense, the term can describe the history of anything, from the entire cosmos to a political system or an idea. In modern biology, evolution can simply mean a change in gene (i.e., allele) frequencies, or it can mean DARWINISM.

**evolutionary informatics** A branch of information theory that studies the informational requirements of evolutionary processes.

**evolvability** The capacity of a system to evolve.

**exon** A region of DNA that is transcribed into mRNA and then translated into protein. *Compare* INTRON.

**fixation** The establishment in a population of one allele of a gene through elimination of alternative alleles.

**fold (protein folding)** The process by which a sequence of L-amino acids joined by peptide bonds organizes itself (or with the help of CHAPERONEs) into a 3-dimensional shape that determines its function.

**founder effect** Reduction of the gene pool and alteration of gene frequencies in a newly formed population. Such a population results when a few organisms (the founders) break away and isolate themselves from a parental population.

**free oxygen** Oxygen uncombined with other substances.

**functional information** (1) Information in the base sequence of a species' DNA that codes for structures capable of performing biological functions; much of this functional information exhibits specified complexity. (2) More generally, patterns embodied in material structures that enable them to perform functions.

**gamete** A germ cell (e.g., sperm or egg) from a sexually reproducing organism that participates in fertilization. *Compare* ZYGOTE.

**gene** A unit of heredity located within a chromosome and composed of DNA. Often used to refer to a DNA sequence that encodes a specific protein.

**gene flow** A loss or gain of alleles between two populations of a species.

**gene pool** The total genetic material in the population of a species at a given time.

**genetic drift** A change in gene frequencies in a population that results from random mating rather than natural selection.

**genetic information** The specific linear sequences of subunits in DNA and RNA molecules that are required to sustain the living state by specifying the sequences of amino acids in proteins. Genetic information can be both complex and specified.

**genome** The total DNA for a given organism.

**genotype** The combination of alleles inherited for a particular trait.

**god-of-the-gaps fallacy** The fallacy of invoking a nonmaterial cause (e.g., God) where a material cause, though for the present unknown, in fact operates. Logically speaking, however, there is no guarantee that nature is an interlocking nexus for which every apparent gap can in fact be filled by a material cause. If no material cause exists, it is not a fallacy to invoke a nonmaterial cause.

**gradualism** The view that evolution occurred gradually over time, with transitional forms smoothly connecting ancestors and descendants. *Compare* SALTATIONISM.

**hemophilia** A genetic disorder of the human blood clotting mechanism produced by the absence of either of two proteins essential to the formation of blood clots. The result is uncontrolled bleeding from even minor cuts or skin punctures, which may cause death.

**heterotroph** An organism that depends on other organisms for its source of nutrients. *Compare* AUTOTROPH.

**heterozygous** Having a mixed gene pair in which one allele is dominant and the other recessive.

**homeobox gene** A gene containing a 180-base-pair segment (the "homeobox") that encodes a protein domain involved in binding to (and thus regulating the expression of) DNA. Genes with different functions in many different organisms may contain remarkably similar homeoboxes.

**homeotic gene** A gene that affects embryo development by specifying the character of a body segment. The classic example is *Antennapedia*, a gene that, when mutated, can cause a fruit fly to grow a leg in place of an antenna.

**homochirality** Condition of a polymeric molecule in which all the subunits (the monomers) are of the same handedness (mirror-image form), that is, either 100 percent left-handed or 100 percent right-handed.

**homology** Similarity of structure and position, but not necessarily function. For example, the bones of vertebrate forelimbs are homologous even though the limbs can function in widely different ways (e.g., for swimming, flying, running, or grasping). In Darwinian theory, homology is attributed to inheritance from a common ancestor. Indeed, "homology" is often taken to mean "similarity due to common ancestry"—though this conflates the explanation with what needs to be explained.

**homoplasy** Structural similarity that in Darwinian theory is thought not to be due to inheritance from a common ancestor.

**Hox gene** One of a cluster of homeotic genes. Many different types of animals have remarkably similar Hox gene clusters.

**hydropathic signature** The pattern of hydrophobic interactions characteristic of certain classes of proteins.

**hydrophobic core** In a protein, the grouping of amino acids that lack polar side-groups and thus tend to avoid water.

**hypercycles** Theoretical prebiotic cycles of chemical reactions involving the autocatalytic amplification of substances thought to have been required for the spontaneous origin of the first life forms.

**hypothesis** In science (1) an educated guess; (2) a tentative explanation; or (3) a proposition yet to be tested and confirmed.

**improbability** The degree to which a probability is close to zero.

**indirect Darwinian pathway** An evolutionary progression in which a biological structure and its function both change over time. In an indirect Darwinian pathway, biological structures and functions coevolve. *Compare* DIRECT DARWINIAN PATHWAY. *See also* COEVOLUTION.

**inference to the best explanation** A method of reasoning employed in the sciences in which scientists elect that hypothesis that would, if true, best explain the relevant evidence. Recent work in the philosophy of science has shown that those hypotheses that qualify as "best" typically provide simple, coherent, comprehensive, and causally adequate explanations of the evidence or phenomena in question.

**information** Literally, the act of giving form or shape to something. Because giving form to a thing rules out other forms that it might take, information theory characterizes information as the reduction of possibilities or uncertainty. In classical information theory, the amount of information in a string of characters is inversely related to the probability of the occurrence of that string. Hence, the more improbable the string, the more uncertainty is reduced by identifying it and therefore the more information it conveys. Information defined in this way provides only a mathematical measure of improbability or complexity. It does not establish whether a string of characters conveys meaning, performs a function, or is otherwise significant. *See also* SEMANTIC INFORMATION, SYNTACTIC INFORMATION, SHANNON INFORMATION, SPECIFIED COMPLEXITY, and COMPLEX SPECIFIED INFORMATION.

**intelligence** A type of cause, process, or principle that is able to find, select, adapt, and implement the means needed to effectively bring about ends (or achieve goals or realize purposes). Because intelligence is about matching means to ends, it is inherently teleological.

**intelligent design** The study of patterns in nature that are best explained as the product of intelligence. As a theory of biological origins, intelligent design attempts to show that intelligent causation rather than blind material forces are required to adequately explain certain patterns of biological complexity and diversity. Intelligent design needs to be distinguished from APPARENT DESIGN and OPTIMAL DESIGN. Objects that are only apparently design look designed but actually are not. Objects that are optimally design exhibit perfect design that cannot be improved. The adjective "intelligent" in front of "design" stresses that the design in question is actual, but makes no assumption about the goodness, cleverness, or optimality of design.

**intron** A region of DNA that gets spliced out as a gene is transcribed into mRNA. *Compare* EXON.

**irreducible complexity** A system exhibits irreducible complexity provided it is composed of several well-matched, interacting parts, each of which is indispensable to maintaining the system's basic function.

**isomorphism** A relation of essential identity between two things. Thus, two things are isomorphic provided they agree in all essential respects. Isomorphism is a much stronger relation than analogy—two things can be analogous without being isomorphic.

**junk DNA** Genetic material for which no biological function has yet to be identified. Because functions keep being discovered for supposed "junk DNA," this term is falling out of favor. *Compare* PSEUDOGENE.

**Kolmogorov complexity** *See* COMPLEXITY.

**law** (1) An extremely well established universal generalization such as the first law of thermodynamics (i.e., in an isolated system the total energy remains constant). (2) A relation between antecedent conditions or causes and consequent states or events in which the antecedent either determines or renders probable the consequent (the former being a deterministic law, the latter being a nondeterministic law).

**linkage** The tendency of genes located on the same chromosome to be inherited together.

**lipids** Compounds such as fats, oils, or steroids that are insoluble in water.

**macroevolution** Large-scale genetic and structural changes in organisms leading to new and higher levels of complexity. Such changes have not been observed within the time scale of recorded human history.

**material mechanism** A type of cause that brings about its effects through the interplay of matter and energy as constrained by unbroken (autonomous) deterministic and nondeterministic laws (i.e., necessity and chance). Material mechanisms rule out the ongoing activity of intelligence though they themselves might be the result of intelligence. Darwin, for instance, thought that a divine designer had set up the mechanism of the world but then let that mechanism run its course without any real-time intelligent input. In this way natural selection became for him a designer substitute.

**materialism** The philosophical worldview that reality consists of space, time, matter, and energy and that these are governed by unbroken natural laws. Materialism affirms that the material world is all that exists and therefore denies the existence of any intelligence outside or beyond it. Materialism is a metaphysical doctrine, taking a definite stance on the nature and limits of reality. As such, it cannot be justified merely as a scientific inference. Compare METHODOLOGICAL MATERIALISM and SCIENTIFIC MATERIALISM.

**melanism** An increased dark or black coloration of the skin, fur, or feathers due to a high occurrence of the pigment melanin.

**metabolism** The ensemble of chemical processes that occur in living cells, including both biosynthetic and biodegradative pathways.

**metaphysical materialism** *See* MATERIALISM.

**methodological materialism** A methodological principle that some scientists think ought to guide science. Methodological materialism requires that scientists limit themselves to materialistic explanations, accounting for the physical world purely in terms of matter, energy, and forces. On this view of science, explanations that invoke intelligent causes or the actions of intelligent agents do not qualify as scientific unless these can be reduced to materialistic explanations. Since any intelligence responsible for life in the universe could not be reduced in this way, those who hold to methodological materialism reject intelligent design in biology. *Compare* MATERIALISM.

**microevolution** Small-scale genetic and structural changes in organisms. Microevolution (unlike MACROEVOLUTION) has been observed and produced experimentally.

**mimetic butterflies** A species of butterfly whose striking resemblance to another species of butterfly may help defend it against predators.

**minimal complexity** The simplest state, either genetic or metabolic, that is consistent with the viability of an organism.

**monomer** A single, simple molecule that can link with others to form POLYMERs.

**mononucleotide** A single nucleotide.

**monophyletic** Descriptive of a collection of organisms that share a common ancestor.

**morphogenesis** Development of the architectural features of an organism during ONTOGENY.

**morphology** The form or structure of an organism.

**mRNA** Messenger RNA. An RNA sequence that contains the same information as the DNA sequence in a gene and specifies the order of amino acids in a protein. The sequence of mRNA is read by a ribosome to make the actual protein.

**mutation** An alteration of the genetic material of a cell that may be caused either by spontaneous changes or by external forces (such as radiation). Mutations that occur in the gametes (sex cells) of an organism may be heritable.

**natural selection** A process in which organisms with certain characteristics are better able to survive and reproduce than organisms with other characteristics. If the characteristics that favor survival and reproduction are heritable, they become more prevalent in succeeding generations. According to Darwin's theory, this process has been the "most important, but not the exclusive" means by which organisms have evolved.

**neo-Darwinism** The modern version of evolutionary theory (also called the Modern Synthesis), which combines Darwinian evolution with Mendelian and molecular genetics. Neo-Darwinism postulates that natural selection acts on the heritable (genetic) variations of individual organisms within populations and that mutations (random copying errors in DNA) provide the main source of these genetic variations. It holds that the genetic processes responsible for small-scale micro-evolutionary changes can be extrapolated indefinitely to produce large-scale macro-evolutionary changes leading to major innovations in form.

**notochord** A flexible column located between the gut and nerve cord in the embryos of all chordates (a group of animals that includes the vertebrates, or animals with backbones).

**nucleic acid** One of a class of molecules composed of nucleotides that are joined together (such as DNA and RNA).

**nucleotide** The fundamental structural unit of a nucleic acid (DNA or RNA) made up of a nitrogen-carrying base (purine or pyrimidine), a sugar molecule, and a phosphate group.

**nucleus** A membrane-bound organelle that includes most of a EUKARYOTIC cell's genetic material.

**oligomers** Short chains of molecular subunits (e.g., short chains of amino acids in proteins or short chains of nucleotides in DNA and RNA).

**ontogeny** Embryological development of an organism from ZYGOTE to mature form.

**ontological** Relating to what exists. *Compare* EPISTEMIC.

**Oparin Hypothesis** (or **Oparin-Haldane Hypothesis**) The hypothesis that life emerged spontaneously from nonlife over hundreds of millions of years in stages of gradually increasing chemical complexity.

**optimal design** Optimal design is perfect design that cannot be improved. INTELLIGENT DESIGN does not require optimal design. Often optimal design and intelligent design are confused on account of unwarranted philosophical or theological presuppositions about the nature of the designing intelligence responsible for biological complexity. Such presuppositions have no place in the scientific theory of intelligent design.

**order** (1) Simple or repetitive patterns, as in crystals, that are the result of laws and cannot reasonably be used to draw a design inference. (2) More generally, the arrangement of parts into a pattern (that may or may not be reasonably used to draw a design inference). (3) The level of biological classification more general than a family and more specific than a class.

**oxidation** The combination of oxygen with some other substance.

**pananimalian genome** A hypothetical genome proposed by Susumu Ohno in which the ancestor of the animal phyla contained a genome that could code for all the proteins necessary to build each phylum.

**paleontologist** A scientist who studies fossils.

**paleontology** The study of life in past geological periods from fossil remains.

**parallel evolution** The acquisition of similarities in separate evolutionary lineages through the operation of similar selective factors on both lines.

**phylogenetic** Referring to the presumed evolutionary history of a group of organisms from a common ancestor.

**phylogenetic inertia** The tendency of populations to maintain an average morphology as well as a limited degree of variability around the population average. Phylogenetic inertia underscores that traits in a population tend to be faithfully transmitted from one generation to the next and therefore stable.

**phylogeny** The evolutionary history or pattern of evolutionary relationships characterizing a group of organisms.

**phylum** The highest taxonomic rank of biological classification among animals. (Among plants, the equivalent category is called a "division.") A phylum classifies a collection of organisms that share the same distinctive body-plan.

**polymer** A large molecule composed of multiple similar structural subunits called MONOMERS.

**polymerization** The formation of POLYMERS from MONOMERS.

**polymorphic** Descriptive of a population with diverse body forms.

**polymorphism** Phenotypic variation in a species population, not necessarily genetically based.

**polynucleotide** A polymer that combines many mononucleotides (e.g., DNA and RNA).

**polypeptide** A molecule consisting of many amino acids joined by peptide bonds. All proteins are polypeptides but not all polypeptides are proteins because polypeptides, as such, need not perform a biochemical or biological function (as proteins do).

**polyploidy** See SECONDARY SPECIATION.

**prebiotic** Prior to the existence of biological life.

**prebiotic evolution** See CHEMICAL EVOLUTION.

**prebiotic soup** A chemical solution (such as sea water) in which prebiotic or chemical evolution is supposed to have occurred.

**primary speciation** The origin of a new species by the splitting of an existing species, due to variation, selection, and perhaps geographical separation, into two or more branches that then continue to diverge from each other. *Compare* SECONDARY SPECIATION.

**primary structure (of protein)** The sequence of amino acids that make up a protein. Compare SECONDARY STRUCTURE and TERTIARY STRUCTURE.

**probabilistic complexity** A measure of the difficulty for chance-based processes to reproduce an event. *See* COMPLEXITY.

**probabilistic resources** The number of opportunities for an event to occur (replicational resources) or to be specified (specificational resources). An event that seems quite improbable can become highly probable once one factors in all the probabilistic resources relevant to its occurrence.

**prokaryote (prokaryotic)** An organism that lacks membrane-bound organelles and whose chromosomes are therefore not contained within a membrane-bound nucleus.

**protein** A three-dimensional macromolecule (POLYMER) composed of amino acids joined by peptide bonds (and therefore a POLYPEPTIDE) that performs a biological or biochemical function.

**protein synthesis** The production of proteins common to all cells in which ribosomes assemble specific linear chains of amino acids using an mRNA template.

**proteinoids** Chains of amino acids that are synthesized by the removal of water and that do not exhibit the specified sequences or bonding characteristics of proteins.

**proteasomes** Large protein assemblies inside cells that break down unused or defective proteins.

**pseudogene** A stretch of DNA that resembles a gene but is incomplete or defective and therefore unable to code for a functional protein. In evolutionary biology, pseudogenes are presumed to have once been functional but then to have lost their function through the accumulation of genetic errors. Because genetic material can do more than simply code for protein, it is questionable whether pseudogenes in fact serve no biological function. *Compare* JUNK DNA.

**punctuated equilibrium** The theory that PRIMARY SPECIATION occurs relatively quickly, when spurts of rapid genetic change "punctuate" the "equilibrium" of primarily constant morphology (*see* STASIS). This theory was first proposed by Niles Eldredge and Stephen Jay Gould.

**racemic** Referring to molecular mixtures that have roughly equal amounts of left- and right-handed forms of a molecule.

**random variation** Minor differences among the individual organisms in a population that are random and thus occur without regard for the uses to which they may be put in the course of evolution. Random variations are uncorrelated with future benefits to the organisms.

**recessive gene** A gene allele that is masked, or not expressed, in the presence of its dominant allele.

**reducing atmosphere** An atmosphere with little or no free oxygen but rich in hydrogen, which is available for energy-storing combinations with other substances.

**rejection region** A type of SPECIFICATION employed in Ronald Fisher's theory of significance testing in which an experimenter identifies a pattern in advance of an experiment and then runs the experiment by taking a sample. Provided the sample matches the pattern (i.e., "falls within the rejection region"), the experimenter then rejects or eliminates the hypothesized probability distribution originally thought to have been responsible for the sample (i.e., "eliminates the null hypothesis").

**replicational resources** *See* PROBABILISTIC RESOURCES.

**residue** A molecule incorporated into a larger molecule. For instance, within a protein an amino-acid residue is one amino acid joined to others by peptide bonds.

**ribosome** An organelle composed of at least 50 proteins and ribonucleic acids that reads mRNA and thereby synthesizes proteins. Ribosomes are the workbenches at which all cells manufacture proteins.

**saltationism** The view that major evolutionary change occurs in sharp jumps. On a saltationist view, TRANSITIONAL FORMs need not smoothly connect ancestors and descendants. *Compare* GRADUALISM.

**scientific creationism** *See* CREATION SCIENCE.

**scientific materialism** The view that MATERIALISM is true and that science is capable of justifying or confirming it.

**Scopes trial** The trial of John Scopes in 1925 that focused attention on whether a teacher had academic freedom to teach evolution. Specifically, the trial challenged Scopes's right to teach that humans evolved from a lower form of life. While Scopes lost the case, public sentiment turned against attempts to censor the teaching of evolution. Teachers today are more likely to be censored if they try to present evidence that challenges the Darwinian theory of evolution.

**secondary speciation** Speciation due to an increase in the number of chromosomes. It is also known as *polyploidy* and often follows hybridization. *Compare* PRIMARY SPECIATION.

**secondary structure (of protein)** Patterns such as alpha helices and beta strands that are produced by local interactions among amino acids in a protein. *Compare* PRIMARY STRUCTURE and TERTIARY STRUCTURE.

**selection pressure** The intensity with which an environment tends to eliminate an organism, and thus its genes, or to give it an adaptive advantage.

**selective advantage** A genetic advantage of one organism over its competitors that raises its relative likelihood of survival and reproduction.

**self-organizational theories** Theories ascribing to unassisted matter and energy the ability to organize themselves (e.g., on the prebiotic Earth) into systems of increasing complexity culminating in the EMERGENCE of living cells.

**semantic information** Information that conveys meaning.

**Shannon information** The concept of information developed by Claude Shannon and Warren Weaver in the 1940s. Shannon information is concerned with quantifying information (usually in terms of number of bits) to keep track of alphanumeric characters as they are communicated sequentially from a source to a receiver. The amount of Shannon information contained in a string of characters is inversely related to the probability of its occurrence. Unlike SPECIFIED COMPLEXITY, Shannon information is solely concerned with the improbability or complexity of a string of characters and not with its patterning or significance.

**sibling species** Two or more species that are very close in physical appearance (morphology) but not necessarily in ancestry (genealogy).

**speciation** The origination of a new species.

**species** The taxonomic rank below genus. There are many definitions of species, the most common of which is "a population of interbreeding organisms that is reproductively isolated from other such populations." This definition cannot be used with organisms that do not interbreed, such as bacteria. Nor can it be rigorously applied to fossilized organisms, whose ability or inability to interbreed cannot be verified.

**specification** A pattern that in the presence of COMPLEXITY can be employed to draw a design inference. Such a pattern exhibits low DESCRIPTIVE COMPLEXITY.

**specificational resources** See PROBABILISTIC RESOURCES.

**specified complexity** An event or object exhibits specified complexity provided that (1) the pattern to which it conforms identifies a highly improbable event (i.e., has high PROBABILISTIC COMPLEXITY) and (2) the pattern itself is easily described (i.e., has low DESCRIPTIVE COMPLEXITY). Specified complexity is a type of INFORMATION.

**spontaneous generation** The unaided (and usually rapid) origination of living organisms from nonliving materials based solely on the capabilities inherent in those materials. Compare the OPARIN HYPOTHESIS.

**stabilizing selection** Natural selection operating to eliminate extreme members of a population, thereby reducing variation in a population.

**stasis** The persistence of unchanging morphology in a species over a long period of geologic time. *See* PUNCTUATED EQUILIBRIUM.

**sympatric speciation** The origin of a new species through reproductive isolation from its ancestral population where the reproductive isolation occurs without the two populations being geographically separated. *Compare* ALLOPATRIC SPECIATION.

**syntactic information** Information that is expressed in strings taken from a fixed set of characters (e.g., alphanumeric characters). Bits, as represented on a computer, and the four-character DNA code are examples of syntactic information. Syntactic information may or may not be specified. See SPECIFIED COMPLEXITY and SPECIFICATION.

**taxon (plural taxa)** A category for classifying biological organisms.

**taxonomy** The science of biological classification. Traditionally, living things have been classified as members of a species, then genus, family, order, class, phylum, and finally kingdom. Each of these categories is referred to as a TAXON.

**teleological** Directed toward an end or goal; marked by design or purpose.

**teleology** The study of end-directed processes, especially in nature or history.

**tertiary structure (of protein)** The three-dimensional shape of a protein, which is determined by how its chain of amino acids folds and which determines its function. *Compare* PRIMARY STRUCTURE and SECONDARY STRUCTURE.

**theory** A scientific explanation that accounts for a wide range of observations. Darwinists often add to this definition that theories are "well established" or "well supported," thus ascribing to Darwin's theory a credibility it may not deserve. Many "well established" theories of the past have now been thoroughly discredited.

**therapsids** An extinct group of reptiles from which mammals are thought to have evolved.

**transcription** Synthesis of mRNA using a DNA template. *Compare* TRANSLATION.

**transitional form** An organism possessing features that it shares with a presumed ancestor and with a presumed descendant but that neither of these shares with each other. For instance, *Archaeopteryx* has feathers in common with birds and scales in common with reptiles, but neither of these features belongs to both birds and reptiles.

**translation** Synthesis of a protein using an mRNA template. Compare TRANSCRIPTION.

**tRNA** Transfer RNA. Each tRNA attaches to a specific amino acid. As a protein is being assembled by a ribosome, the tRNA positions its associated amino acid in the proper sequential location specified by mRNA.

**unit of selection** The level of organization of an evolving system at which selection is effective.

**unit of variation** The level of organization of an evolving system at which variations occur.

**universal probability bound** A degree of improbability for which a specified event of probability less than it cannot reasonably be attributed to chance regardless of whatever probabilistic resources from the known universe are factored in. Universal probability bounds have been estimated between $10^{-50}$ (Émile Borel) and $10^{-150}$ (William Dembski).

**vertebrates** Chordates (organisms belonging to the phylum Chordata) with a backbone. Vertebrates include cartilaginous and bony fishes, amphibians, reptiles, birds, and mammals.

**vestigial structures** Features that apparently serve no function in an organism and are allegedly holdovers from an evolutionary past. Such features, though no longer useful (as far as we know), are presumed to have been useful in ancestral species.

**zygote** A cell that results from the union of gametes (e.g., sperm and egg) during fertilization.

# Index

(Note: *Italicized* page references identify illustrations of the subject; [GN] page references are found in the General Notes on the CD)

# ABOUT THE Authors

**William A. Dembski,** *Senior Fellow*
Discovery Institute's Center for Science and Culture

William Dembski authored the first book on intelligent design to be published by a major university press, *The Design Inference: Eliminating Chance Through Small Probabilities* (Cambridge University Press, 1998). In it he lays out a rigorous, scientific method for detecting design. Dembski's work has been featured on the front page of the *New York Times* and in many other publications. He has debated top Darwinists at the American Museum of Natural History, and he has appeared on numerous radio and television broadcasts, including Jon Stewart's "The Daily Show" and ABC's "Nightline." He lectures around the globe on the topic of intelligent design (e.g., the Niels Bohr Institute in Denmark, Cambridge and Oxford Universities, U.C. Berkeley, UCLA, Princeton, Yale, MIT). He is the author or editor of more than ten books, including *Darwin's Nemesis* (IVP), a Festschrift volume in honor of Phillip E. Johnson. *Christianity Today* calls Dembski "Johnson's successor as the informal leader of the intelligent design community."

**Jonathan Wells,** *Senior Fellow*
Discovery Institute's Center for Science and Culture

Jonathan Wells holds two doctorates, one in molecular and cell biology from the University of California at Berkeley, the other in religious studies from Yale University. He has worked as a postdoctoral research biologist at the University of California at Berkeley, supervised a medical laboratory in Fairfield, California, and taught biology at California State University in Hayward. He has published articles in *Development, Proceedings of the National Academy of Sciences USA, BioSystems, Rivista di Biologia, The Scientist,* and *The American Biology Teacher.* As the author of *Icons of Evolution: Why Much of What We Teach about Evolution Is Wrong* (Regnery, 2000), Wells has emerged as one of the key figures for reforming the teaching of evolution by correcting textbook errors and by insisting that the evidence that both confirms and disconfirms Darwinism be taught. He is a widely acclaimed lecturer and debater on the topic of intelligent design. He has inspired many younger scholars to develop intelligent design as a fruitful scientific research program.

# Credits

**Cover Image** 3D Clinic/Collection Mix: Subjects/Getty Images.

**Preface Collage**
*Icons of Evolution*, courtesy of Regnery Publishing, Washington, DC; *Darwin's Black Box:* Man and Chimpanzee: James Balog Photography/Stone/Getty Images, Solar Eclipse: Amana Inc./Amana Images/ Getty Images; ABC screen capture, ABC News/Nightline; *The New York Times* front page, August 21, 2005, ©The New York Times, Inc. All rights reserved. Used by permission and protected by the Copyright Laws of the United States. The printing, copying, redistribution, or transmission of the Material without express written permission is prohibited.; **Image of** American Museum of Natural History ©Michael S. Yamashita/Corbis; CNN screen capture, courtesy of CNN.

**Chapter 1**
**Image of** Dr. Harold Morowitz, courtesy of Dr. Harold Morowitz; **Image of** Charles Darwin, FPG/Taxi/ Getty Images; **Image of** Dr. Francisco Ayala, courtesy of Dr. Francisco Ayala; **Image of** Arecibo, courtesy of the NAIC-Arecibo Observatory, a facility of the NSF, Photo by: Tony Acevedo; **Image of** Francis Crick, ©Barrington Brown/Photo Researchers, Inc.; **Image of** Dr. Leslie Orgel, courtesy of Dr. Leslie Orgel; An African Grey Parrot, ©Henry Horenstein/Corbis; **Image of** Louis Pasteur, ©Pixtal/SuperStock; **Image of** Michael Ruse, courtesy of Michael Ruse; **Image of** Dutch Rescuers, ©United States Holocaust Memorial Museum, courtesy of Sytske Huisman-Bakker. "The views or opinions expressed in this book, and the context in which the images are used, do not necessarily reflect the views or policy of, nor imply approval or endorsement by, the United States Holocaust Memorial Museum."; **Image of** Mother Teresa, AP Images/Tim Graham.

**Chapter 2**
**Image of** Charles Darwin, FPG/Taxi/Getty Images; **2.1** mood board/PunchStock; **2.2** Both from Corbis/ PunchStock; **Image of** Gregor Mendel, ©Science Source/Photo Researchers, Inc.; **2.3** Cobis/PunchStock; **2.5** ©Vasil Ishmatov/BigStockPhoto.com, ©John Richbourg/BigStockPhoto.com, Corbis/PunchStock; **Image of** DNA, ©SPL/Photo Researchers, Inc.; **2.7** photodisc/PunchStock; **2.10** ©Darwin Dale/Photo Researchers, Inc.; **2.11** ©SPL/Photo Researchers, Inc., ©Phototake Inc./Oxford Scientific Images; **2.12** courtesy of F. R. Turner, Ph.D, Indiana University.

**Chapter 3**
**3.1** ©imv/BigStockPhoto.com; **3.4** ©Tom McHugh/Photo Researchers, Inc.; **3.6** ©Chase Studio Inc./ Photo Researchers, Inc.; **3.10** ©Jim Corwin/Photo Researchers, Inc.

**Chapter 3 General Notes**
**3.14** ©Stephen J. Krasemann/Photo Researchers, Inc.

**Chapter 4**
**4.3** Stockbyte/PunchStock, courtesy of Mary Jane Cooper; **4.6** ©David Phillips/Photo Researchers, Inc.; **4.7** Photo of Hemignathus munroi, courtesy of Jack Jeffrey.

**Chapter 4 General Notes**
**4.9** Corbis/PunchStock.

**Chapter 5**
**5.2** ©Tom McHugh/Photo Researchers, Inc., ©N. Bergkessel Jr./Photo Researchers, Inc., ©SPL/Photo Researchers, Inc., ©Petr Mašek/BigStockPhoto.com; **5.6** Dorling Kindersley/PunchStock; **5.7** ©Leonid Smirnov/BigStockPhoto.com, ©Lori Carpenter/BigStockPhoto.com, Sergei Belski; **5.9** PunchStock; **5.10** ©Inga Spence/Photo Researchers, Inc.; **5.12** ©Dante Fenolio/Photo Researchers, Inc., ©Dante Fenolio/Photo Researchers, Inc.; **5.15** ©Marcel Nijhuis/BigStockPhotos.com.

**Chapter 5 General Notes**
**5.16** Photos by Jerry Heasley.

**Chapter 6**
**6.1** ©Suprijono Suharjoto/BigStockPhoto.com; **6.2** ©Pete Bax/BigStockPhoto.com, ©Winston Luzier/Transtock/Corbis; **6.3** *Cellular Microbiology*, 2e by Cossart, BoQuet, Normark and Rappuli, a publication of ASM Press, Washington D.C., 2005, *Science*:280: 602-605 1998 w/permission from AAAS-Kubori et al., courtesy of Shin-Ichi Aizawa, prof. PhD., courtesy of Jorge E. Galan, Ph.D.

**Chapter 6 General Notes**
**6.4** ©Ellen Henneke/BigStockPhoto.com.

**Chapter 7**
**7.1** "WALL STREET"©1987 Twentieth Century Fox. All rights reserved.; **7.3** AP Images/Francois Duhamel; **7.4** courtesy of the NAIC-Arecibo Observatory, a facility of the NSF, Photo by: Tony Acevedo; **Image of** Seth Shostak, courtesy of Seth Shostak; **7.5** courtesy of Robert M. Elowitz; **7.7** ©Joe Gough/BigStockPhoto.com.

**Chapter 7 General Notes**
**7.8** ©Raymond Kasprzak/BigStockPhoto.com; **7.9** AP Images; **7.10** AP Images; **7.11** ©David Biagi/BigStockPhoto.com; **7.12** AP Images; **7.13** DigitalGlobe; **7.14** Photodisc/PunchStock, ©Lit Liu/BigStockPhoto.com, ©Andre Klaassen/BigStockPhoto.com, Holger Wulschlaeger/BigStockPhoto.com, Tom Oliveira/BigStockPhoto.com; **7.15** ©Bobby Aback/BigStockPhoto.com, ©David Davis/BigStockPhoto.com, ©The Gallery Collection/Corbis; **7.16** Corbis/PunchStock, ©Tyler Olson/BigStockPhoto.com.

**Chapter 8**
**8.1** courtesy of Illustra Media; **8.2** ©SPL/Photo Researchers, Inc.; **8.3** ©Andreas Guskos/BigStockPhoto.com; **8.4** courtesy of John Wiester; **8.5** ©SPL/Photo Researchers, Inc.; **8.11** ©Laguna Design/Photo Researchers, Inc., digital vision/PunchStock; **8.12** ©STEVEN BROOKE STUDIOS.

**Chapter 8 General Notes**
**Image of** Thomas Henry Huxley ©Mary Evans/Photo Researchers, Inc.; **Image of** Ernst Haeckel ©SPL/Photo Researchers, Inc.; **8.14** ©Omikron/Photo Researchers, Inc.